T-LEVELS
THE NEXT LEVEL QUALIFICATION

T0173325

ENGINEERING AND MANUFACTURING

CORE

Paul Anderson
David Hills-Taylor
Andrew Buckenham
Polly Booker
Andrew Topliss
Himanshu Jadav (Contributor)

Orders: please contact Hachette UK Distribution, Hely Hutchinson Centre, Milton Road, Didcot, Oxfordshire, OX11 7HH. Telephone: +44 (0)1235 827827. Email education@hachette.co.uk Lines are open from 9 a.m. to 5 p.m., Monday to Friday. You can also order through our website: www.hoddereducation.co.uk

ISBN: 978 1 3983 6092 1

© Paul Anderson, David Hills-Taylor, Andrew Buckenham, Polly Booker and Andrew Topliss 2023

First published in 2023 by
Hodder Education,
An Hachette UK Company
Carmelite House
50 Victoria Embankment
London EC4Y 0DZ

www.hoddereducation.co.uk

Impression number 10 9 8 7 6 5 4 3

Year   2027 2026 2025 2024

Cover photo © PARILOV EVGENIY - stock.adobe.com

Illustrations by Integra

Typeset in India by Integra

Printed by CPI Group UK

A catalogue record for this title is available from the British Library.

Contents

Acknowledgements

The Publishers would like to thank the following for permission to reproduce copyright material.

Photo credits

p.1 © Littlewolf1989/stock.adobe.com; p.2 © Tatomm/stock.adobe.com; p.10 © Touch The Skies/Alamy Stock Photo; p.11 © Archivist/stock.adobe.com; p.12 © Salman2/stock.adobe.com; p.13 © Sittinan/stock.adobe.com; p.17 © Blue Planet Studio/stock.adobe.com; p.18 © Poco_bw/stock.adobe.com; p.20 © ABCDstock/stock.adobe.com; p.22 © Fotomatrix/stock.adobe.com; p.23 © Scanrail/stock.adobe.com; p.24 © Belogorodov/stock.adobe.com; p.25 © Gorodenkoff/stock.adobe.com; p.27 © Deatonphotos/stock.adobe.com; p.31 © DC Studio/stock.adobe.com; p.32 © Justanotherspare/stock.adobe.com; p.35 (top) © Ratmaner/stock.adobe.com, (bottom) © Cosmin Sava/Shutterstock.com; p.36 © Vector FX/stock.adobe.com; p.37 (top) © Sergey Ryzhov/stock.adobe.com, (bottom) © National Motor Museum/Heritage Images/Alamy Stock Photo; p.46 © Sergey Ryzhov/stock.adobe.com; p.51 © I Believe I Can Fly/stock.adobe.com; p.75© Visionsi/stock.adobe.com; p.80 © Michelangeloop/stock.adobe.com; p.86 © Minet/stock.adobe.com; p.102 (top) © Vangelis_Vassalakis/Shutterstock.com, (bottom) © Coprid/stock.adobe.com; p.107 © Hor/stock.adobe.com; p.130 © ArtEvent ET/stock.adobe.com; p.133 © Denklim/stock.adobe.com; p.135 © Artsiom P/stock.adobe.com; p.136 © Andrea Izzotti/stock.adobe.com; p.139 © Hrui/stock.adobe.com; p.141 © SKW/stock.adobe.com; p.144 © Pioneer111/stock.adobe.com; p.147 (left) © Alexlmx/stock.adobe.com, (right) © Maksym Yemelyanov/stock.adobe.com; p.151 © Phonlamaiphoto/stock.adobe.com; p.154 © Cristianstorto/stock.adobe.com; p.155 (left) © Milan Lipowski/stock.adobe.com, (right) © Timothyh /stock.adobe.com; p. 156 © Sorapolujjin /stock.adobe.com; p.161 © Gumpapa/stock.adobe.com; p.166 © Nomad_Soul/stock.adobe.com; p.167 © Juergen Wallstabe /stock.adobe.com; p.169 © Chitsanupong/stock.adobe.com; p.172 © Gorodenkoff/stock.adobe.com; p.174 © KPixMining/stock.adobe.com; p.175 (left) © Joy_Studio/Shutterstock.com, (right) © Aapsky /stock.adobe.com; p.178 © Insta_photos /stock.adobe.com; p.180 © Björn Wylezich/stock.adobe.com; p.181 © Monkey Business/stock.adobe.com; p.183 © Littlewolf1989/stock.adobe.com; p.184 © Kokliang1981 /stock.adobe.com; p.185 © Andrey Burmakin /stock.adobe.com; p.186 © Tongpatong /stock.adobe.com; p.187 (top) © Andrey Popov /stock.adobe.com, (bottom) © Auremar/stock.adobe.com; p.188 © Antoine2k /stock.adobe.com; p.190 © Amorn /stock.adobe.com; p.191 © Sebastian Duda/stock.adobe.com; p.192 © NVB Stocker /stock.adobe.com; p.193 (top) © StevenK/Shutterstock.com, (bottom) © Kokliang1981 /stock.adobe.com; p.195 © SpeedShutter /stock.adobe.com; p.198 © Warut /stock.adobe.com; p.199 © TheDirector/Shutterstock.com; p.200 (left) © Dmitry/stock.adobe.com, (right) © Tomas /stock.adobe.com; p.201 © Surapong /stock.adobe.com; p.203 © Gorodenkoff/stock.adobe.com; p.204 © Wutzkoh/stock.adobe.com; p.205 © Highwaystarz/stock.adobe.com; p.206 (top left) © NVB Stocker/stock.adobe.com, (bottom left) © Kadmy/stock.adobe.com, (right) © Sdecoret/stock.adobe.com; p.207 © St.Marco/Shutterstock.com; p.208 © Rainer/stock.adobe.com; p.210 © Fizkes/stock.adobe.com; p.211 © Pressmaster/stock.adobe.com; p.214 © Insta_photos/stock.adobe.com; p.217 © Drazen/stock.adobe.com; p.218 © Gorodenkoff/stock.adobe.com; p.219 © Sanpom/stock.adobe.com; p.221 © Rawpixel.com/stock.adobe.com; p.223 © Industrieblick/stock.adobe.com; p.225 © Gorodenkoff/stock.adobe.com; p.226 © Bannafarsai/stock.adobe.com; p.227 © MrSegui/Shutterstock.com; p.229 © Cagri Kilicci/Shutterstock.com; p.230 (left) © Snapper8S8/Shutterstock.com, (right) © NDABCREATIVITY/stock.adobe.com; p.231 (left) © Es sarawuth/stock.adobe.com, (right) © Zapp2photo/stock.adobe.com; p.234 © Prostock-studio/stock.adobe.com; p.237 © Firma V/stock.adobe.com; p.238 © Pressmaster/stock.adobe.com; p.242 © Teptong/stock.adobe.com; p.244 © Seventyfour/stock.adobe.com; p.245 © Andrey Popov/stock.adobe.com; p.247 © Boonchok/adobe.stock.com; p.250 (left) © Bobo1980/stock.adobe.com, (right) © NVB Stocker/stock.adobe.com; p.252 © NicoElNino/stock.adobe.com; p.253 © Monkey Business/stock.adobe.com.

Acknowledgements

p.98 Figure 5.36 © Nasa; p.189 Table 12.2 Contains public sector information published by the Health and Safety Executive and licensed under the Open Government Licence; p.190 Table 12.3 Contains public sector information published by the Health and Safety Executive and licensed under the Open Government Licence.

About the authors

Paul Anderson

Paul Anderson is an experienced teacher of Design, Technology and Engineering, having worked across a number of teaching roles, including as head of department for many years. He has experience as a consultant and teacher adviser, working with departments and schools to improve teacher attainment. Notably, he was awarded a Gatsby fellowship for implementing Assessment for Learning in Design and Technology. Paul has significant examining experience, having worked with a number of major exam boards to develop a range of qualifications across Design and Technology and Engineering, and he has held a range of examining roles, including Chief Examiner, Principal Moderator and Chair of Examiners. He is a bestselling author of numerous textbooks and high-quality STEM resources for teachers, schools and colleges.

David Hills-Taylor

David Hills-Taylor is a qualified teacher, experienced senior examiner and freelance consultant for engineering subjects from secondary to degree level. He has 14 years of teaching and leadership experience in technology and engineering, with specialisms including engineering design, electrical and electronic engineering, and control and instrumentation. He is a previous winner of the IET Digital D&T Teacher of the Year Award and a co-author of Hodder Education's *AQA GCSE (9–1) Engineering*, among several other publications.

Andrew Buckenham

Andrew Buckenham graduated in mechanical engineering prior to a career in design and manufacturing at sites worldwide. A move into teaching later saw him head the engineering department of a large comprehensive school. He has written several textbooks and revision guides for vocational engineering subjects. He has also worked with a range of awarding bodies to develop engineering qualifications, as an examiner and external verifier and to deliver teacher training. He currently lives in the Scottish Borders and lectures in Engineering at Borders College.

Polly Booker

After graduating from Salford University, Polly Booker worked as a hardware and software engineer in satellite communications and scientific instrumentation. A chance for a career change in 2003 took her to teach in Design and Technology. With engineering, and systems in particular, she has focussed on introducing learners to the possibilities, challenges and opportunities in technological careers. For relaxation she gardens, drives her Triumph Dolomite Sprint, plays with the grandkids and makes stuff.

Andrew Topliss

Andrew Topliss is a former head of faculty for Art, Design, Engineering and Technology with over 20 years of teaching experience. He has taught in both Japan and New Zealand. Before teaching he was an engineer in the British Army and later a software consultant. He works with a number of examination boards as an author, examiner, moderator, editor and course writer. He is also a registered health and safety consultant.

Himanshu Jadav (Contributor)

Himanshu Jadav is a highly motivated and accomplished engineer with a diverse background in the field. With a passion for technology and innovation, he excels in challenging environments. He holds a bachelor's degree in mechanical engineering and a master's degree in mechanical engineering with specialisation in CAD/CAM. He has earned recognition for his exceptional academic performance. His experience spans CAE, automation and advanced manufacturing. Himanshu possesses technical skills in engineering design, CNC programming and project management, utilising cutting-edge software to solve complex problems. A natural leader and team player, he thrives in collaborative environments. With a proven track record, Himanshu drives innovation and contributes significantly to the field.

Introduction

This book is based on the authors' extensive knowledge and experience of teaching Engineering and Manufacturing and is designed to be accessible. This is reflected in both the clear and concise content and the simple, clearly explained and purposeful learning features.

The chapters and section headings follow the structure of the specification. Key term boxes define important terms and phrases, and these are compiled in a glossary at the back of the book for easy reference.

Accessibility has also informed the page design, with numerous artworks and photographs providing visual references for the key concepts covered in the text.

Engineering common core component

The book has been designed to help learners develop the knowledge, understanding and practical skills required for the Engineering common core component of the following qualifications:
▶ T Level Technical Qualification in Maintenance, Installation and Repair for Engineering and Manufacturing (8712)
▶ T Level Technical Qualification in Engineering, Manufacturing, Processing and Control (8713)
▶ T Level Technical Qualification in Design and Development for Engineering and Manufacturing (8714).

It provides complete coverage of the specification's content, range and assessment objectives for the common core component, and learners will gain insight into a broad range of principles and practices that are relevant to the engineering sector. Specifically, they will learn:
▶ the job roles, responsibilities, attitudes and behaviours needed to work in the engineering and manufacturing sector
▶ essential maths and science for engineering and manufacturing
▶ materials and their properties
▶ mechanical, electrical, electronic and mechatronic principles
▶ control systems, quality management and health and safety within the industry
▶ business management skills, including stock management, project management and continuous improvement.

Occupational Specialisms

In order to achieve any of the Technical Qualifications listed above, learners need to complete any standalone Occupational Specialism from their chosen pathway in addition to the Engineering common core component.

Occupational Specialisms are designed to develop the knowledge, skills and behaviours required to enter a particular occupation within the engineering and manufacturing industry. **They are not covered in this book**.

Assessment

For the Engineering common core component, learners need to complete:
▶ two externally set exams covering knowledge from the Engineering common core
▶ an employer-set project covering knowledge and core skills from the Engineering common core.

Exams

Both exams include a mixture of short-answer and extended-response questions.

Each exam lasts 2.5 hours, is worth 100 marks and makes up 35% of the total assessment for the core component of the Technical Qualification (so 70% across the two papers).

Employer-set project

The employer-set project is based on a real industry brief. It is worth 90 marks and makes up 30% of the total assessment for the core component of the Technical Qualification.

Supporting the assessment

The book is designed to support assessment using a range of features:
▶ Test yourself: a range of knowledge-recall questions throughout the book to check learners' understanding
▶ Assessment practice: practice questions that allow learners to test their knowledge and understanding at the end of each chapter in preparation for the written exam
▶ Project practice: a range of assessment scenarios and focused tasks that allow learners to apply the skills and knowledge they have learned in each chapter to support their preparation for the employer-set project.

How to use this book

The following features can be found in this book.

Learning outcomes

Summaries of the core knowledge outcomes that you must understand and learn

Key terms

Definitions of important terms that you need to understand

Industry tips

Tips and advice to help you in the workplace

Research

Research-based activities: either stretch and challenge activities enabling you to go beyond the course, or industry placement-based activities encouraging you to discover more about your placement

Case study

Scenarios that place knowledge into a real-life context in order to introduce problem solving and dilemmas

Example

Worked examples to show you how to perform key calculations

Test yourself

A knowledge-consolidation feature containing questions and tasks to aid understanding and guide you to think about a topic in detail

Health and safety

Important points to ensure safety in the workplace

Improve your maths

Short activities that encourage you to apply and develop your functional maths skills in context

Improve your English

Short activities that encourage you to apply and develop your functional English skills in context

Assessment practice

Knowledge-based practice questions to help prepare you for the exam

Project practice

Short scenarios and focused activities reflecting one or more of the tasks that you will need to undertake during completion of the employer-set project

1 Working within the engineering and manufacturing sectors

Introduction

Modern life as we know it would be impossible without the efforts of millions of engineers and engineering technicians who apply scientific principles to find practical solutions to real-world problems. They are responsible for designing and manufacturing a staggeringly broad range of products across multiple sectors, from chemical engineering and pharmaceuticals to aviation and space exploration.

This chapter provides an overview of the engineering and manufacturing industry and explores the activities of engineers working in these sectors.

Learning outcomes

By the end of this chapter, you will understand:

1   how different types of manufacturing processes influence the design of engineered products
2   how different requirements affect the user and designs related to the manufacture of products
3   the steps of the linear and iterative design processes, and the contribution that testing makes to achieve a suitable and effective design
4   how to interpret anthropometric data
5   the role and purpose of maintenance, repair and installation
6   the advantages and disadvantages of different approaches to maintenance
7   the responsibilities of the different roles involved in maintenance
8   approaches to monitoring and the reasons for carrying out monitoring
9   the reasons for, and implications of, shutdown and servicing
10 an overview of the types of tools and equipment used
11 the reasons for commissioning activities
12 how effective maintenance reduces impact on the environment, and the safe and environmentally friendly disposal of waste
13 how the scale of manufacture affects the level of automation
14 examples of products made at different scales of manufacture
15 different types of manufacturing infrastructure, their purpose and relative advantages and limitations
16 the purpose and application of CAM systems and software
17 the advantages and limitations of different levels of automation.

1.1 Key principles and methodologies in engineering and manufacturing design

Types of manufacturing process

A wide range of processes are used in the manufacturing industry:

▶ **Wasting** involves the removal of material from a workpiece to produce a component of the required size and shape. For example, when drilling, the drill bit cuts material away to form a hole. The waste material generated is carried away from the cutting edges in the form of chips or swarf.

▶ **Forming** uses the application of force to change the shape of a material. For example, using a press tool to bend a length of flat steel bar into an L-shaped bracket.

▶ **Shaping** involves pouring or injecting liquid material into a mould where it then solidifies into the required shape.

▶ **Casting** is a shaping process which involves heating a material above its melting point and pouring the resulting liquid into a mould, where it then cools and solidifies into the required shape.

▶ **Joining** processes are used to fix two or more components together into a larger assembly. Examples include welding, where components are fused together permanently, or the use of nuts and bolts that allow disassembly for maintenance or repair.

▶ **Finishing** adds a coating to the outer surface of a component that protects it from corrosion. It can also improve aesthetics.

▶ **Additive manufacturing** describes a range of advanced processes able to create complex components directly from three-dimensional computer models. Often referred to as 3D printing, it involves the direct addition of material, usually in a series of thin layers which are gradually built up into the required shape. This differs from conventional wasting processes, which remove material from a workpiece to create the feature required.

Key terms

Wasting: removal of material from a workpiece to produce a component of the required size and shape.

Forming: application of force to change the shape of a material.

Shaping: pouring or injecting liquid material into a mould where it solidifies into the required shape.

Casting: a shaping process which involves heating a material above its melting point and pouring the resulting liquid into a mould, where it then cools and solidifies into the required shape.

Joining: a process used to fix two or more components together into a larger assembly.

Finishing: adding a coating to the outer surface of a component that protects it from corrosion. It can also improve aesthetics.

Additive manufacturing: creation of complex components using 3D computer models; material is added in a series of thin layers which are gradually built up into the required shape.

▲ Figure 1.1 In sand casting, molten metal is poured into a sand mould

Test yourself

State the type of manufacturing process for each of the following:
▶ 3D printing
▶ MIG welding
▶ drilling.

The influence of manufacturing process on product design

Engineered products and their components are designed to make their manufacture as easy and cost effective as possible. As such, the preferred manufacturing process or processes can have a considerable influence on design.

For example, a company that specialises in welding assemblies from flat steel components will naturally prefer this approach when asked to design a garden bench. In contrast, an iron foundry would be able to produce a far more decorative, flowing design in cast iron.

Often, design features are included that serve no other purpose than making manufacturing easier, or in some cases, making manufacturing even possible. For example, when casting into a reusable mould, it is essential that the component is designed with a slight taper or draft angle. Without this, the component would become stuck in the mould once it cooled and solidified.

On many sheet-metal components, you will find small holes placed along edges or in corners. Often these are hanging holes, which allow the bare metal components to be suspended on hooks for painting or powder coating during manufacture.

The design process

The first step in the design of a new product is for the client to write a **design brief**. This defines the purpose of the product and outlines all the needs that must be met in order for it to be **fit for purpose**. These will include the key **functional requirements** of the product, as well as other important factors, such as its impact on the environment, its aesthetic appeal or its cost.

There are several design methodologies that can be used to structure the development of a final design proposal that will meet this brief.

Linear design

A **linear design methodology** follows a series of sequential steps in order to generate a single final design solution.

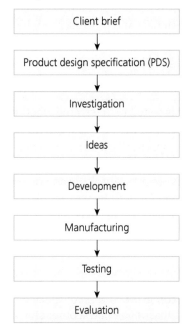

▲ Figure 1.2 The sequential steps of the linear design process

▶ Client brief – the client outlines their requirements and all the needs that must be met by the final design solution.

▶ **Product design specification (PDS)** – design engineers analyse the contents of the client brief and translate each requirement into specific and measurable criteria. This PDS is discussed and agreed with the client, and it describes in detail how designers will fulfil the needs laid out in the client brief. The PDS is used alongside the brief to determine whether the final design proposal is fit for purpose.

▶ Investigation – design engineers investigate ways of achieving each of the criteria in the PDS. This includes research into materials, processes, existing related products and applicable technologies.

▶ Ideas – designers generate many ideas to address each requirement of the PDS. This can include the use of thumbnail sketching, brainstorming and other ideation techniques.

▶ Development – designers take the best elements of the generated ideas and develop these into a design solution that will address the requirements of the PDS. At this stage, materials and processes will be defined and design drawings will be finalised for the completion of the product.

▶ Manufacturing – the physical product is manufactured.

▶ Testing – the product is tested to determine whether it fulfils the requirements of the PDS and the client brief. This is the first opportunity to see how a new product works in the real world.

▶ Evaluation – this is an opportunity at the end of the linear design process for designers to assess how well their solution performed against the requirements of the PDS and the client brief.

Key terms

Product design specification (PDS): a list of specific and measurable criteria that define how a product will meet the requirements of a design brief.

Test yourself

Name all the steps in the linear design process.

The principal weakness of the linear design process is that the first fully developed solution is adopted and used to manufacture the finished product. In practice, it is extremely unlikely that this initial solution will offer optimal performance, and the evaluation stage will uncover weaknesses that could and should be addressed.

Iterative design

An **iterative design methodology** uses similar steps to linear design but with two significant differences:

▶ Instead of development leading straight into product manufacture, a model or prototype is made and tested first.

▶ After testing, the evaluation stage generates suggestions for improvements that will bring the performance of the product closer to the needs identified in the PDS and the client brief.

Key term

Iterative design methodology: an iterative approach to the design process, with structured opportunities to improve the design solution over several design cycles.

The suggested improvements are then fed back into the investigation stage and the process is repeated in order to realise an improved design solution. This iterative cycle is repeated until the final design fulfils all the requirements of the PDS and the client brief, and no further improvement is necessary. This approach enables and encourages a culture of continuous improvement, which ultimately leads to better products and higher sales.

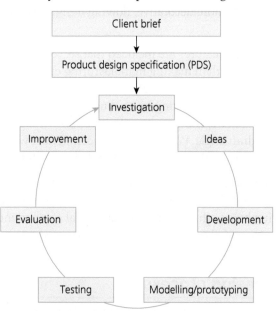

▲ Figure 1.3 The iterative design cycle

Test yourself

Explain the difference between a linear and an iterative design process.

Other approaches to design

There are several other approaches to the design process which emphasise a range of wider social, economic and environmental priorities:

▶ **Inclusive design** focuses on making design solutions that are accessible to the widest possible proportion of society, including those with physical or mental impairments.
▶ **Ergonomic design** ensures a product is safe and comfortable to use by taking into consideration both human behaviour and the size, strength and limitations of the body. This includes referring to **anthropometric data**, which provides guidance on typical measurements of different parts of the human body.
▶ **User-centred design** is an iterative approach that focuses on feedback from the end user of a product during its design and development.
▶ **Design for manufacture** ensures products incorporate features that support the use of a particular process in their manufacture. For example, the use of draft angles on injection-moulded components allows them to be easily released from the mould tooling during manufacture.
▶ **Design for assembly** ensures that assemblies can be put together easily, with intuitive positioning and alignment of each individual component. For example, asymmetric mounting holes on symmetrical components so they can only be fitted in the correct orientation, self-jigging features such as slots, and locating holes to allow quick and easy positioning of components.
▶ **Sustainable design** embeds sustainable principles in the design process to ensure that environmental impact is minimised through reduced use of energy, natural resources and materials.

Key terms

Inclusive design: design approach that focuses on making design solutions that are accessible to the widest possible proportion of society.

Ergonomic design: design approach that ensures a product is safe and comfortable to use by taking into consideration both human behaviour and the size, strength and limitations of the body.

Anthropometric data: research-based data that provides guidance on typical measurements for different parts of the human body.

User-centred design: iterative design approach that focuses on feedback from the end user of a product during its design and development.

Design for manufacture: design approach that ensures products incorporate features that support the use of a particular process in their manufacture.

Design for assembly: design approach that ensures assemblies can be put together easily.

Sustainable design: design approach that ensures environmental impact is minimised through reduced use of energy, natural resources and materials.

▼ Table 1.1 The 6 Rs of sustainable design

Reduce	Reduce the amount of material, energy and waste involved in manufacturing and using a product
Refuse	Refuse to use harmful or polluting materials or processes
Rethink	Rethink conventional, unsustainable approaches, for example the choice of materials or production processes
Repair	Design products that enable straightforward repair and refurbishment in order to extend their useful lives
Reuse	Design products that can be reused and avoid single-use disposable products
	Repurpose existing components, products or waste materials
Recycle	Use recycled materials and/or ensure materials are recyclable
	The design should allow easy separation of different material types when the product reaches the end of its useful life

Research

Research is an important way of gathering information that will inform the design of a new product through all stages of its development.

At an early stage, market research is carried out to establish if there is sufficient interest in an idea for a new product to justify its development. The use of interviews and questionnaires can also help to gauge the relative importance of the features and capabilities of the product among potential customers. Useful insights from this process will be incorporated into the client brief.

When a prototype or early production model is available, further market research and user experience testing are used to determine the strengths and weaknesses of the design from the point of view of the end user. Responses can be fed back into the iterative design cycle and used to inform improvements that will make the product more attractive to potential customers. This process might be repeated several times and helps to maximise the commercial success of a new product by ensuring it fulfils the expectations of its target market.

Using anthropometric data

Research on body sizes provides anthropometric data that plays a key role in ergonomic design. It enables designers to understand the different shapes and sizes of the human body, which is vital information for a product which physically interacts with the user.

When summarised in frequency distribution graphs, anthropometric data shows how common specific body measurements are across a population. Anthropometric data is normally distributed, and graphs of measurement versus frequency show the classic bell-shaped curve.

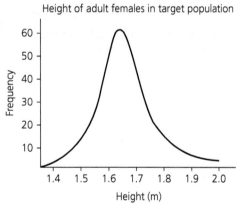

▲ Figure 1.4 Graph showing frequency distribution

Clothing manufacturers rely on accurate anthropometric data to determine the size ranges for their products. Each size is designed to accommodate a narrow range of body measurements. The full range of sizes, typically from extra small (XS) to extra large (XL), accommodate the majority of people in the anthropometric data distribution.

The population at the smallest and largest extremes of the data distribution are usually outside the range of standard sizes supplied by manufacturers. The demand for sizes smaller than XS or larger than XL is low, so it may not be commercially viable to manufacture clothes in these sizes.

Other products, such as chairs, adopt a one-size-fits-all approach. Here, the shape, strength and height of the product are designed to accommodate a wide range of body shapes, sizes and weights. For example, if we consider the design of a chair, then it is reasonable to make it strong enough to accommodate bodyweight at the 99th percentile of the anthropometric data distribution. That way, 99 per cent of the population will be able to use it and the number of potential customers will be maximised. If the chair was designed to accommodate bodyweight at the 50th percentile, or median, then it would be less expensive to make but only half of the population could safely use it.

> **Test yourself**
>
> Explain how anthropometric data might be used in the design of an office desk.

Testing

Testing is a vital part of the design process and ensures that design solutions are suitable, effective and safe.

Tests are designed to check that each element of the client brief and the PDS has been fulfilled. They are conducted using a range of methodologies:

- ▶ Functional testing confirms that each of the functional requirements of the product is fulfilled. This ensures that the product's main purpose is achieved and that it can perform the actions it was designed to complete. Functional tests are performed by engineers who are comfortable with the technical aspects of the design and who are able to conduct complex testing and interpret the results.
- ▶ User-experience testing is based on field tests, where products are used in their normal working environment by real end users. This provides feedback on more human-focused reactions to a product: whether it is attractive, comfortable, useful, intuitive to operate, and whether it lives up to the expectations of the user. Users are generally non-technical, with little or no knowledge of the technical aspects of the design. As such, the tests they are required to complete must be explained clearly and any questions they are asked must avoid using jargon or complex technical terms.
- ▶ Safety testing is an essential part of product design and manufacture. The tests range from basic visual checks for sharp edges on individual components to full destructive crash tests on new car designs.

▶ Compliance testing ensures that a product complies with any applicable safety or quality assurance standards that might be required for a product to be sold in a particular market. For example, CE marking is used to indicate goods sold in the European Economic Area (EEA) that have been tested and comply with basic safety standards.

Communicating with different audiences

Market research, testing and other design activities require input from both technical and non-technical audiences. Different approaches must be taken when communicating with these two groups.

For example, technical audiences are likely to be interested in exactly how certain aspects of a complex product, for example a mobile phone or wireless earbuds, function in granular detail. They will want to drill down into performance data and technical specifications, and will be able to interpret technical language.

In contrast, most end users of complex products will have little or no insight into the technical aspects of how these products work. However, they still need to be engaged in order to provide essential feedback on how a product looks, feels and operates in real-world conditions. For non-technical audiences, complex information needs to be summarised and simplified. Often this information is made more visually engaging by expressing it graphically using images, charts or diagrams. Where explanatory text is needed, this should be written in plain language without the use of jargon or complex technical terms.

> **Test yourself**
>
> Name two things that should be avoided when communicating with non-technical audiences.

1.2 The role of maintenance, repair and installation in engineering

Installation and commissioning

Before new equipment arrives on site, preparations must be made by the local maintenance team to ensure that:
▶ Any disruption to the normal running of the factory is minimised.
▶ Access is large enough for the equipment to enter the factory.

▶ The site for the equipment is large enough and the floor is strong enough to support it.
▶ **Provision of services** has been organised and prepared for connection to the new equipment.

Installation is usually completed by specialist teams trained by the equipment manufacturer. This ensures that the equipment is transported, positioned, mounted, assembled and connected to services safely, efficiently and according to the manufacturer's recommendations.

Following installation, the equipment is commissioned. **Commissioning** is the process of getting everything up and running, and training staff to use and maintain the equipment safely and effectively. This will help to prevent accidents, minimise breakdowns and extend the service life of the equipment.

Once everything is operating smoothly and the client is satisfied with the performance of the equipment, they will sign off the installation and accept responsibility for the equipment from the installers.

> **Key terms**
>
> **Provision of services:** supply and connection of electricity, gas, ventilation, compressed air or water to equipment or machinery.
>
> **Commissioning:** process of getting a newly installed machine or piece of equipment up and running, ready for handover to the customer.

> **Health and safety**
>
> An important aspect of installing and commissioning a new piece of equipment is to review and update the relevant health and safety risk assessments to ensure that any new hazards or increased risks are managed effectively.

Maintenance

Maintenance encompasses all the activities involved in the installation, commissioning and ongoing care of machines and equipment that ensure they are kept in good working order.

Effective maintenance:
▶ maximises the service life of machinery and equipment so that it lasts longer and needs to be replaced less frequently

- minimises downtime caused by unplanned breakdowns, thereby avoiding costly delays in production
- ensures equipment runs correctly and efficiently
- reduces impact on the environment by ensuring waste materials are managed safely and disposed of responsibly.

> **Test yourself**
>
> Explain the importance of effective maintenance.

Types of maintenance

Planning maintenance activities is essential in order to minimise equipment downtime and any disruption that this may cause. Ways of organising ongoing planned maintenance include:

- **reactive maintenance**
- **preventative maintenance**
- **condition-based maintenance**.

> **Key terms**
>
> *Reactive maintenance:* maintenance carried out when there is an equipment failure.
>
> *Preventative maintenance:* maintenance that uses manufacturer-recommended fixed service schedules to regularly repair or replace components prone to wear, before they fail.
>
> *Condition-based maintenance:* maintenance enabled by modern sensor and communication technology that allows real-time monitoring of the condition of complex equipment.

Reactive maintenance

Reactive maintenance is carried out when there is an equipment failure. This requires contingency planning to ensure that equipment can be replaced or repaired as quickly as possible when it fails. This makes sense in situations where replacement or repair is fast, simple and inexpensive. For example, when drilling holes, drill bits occasionally break. Having a stock of drill bits on hand and quickly replacing the broken bit is an example of reactive maintenance.

Preventative maintenance

Preventative maintenance is generally used on large complex systems with many moving parts. It uses manufacturer-recommended fixed service schedules to regularly repair or replace components that are prone to wear, before they fail. It also ensures that consumable items, such as lubricants or filters, are replaced regularly. This minimises the likelihood of serious equipment failure and any associated unscheduled downtime required to carry out repairs.

For example, trains are complex systems that include dozens of mechanical and electrical systems. Train breakdowns can cause massive disruption to rail services and so must be avoided. In this case, it is cost effective and convenient to schedule regular preventative maintenance tasks to be carried out quickly in specialist facilities, rather than deal with an increase in in-service breakdowns.

Condition-based maintenance

Condition-based maintenance is enabled by modern sensor and communication technology that allows real-time monitoring of the condition of complex equipment. Data on vibration, temperature and operating conditions is recorded and monitored over time. Analysis of this data can indicate exactly when servicing or component replacement is required. This means that parts in good working order, and which show no signs of wear, are not replaced, which they may have been as a matter of routine in basic preventative maintenance schedules. It also means that parts that have worn prematurely between scheduled maintenance cycles are identified and replaced.

For example, Rolls Royce monitors real-time data from its jet engines in flight around the world. Analysis of this data is used to identify early signs of potential problems and indicates when maintenance procedures need to be carried out on specific engines. This approach minimises any chance of engine failure and increases the efficiency and effectiveness of maintenance activities.

> **Test yourself**
>
> Explain the difference between reactive and preventative maintenance.

Maintenance operations

Maintenance operations are practical actions carried out by maintenance staff and machine operators to keep equipment and machinery in good working condition. They include the following:

▶ Monitoring the condition of equipment and machinery: this can be as simple as the operator performing a basic visual inspection of their machine at the start of each shift. More complex monitoring strategies involve using sensors to monitor and record key operating characteristics, such as temperature, vibration or oil pressure. Any change in these might be indicative of an emerging problem that warrants investigation.

▶ Shutdown: this involves taking a machine or piece of equipment out of service so that it can be worked on. It is more than just switching it off. To ensure the safety of maintenance engineers, all the services connected to the equipment must be isolated and locked off, including electricity, gas, water or compressed air. Next, any stored energy in the equipment itself must be released. This ensures that everything is completely de-energised and can be dismantled and worked on safely.

▶ Servicing: this is usually a scheduled set of procedures required to care for a machine and keep it in good working order. It can include replacing consumable items like filters, topping up fluid levels and lubricating working parts.

▶ Repair: this is often an unplanned response to a breakdown that requires the replacement of worn or broken parts.

Key term

Maintenance operations: practical actions carried out by maintenance staff and machine operators in order to keep equipment and machinery in good working condition.

Health and safety

When isolating machinery from the electricity supply, maintenance engineers will use a lock out/ tag out system. This involves fitting padlocks to the isolator switches so that they cannot be turned back on by mistake while the maintenance engineers are working.

Test yourself

Explain the importance of safe shutdown procedures when carrying out maintenance tasks.

Roles and functions in maintenance operations

A range of roles are involved in maintenance operations:

▶ Machine operators are responsible for the day-to-day operation of machines and equipment. They conduct basic daily maintenance routines, such as lubrication, monitoring fluid levels and checking for damage or wear. Operators familiar with their machines and equipment will often be the first to notice signs of potential problems. They report any immediate concerns they have about the condition, operation or safety of the equipment to their supervisors or the maintenance manager.

▶ Maintenance engineers conduct the maintenance and repair tasks required by planned maintenance schedules or in response to breakdowns. This includes the safe shutdown of complex systems so that they are safe to work on.

▶ Maintenance managers are responsible for planning and organising maintenance activities, and for putting appropriate systems in place to ensure time lost to planned maintenance and unplanned, disruptive breakdowns is minimised. Many management decisions are driven by data collected during condition monitoring.

Maintenance tools and equipment

A wide range of general-purpose tools and equipment is required when conducting maintenance operations. These tools and equipment must cover both mechanical and electrical tasks, and include:

▶ portable power tools (for example drills, angle grinders, impact drivers and soldering irons)

▶ hand tools (for example spanners, screwdrivers, sockets, combination pliers, hacksaws, wire strippers, wire cutters and hammers)

▶ measuring instruments and gauges (for example tape measures, steel rules, digital callipers, feeler gauges, infrared thermometers, pressure gauges, flow meters and multimeters).

Developments in maintenance

Environmental concerns are increasingly at the forefront of people's minds. Effective maintenance is essential in order to protect the environment by reducing energy usage, minimising emissions and preventing leaks of oil, hydraulic fluid or other chemicals. Even the impact of old equipment can be reduced by regular servicing and changing over to modern, low-toxicity, long-life, synthetic lubricants, coolants and other working fluids. These can reduce the amount of waste produced during routine maintenance and lessen the environmental impact following disposal.

As mentioned earlier, the development of internet-enabled sensor technology allows remote condition monitoring of complex systems. This supports real-time analysis of a wide range of key performance parameters, enabling greater efficiency and effectiveness of condition-based maintenance.

Manufacturers of complex machines can now use artificial intelligence to analyse vast amounts of performance monitoring data, allowing them to predict with ever greater accuracy when maintenance procedures are required. This approach further minimises costs while maintaining reliability, optimising maintenance activities and reducing downtime.

Case study

Rolls Royce monitors the performance and condition of its jet engines in real time around the world. Its latest aircraft engines are equipped with engine health monitoring systems (EHMS) – advanced systems of onboard sensors and communications equipment that monitor the condition of each engine and relay that information to a ground-based engineering centre. Engineers on the ground are even able to reconfigure the engine sensors remotely and run diagnostic tests with the engine still in service. All this is aimed at maximising aircraft availability by minimising time spent on the ground for maintenance and repair.

Questions

1 Why is effective maintenance vital in the aerospace industry?
2 What new technologies does Rolls Royce rely on to carry out the condition-based monitoring of its jet engines?

▲ Figure 1.5 Rolls Royce jet engine

1.3 Approaches to manufacturing, processing and control

general purpose tools and very little automation. In contrast, mass production, where thousands of the same item are made, requires low-skilled workers, highly specialised tools and high levels of automation.

Scale of manufacture

The number of products being made has a strong influence on the level of automation used in their manufacture. Making one-off products, where each piece is different, requires highly skilled workers,

▼ Table 1.2 Comparison of different scales of manufacture

	One-off	Batch	Mass	Continuous
Batch size	1	2 to 1000	1000 to 100000	100000+
Set-up cost	Low	Medium	High	High
Specialist tools	None	Some	Most	All
Skill	High	High/Medium	Low	Low
Automation	None	Some	Most	All
Automation technology	Manual	Manual, CAD/CAM	Robotic	Fully automated
Cost per product	High	Medium	Low	Very low
Flexibility	High	High/Medium	Very low	None
Lead time	High	Medium	Low	Very low
Variation	High	High/Medium	Very low	None
Examples	Bespoke furniture	Bread rolls	Washing machines	Nails
	Custom car exhaust	Classic car replacement chassis	Mobile phones	Wood screws
				Chemicals

Manufacturing infrastructure

The layout of a factory will depend on the scale of manufacture and manufacturing methodologies being employed.

Product layout (or production line)

The factory layout that most people are familiar with is the production line used in mass production. This approach, pioneered by Henry Ford, breaks down the complex task of building a sophisticated product into a series of simple steps.

Materials, tools and equipment required for each step are arranged in a long line of workstations. The product itself moves slowly along this line and is assembled piece by piece as it progresses. At each workstation, a semi-skilled worker uses a small number of specialist tools and/or automated equipment to carry out the same task on each passing product.

Specialising in manufacturing a single product in this way results in a low manufacturing cost per item, due

to low labour costs and high production rates. This means products can be produced quickly and cheaply. However, the facilities, specialist tools and high levels of automation required are expensive, time consuming to set up and lack the flexibility needed to cope with any variation in the product being manufactured.

▲ Figure 1.6 Production line at the Ford Motor Company, 1929

▲ Figure 1.7 A modern Ford production line

Functional layout

In a functional layout, similar equipment and skilled workers are arranged into workstations based on process type. For example, welders and welding equipment are in one place, and machinists and lathes are in another. In this model, workflow is routed between workstations until all the required operations are completed.

This is a flexible system, using general-purpose equipment with little automation, that works well in batch production where there is a high degree of variation in the products being manufactured. However, work-in-progress spends significant time being moved between workstations and waiting to be processed. The highly skilled workers required also drives up costs in comparison to a high-volume, low-skilled product layout.

Cellular layout

Cellular layouts use small, self-contained manufacturing cells for the batch production of variants within a family of similar components or products.

Sequential workstations are usually arranged in a horseshoe shape to enable workers to move around the cell quickly and easily to complete multiple tasks. This eliminates much of the movement of work-in-progress and waiting inherent in a functional layout.

This approach maintains flexibility by using highly trained workers who are familiar with all the product variants that they might be asked to manufacture and who also know how to complete any of the operations needed. This ensures that changeovers from making one variant to another are quick and easy.

So, partial specialisation enables some of the advantages of a product layout, while multi-skilling and rapid changeovers enable some of the advantages of a flexible functional layout. This means that customers can buy a wide range of product variants with short lead times, without the manufacturer having the expense of maintaining a stock of finished goods for each type.

Matrix arrangement

Matrix arrangements provide additional flexibility and capability to a cellular layout by allowing some products to be routed between individual processes or machines in different product cells. This is sometimes necessary in order to allow better utilisation of high-cost equipment or in cases where complex or potentially hazardous processes are used.

Levels of automation

Manual systems

Completely manual systems do not use automation. They provide a very high level of flexibility but also demand a high level of skill from workers and tend to be slow and expensive.

CAD/CAM systems

CAD/CAM systems integrate computer-aided design (CAD) and computer-aided manufacturing (CAM) and can be used to automate a manufacturing process.

Most engineered products are designed using CAD software to create three-dimensional models and associated engineering drawings of their components. This information is then used to generate the necessary programming code to instruct a computer numerical control (CNC) machine to manufacture a component.

CAD/CAM systems are often used in batch production in order to machine mechanical components on a CNC lathe or milling machine, and they have numerous advantages over manual systems:
- Accuracy is achieved consistently for every component, as it is not dependent on the skill of an individual machinist.
- Speed of manufacture is increased, as tool repositioning and changeover is automated.
- Capacity is increased, as CNC machines can operate 24 hours a day without breaks or holidays.

Robotic systems

Robotic systems are used where a repeated sequence of movements is required when manufacturing a large batch of products. Movements are carefully choreographed, and programming these systems can be complex and time consuming.

During the lifetime of a product, there are often small changes in design or manufacturing processes. Robotic systems are flexible and programs can be updated to accommodate change. This flexibility also means that robotic systems can be completely repurposed and reprogrammed to manufacture a different product when required.

A robotic arm can position an attached accessory anywhere within its **operational envelope** quickly and with pinpoint precision to complete a wide range of operations. Attachments include manipulators, spray-painting guns, welders and measuring equipment. Applications include relatively simple tasks such as picking up finished components and placing them in packing boxes, or much more complex tasks like spot welding pressed-steel components together into car bodies.

Key term

Operational envelope: area of three-dimensional space representing the maximum extent or reach of a robotic arm.

▲ Figure 1.8 Welding robots in the automotive industry

Health and safety

Robotic systems are able to operate in hazardous environments that would be far too dangerous for a human to work in.

Test yourself

Explain why robotic vehicles are widely used by bomb disposal teams.

Fully automated systems

Fully automated systems are used in continuous manufacturing, where a system or machine is built around the manufacture of a single product. The same product is manufactured for the entire service life of the system or machine. For example:

▶ Chemical products such as pharmaceuticals and petrochemicals are manufactured using dedicated automatic systems that are never switched off.
▶ High-volume consumable products such as nails, machine screws and springs are manufactured on dedicated automatic machines.

Control

Complex manufacturing systems demand a high degree of control to ensure they function efficiently and manufacture products and components of the right quality, in the required quantity and at the right time.

Performance monitoring

Essential to the control of manufacturing systems is the collection of key performance data, such as production rates or the number of product defects identified over time. This data forms the basis of performance management systems such as statistical process control (SPC).

SPC uses statistical tools to analyse data and reveal trends that cannot be explained by the random small variations present in any complex system. An unexplained rise in defects in the components made on a particular machine is an early warning sign and will prompt an investigation to identify, and then correct, the cause.

This proactive approach helps identify and address emerging problems before they can have a significant impact on manufacturing output or quality.

Quality control (QC)

At several points during the manufacture of a product, quality control (QC) checks are carried out to ensure compliance with the required specification. The earlier a defect in a product is identified, the easier and less expensive it is to put right.

Defective products reaching the customer must be avoided. Replacing or recalling faulty goods has significant immediate cost implications, but the long-term financial damage caused by gaining a reputation for poor quality can prove fatal to a business.

Test yourself

Explain why a faulty product that reaches a customer is much worse than one identified at the factory.

QC checks are carried out at the following points:
- Goods-inwards inspections are carried out on components and raw materials as they arrive in the factory. These checks are vital to ensure that goods coming in comply with their purchasing specifications, are fit for purpose and can be used in the manufacture of products.
- In-process inspections are carried out after each manufacturing operation. These ensure that any manufacturing error is identified as quickly as possible and that any defects are quarantined and not used in subsequent processes.
- Finished-goods inspections are the final checks made before a product goes to the customer. It is the last opportunity for any problems to be identified and corrected.

There are different types of QC check:
- Visual inspection involves carefully looking over a product or component. This simplest of checks can pick up a range of issues, from scratches and other forms of damage, to missing or poorly fitted components.
- Dimensional checks are measurements of the physical size of components to ensure they have been manufactured to within required tolerances.

- Functional checks are carried out to ensure that component assemblies function as intended. This could be a mechanical check (for example measuring the loading characteristics of a completed car shock absorber) or an electrical check (for example ensuring a PCB powers up and functions as expected).

Industry tip

Maintaining the quality of products and processes is a collective responsibility for all the employees in an organisation. Remain vigilant, and always report any problems or quality issues that you encounter in the workplace.

Test yourself

State at least five quality checks that will be carried out on a washing machine before it leaves the factory.

Quality assurance (QA)

Quality assurance (QA) involves much more than just carrying out QC checks. QA is an integrated approach to monitoring, maintaining and improving the quality of manufactured products.

Typical activities undertaken in an effective QA system include:
- performance monitoring
- QC checks
- standardisation of work methods to ensure all staff carry out manufacturing operations in the same way to avoid inconsistencies in the product
- planning what, when, where and how often QC inspection measurements are taken
- analysis of QC data using systems like SPC to identify areas of concern
- identification of the root cause(s) of quality issues and the redesign of products or processes to eliminate those issues.

Assessment practice

1 Powder coating is a common finishing process. Give two reasons why powder coating might be used on a steel bicycle frame.

2 Explain the differences between a linear and an iterative approach to design.

3 An external staircase used as a fire escape is manufactured by a company specialising in welding and fabrication. Explain one way in which design for manufacture and assembly may have influenced the design of the staircase.

4 A company manufactures mechanical assemblies for use on offshore wind turbines. Identify one way in which design for assembly techniques can be applied to reduce errors during production.

5 A company has decided to review the sustainability of its product range. Explain how considering each of the 6 Rs of sustainable design might help it to come up with new ideas.

6 Explain two ways in which the effective maintenance of machinery and equipment reduces negative impacts on the environment.

7 Explain the advantages of using automated processes when manufacturing large batches of the same product in mass production.

8 Explain why complex prototype components are often manufactured in a polymer material using an additive manufacturing process such as 3D printing.

9 Quality control (QC) inspections are carried out at several points during the manufacture of a product. Explain why goods-inwards inspection of purchased components and raw materials is important.

10 Explain the difference between quality control (QC) and quality assurance (QA).

Project practice

You work as a maintenance engineer for a manufacturing company. The company has just installed and commissioned five new centre lathes in its apprentice training centre. These machines will require regular maintenance to ensure they remain in good working order.

Write a short report to explain how these machines will be maintained. Your report should include:

▶ an explanation of the different types of maintenance
▶ a fully justified recommendation of the most appropriate type of maintenance to use for the lathes
▶ a description of how the system of maintenance will be organised
▶ a description of the roles and responsibilities of those involved.

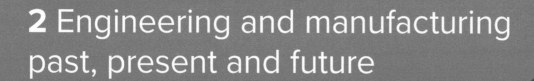

2 Engineering and manufacturing past, present and future

Introduction

In the 300 years since mechanisation brought about the first industrial revolution, engineering has underpinned every aspect of human endeavour. Today, industry and wider society are adopting and adapting to technological innovation at unprecedented speed.

In this chapter, you will learn about the activities undertaken in a wide range of engineering sectors. You will also reflect on some of the key developments that brought us to this point and how new and emerging technologies might shape the future.

Learning outcomes

By the end of this chapter, you will understand:
1 the main activities, products and/or services relating to different sectors of the engineering industry
2 how technological advances and their operations have evolved and contributed to engineering and social and economic development, in areas such as transportation, healthcare, housing, employment and sustainability
3 how innovation and emerging trends are evolving and could influence manufacturing, environmental considerations and social and economic development.

2.1 Sectors of the engineering industry

The engineering industry encompasses a diverse range of specialist sectors. The activities, goods and/or services relating to some of these sectors are outlined in Table 2.1.

▲ Figure 2.1 Aircraft maintenance is conducted by the aerospace sector

▼ Table 2.1 Sectors of the engineering industry

Engineering sector	Main activities, products and/or services
Aerospace	Research and development, design, manufacture and maintenance of manned and unmanned aircraft, rockets, missiles, satellites and spacecraft
Rail	Research and development, design, manufacture and maintenance of rolling stock and rail infrastructure
Agriculture	Research and development, design, manufacture and maintenance of specialist equipment and machinery used in forestry, horticulture and farming
Automotive	Research and development, design, manufacture and maintenance of cars, motorbikes, trucks and other road-going vehicles
Chemical	Research and development, design, manufacture and maintenance of industrial chemical processes, processing plants and equipment

Applications across multiple sectors, including chemical manufacturing, petrochemicals, electronics, pharmaceuticals, food processing and healthcare |

Engineering sector	Main activities, products and/or services
Structural	Research, development and design of load-bearing structures, such as buildings and bridges
Materials	Research, development and manufacture of metals, ceramics, polymers and composite materials used across engineering and a wide range of other sectors
Logistics	Research, development and design of systems to optimise the processes and activities involved in supply-chain management, such as purchasing, storage, warehousing and distribution
Defence	Research, development, design, manufacture and maintenance of a wide range of equipment and technologies involved in maintaining national security and equipping the armed forces

Includes everything from communications, munitions and body armour, to tanks, submarines and combat aircraft |
Electrical and electronic control	Design, manufacture and maintenance of control systems, instrumentation, monitoring and automation for electro-mechanical engineering systems used in a range of sectors
Medical	Research and development, design and manufacture of instruments, equipment, machines and devices for use in healthcare, including heart monitors, MRI machines, surgical instruments and prosthetics
Manufacturing	Research and development, design, manufacture and maintenance of processes, systems, equipment and machinery involved in manufacturing a wide range of products in a range of sectors
Marine	Research, development, design, manufacture and maintenance of oil rigs, offshore wind installations, ships, submarines and other sea-going vessels
Petrochemical	Design, manufacture and maintenance of machinery and equipment used for oil exploration, extraction, transport, processing and refining

Engineering sector	Main activities, products and/or services
Power generation (renewables)	Research, development, design, manufacture and maintenance of infrastructure, machinery and equipment used in generating electricity from renewable resources, including photovoltaic solar, wind turbine, hydroelectric, tidal and geothermal systems
Power generation (non-renewables)	Research, development, design, manufacture and maintenance of infrastructure, machinery and equipment used in generating electricity from non-renewable fossil fuels, such as coal, gas and oil
Power generation (nuclear)	Research, development, design, manufacture and maintenance of infrastructure, machinery, equipment and waste-management technologies used in generating electricity from nuclear fuel
Telecommunications	Research, development, design, manufacture and maintenance of telecommunications infrastructure and equipment, including fixed and mobile telephone networks, fibre-optic broadband networks, satellite communications, television and radio
Water and waste management	Research, development, design, manufacture and maintenance of water and wastewater treatment infrastructure, machines and equipment, including the water distribution network and the collection, treatment and reuse of wastewater

Industry tip

Many engineering sectors overlap or are interlinked. Whatever sector you work in, you will deal with a wide range of products, services and colleagues from across the engineering industry.

Test yourself

A company designs control systems and instrumentation to be installed in offshore wind turbines. What engineering sector does it belong to?

Research

Choose an engineering sector you are interested in and conduct your own detailed research into the products and services it provides. Find out about potential employment opportunities in that sector and the qualifications you would need in order to gain employment.

2.2 Significant technological advances in engineering from a historical perspective

Development of materials

Throughout history, human progress has been shaped by taking advantage of the materials available to us. Early humans spent hundreds of thousands of years reliant on just the natural materials they found in their local environment. In contrast, we are now in a period where our understanding of material science, processing and manufacturing means that we can call upon thousands of materials with a massive range of properties.

Some materials have had a significant impact on the development of modern society:

▶ In 1855, the development of the **Bessemer process** allowed the cost-effective mass production of steel. Steel is the most widely used metal in engineering and has perhaps had the biggest impact on industrial development. It has allowed us to build railways, bridges, dams and other structures that form essential infrastructure and dominate the built environment. Steel is essential in the manufacture of machine tools, cars, ships and any product or structure held together with screws, nails or nuts and bolts.

▶ In 1907, Bakelite became the world's first synthetic polymer. However, it was not until the 1960s that the use of polymers became preferred over traditional materials and they became ubiquitous. Today, polymers can be found in a massive array of applications across all engineering sectors.

Key term

Bessemer process: method of producing steel by burning off carbon and other impurities in pig iron by blasting air through the molten metal.

▶ Ceramics are among the oldest types of materials used by humans. They include glass and ceramic pottery, both of which have had important applications in cookware and food storage for centuries. Modern ceramic materials, such as tungsten carbide, have important applications in manufacturing, where they are used in cutting tools and abrasives.

Electrical power

The harnessing of electrical power underpins much of the scientific progress and societal and technological changes that have occurred in the last hundred years.

Today, we tend to take the supply of electricity for granted, and it is easy to forget that the widespread use of electrical power is very recent. The world's first large-scale, coal-fired power station was opened in London in 1882, but extensive availability of electrical power did not happen until well into the twentieth century. It was not until 1935 that the world's first integrated national grid was completed, connecting power stations and electricity customers, and enabling widespread distribution of electricity in the UK. By 1940, two thirds of homes in the UK, mostly concentrated in urban areas, had a supply of electricity.

One of the first practical applications for electricity was to provide a safe and clean form of artificial light (this is explored further in the next section). Soon after, a wide range of domestic labour-saving appliances were invented, designed to automate or assist in an array of everyday household chores. These included washing machines, electric irons and vacuum cleaners.

Other domestic appliances, such as refrigerators and freezers, allowed food to be preserved for longer. This led to changes in shopping habits, as it was no longer necessary to shop every day for fresh ingredients. Food from larger weekly shops could be refrigerated until needed.

Refrigerated transport led to more fresh fruit, vegetables and fish being available in large cities, which in turn led to improvements in diet and general health.

Electricity also made consumer electronics possible, for example radios, televisions and, more recently, computers and the internet.

The impact of these technologies is explored later, but their importance is such that all technologically advanced societies have become entirely dependent on a reliable and accessible supply of electrical power. Modern life as we know it could not exist without it.

▲ Figure 2.2 The National Grid distributes electricity throughout the UK

Electrical sources of artificial lighting

Prior to the availability of reliable electric lights in 1878, street lighting in major cities and large towns ran on gas. Gas lighting was also used in commercial buildings and some large homes, but most of the population still relied on candles and oil lamps for lighting, and on open coal fires for warmth.

Domestic electric lighting began to replace gas as more homes were connected to the electricity grid. This revolutionised home life, which was no longer constrained by seasonal daylight hours.

In factories, electric lighting allowed manufacturing to continue throughout the night, enabled the introduction of shift work, and increased both capacity and industrial output.

The internal combustion engine

From its invention in the 1860s, the **internal combustion engine** began to replace the steam engines used to drive machinery and run electrical generators in factories and light industry. By the 1890s, the technology had matured sufficiently to allow Carl Benz to launch the first commercially successful motor car powered by an internal combustion engine. However, it was not until Henry Ford introduced the affordable, dependable and easy-to-drive Model T in 1908 that motoring became widely accessible to ordinary families.

> ### Key term
>
> **Internal combustion engine:** engine where fuel is burned inside the engine itself.

In the first half of the twentieth century, the socio-economic impact of the motor car was enormous. It provided ordinary people with unprecedented freedom to travel wherever and whenever they wanted, without reliance on public transport. People were able to travel for work, visit friends and relatives out of town, or take trips to popular tourist destinations, helping to grow the economy in the process.

Roads and transport infrastructure spread across the country, allowing freight to be transported quickly and easily in vans and trucks, thereby improving access to goods and services.

In the fields, tractors and farm machinery using internal combustion engines soon replaced horses and steam-driven traction engines to power agriculture, leading to greater efficiency and increases in food production.

An enormous industry emerged to satisfy the demand for cars, which itself helped to grow the economy and provided well-paid jobs for thousands of factory workers. The need for cheap and plentiful petrol led to the development of the worldwide oil industry, which still dominates many economies. Demand increased for steel, rubber and other raw materials that were essential for car manufacturing, and the size of the industries that supplied these materials increased too.

However, the unfortunate legacy of the widespread use of internal combustion engines is that they are responsible for around 20 per cent of global carbon-dioxide emissions. That makes them a major contributor to climate change and will ultimately lead to their decline as they are replaced with electric vehicles.

> **Test yourself**
>
> What effect did the internal combustion engine have on agriculture?

Electric motors

The first electric motors were developed as early as the 1830s. However, it was not until the 1890s that practical and powerful electric motors found applications in electrically operated trams and lifts.

As the electricity network began to spread in the first half of the twentieth century, electric motors began to replace steam engines in factories. Up until then, machinery was driven by belts running on a complex system of overhead line shafts powered by a single steam engine. These needed constant care and maintenance and suffered large losses due to friction in the shaft bearings and belts.

The transition to electric motors ended the need for line shafts and belts, and meant that each piece of equipment could be operated independently. Electric motors required much less maintenance and were more efficient and reliable, which enabled a significant increase in manufacturing output.

Today, electric motors find applications in an enormous range of products, for example providing haptic feedback in game controllers, powering electric vehicles and running the compressors in refrigeration equipment.

Replaceable parts and mass production

At the start of the industrial revolution, tools, equipment and machinery were manufactured in workshops that produced every component part, right down to individual nuts and bolts.

The development of machine tools, measuring equipment and agreed standards for screw threads enabled the standardisation of mechanical components in the early 1800s.

One of the first examples of mass production using replaceable, interchangeable parts was in the manufacture of muskets. These needed to be inexpensive and easy to manufacture, maintain and repair. Previously, muskets were made individually by skilled gunsmiths, with each part made to fit a specific gun. However, this meant that they could only be maintained and repaired by those gunsmiths, who would have to make and fit any replacement parts needed.

By the 1830s, several manufacturers had succeeded in achieving 100 per cent interchangeability of all the parts used to make their muskets. This resulted in higher production rates, lower costs and much easier maintenance and repair, so that more muskets were available and easy to keep serviceable by conducting field repairs using interchangeable spare parts.

These ideas soon spread to other industries. The mass production of complex products, such as domestic sewing machines and motor cars, was heavily dependent on the interchangeability of parts.

Television

Following on from its pioneering work with radio, the BBC began broadcasting regular black and white television programs in 1936.

After the Second World War (during which television broadcasts were suspended in the UK), television became a popular alternative to radio for affluent and middle-class families. However, the expense of new television sets meant that many families still relied on their radios for news and entertainment.

Television radically changed how people spent their leisure time. For young people especially, time in front of the television soon took the place of reading or playing outdoors as their favoured pastime.

As well as programming aimed at children and to entertain family audiences, television brought educational documentaries, news and current affairs into the home. Indeed, the declared mission of the BBC remains to inform, educate and entertain.

By the 1960s, most households had a television set, and in 1969, colour television was available on all three UK terrestrial channels (BBC1, BBC2 and ITV).

Since then, there have been other key milestones in the history of television:

- Satellite television was first available in 1989 with the launch of Sky.
- Internet-based, on-demand television services were first available in 2006.
- Internet-streaming service Netflix was launched in 2012.

The technology used in televisions has come a long way since the first sets were introduced. Early television was dependent on the **thermionic valve**, an electronic device used to amplify electrical signals and a forerunner to the transistor.

They also used **cathode ray tubes (CRTs)** to create the image on a small screen on the black and white sets of the 1950s. CRTs direct a beam of electrons onto a fluorescent material coating the inside of the screen. This causes the coating to glow and builds up the picture pixel by pixel. Similar CRT technology was still being used in the large colour sets of the late 1990s.

Test yourself

What are thermionic valves?

▲ Figure 2.3 A cathode ray tube of the type used in televisions up until the late 1990s

By the early 2000s, cathode ray tubes had been superseded by liquid crystal displays (LCDs) back lit with **light-emitting diodes (LEDs)**. The liquid crystal was manipulated by an electrical signal to allow it to transmit colour for each pixel and build up the image. By eliminating the large glass cathode ray tube, this technology allowed televisions to be built in the flat-screen format we are now familiar with.

Unlike previous technologies that used fluorescence or light transmission to create an image, individual **organic light-emitting diodes (OLEDs)** can now be used to emit brightly coloured light for each pixel.

OLED technology has allowed the development of curved screens that provide a more immersive viewing experience.

Key terms

Thermionic valve: electronic device that amplifies electrical signals by controlling the flow of electrons in a vacuum tube.

Cathode ray tubes (CRTs): vacuum tubes that direct a beam of electrons onto a fluorescent surface to produce images.

Light-emitting diodes (LEDs): electronic devices that produce light when current flows through them.

Organic light-emitting diodes (OLEDs): special types of LED that contain an organic material layer that produces light when current flows through it.

▲ Figure 2.4 Modern OLED television with a high-definition curved screen

4K or 5K resolution refers to the number of vertical pixels available on a television screen. The higher the number of pixels, the sharper and more realistic the image, and the larger the screen size can be without loss of resolution. High pixel density is made possible by miniaturisation of the OLEDs used in these screens.

Radio

Marconi invented the 'wireless telegraph' in 1895. This enabled simple morse-code messages to be transmitted and received wirelessly using radio waves.

By 1922, the BBC was using radio to transmit news, music and drama across the UK. In 1932, King George V was able to address millions of listeners across the world when the BBC World Service was launched.

Radio revolutionised the mass media, enabling news and political debate to be delivered directly into homes for the first time. During periods of national upheaval, tragedy or celebration, the radio soon became an essential part of everyday life. In wartime, it encouraged a national sense of community and common purpose.

Radio also revolutionised popular culture, bringing a huge variety of music and entertainment into our living rooms. It promoted the success of new music, bands and entertainers and helped cultivate nationally recognised celebrities.

Radio remains popular, with millions of regular listeners in the UK and around the world.

Automated machines

Prior to around 1770, people tended not to travel far from where they were born. They grew food and made anything they needed locally.

During the period between 1770 and 1850, new automated machines driven by steam engines were developed. These were able to manufacture goods quickly, efficiently and in much larger quantities than ever before. This period became known as the industrial revolution and it had a massive effect on the economic development of the UK.

In this period, agricultural machinery began to make food production more efficient but in doing so it reduced the number of jobs available for farm workers. At the same time, workers were needed in the new factories of the towns. These factors prompted the migration of hundreds of thousands of people from rural areas into new industrial towns.

The output from these new factories led to a massive increase in the availability of manufactured goods, and efficient production meant lower prices. The UK become known as the workshop of the world, massively growing its economy by selling factory-made goods across the globe.

Other major economies were swift to realise the advantages of mechanisation to boost industrial output. Notably, the USA embraced industrialisation, mechanisation and automation technologies, and by 1890 became the world's largest economy.

Computers

Early electro-mechanical computing machines were developed in the 1940s and used with great success in the UK to break military codes used by Nazi Germany during the Second World War. These machines were the forerunners to the modern computer. It wasn't until the development of the silicon chip and the microprocessor in the 1970s, however, that practical and affordable personal computers became a reality.

▲ Figure 2.5 IBM Personal Computer XT circa 1983

Today, computers are essential to many aspects of modern living. They are extremely powerful tools for communication and information management, and they have been universally embraced by businesses and government organisations:

▶ Computers are important for controlling the basic functions of business, from communication with customers, to managing sales and financial transactions.

▶ In manufacturing, computers can be used to control stock and create production schedules. They also facilitate computer-aided design (CAD) and computer-aided manufacturing (CAM).

▶ Financial and banking systems are underpinned by secure computer networks, without which business could not function.

▶ In healthcare, computers are used to store medical records, help diagnose disease and control complex medical equipment.

Personal computers, laptops, tablets and smart phones enable easy access to the internet and the advantages that connectivity brings.

Computer processors are embedded in products manufactured in a wide range of sectors. For example, in the aerospace sector, they enable the safe control and operation of complex aircraft systems.

Test yourself

Which key technological developments led to the invention of the modern personal computer?

The internet

In 1989, a British scientist called Tim Berners-Lee invented the World Wide Web (WWW) in order to connect scientists and academics so they could share research. The first web page appeared in 1991. Today, there are estimated to be around two billion websites, containing up to fifty billion pages of content.

The internet has become essential to modern living in many ways:

▶ Email connects businesses worldwide, speeding up internal communications and increasing productivity.

▶ Social media connects huge online communities. Platforms such as Twitter enable millions of people to access news and opinion as it happens and engage in political debate. Other platforms, such as Instagram, TikTok and YouTube, have enabled social-media personalities to attract vast numbers of followers and become extremely influential.

▶ The convenience and low cost of online retailing led to the rapid growth of companies such as Amazon and eBay, while having a detrimental effect on high-street shops.

▶ The availability of online educational resources has revolutionised access to learning.

▶ Advice and guidance on health helps to relieve pressure on face-to-face medical services and educates the public on heath matters.

▶ Remote working is now a reality for many people, with all interaction with colleagues taking place online.

▶ Access to entertainment, music, television and film has been revolutionised. A subscription to an online music service can provide access to millions of songs. On-demand film and television services make huge back catalogues of work available and are now more popular than live television broadcasts.

2.3 Areas of innovation and emerging trends in engineering

Artificial intelligence (AI)

Artificial intelligence (AI) describes the ability of machines to gather information, perform analysis and make autonomous decisions. Current forms of AI rely on complex mathematical algorithms and computing power to allow them to learn behaviours that were previously only possible by animals and humans.

Key term

Artificial intelligence (AI): ability of machines to gather information, perform analysis and make autonomous decisions.

AI has already found applications in healthcare by learning to recognise health conditions such as skin cancers using data from photographic images. This has improved rates of early diagnosis and, in turn, survival rates among patients.

In manufacturing, data from temperature and vibration sensors monitored by AI can be used to accurately identify maintenance issues before they cause breakdowns. This enables machinery and equipment to run more reliably and maintenance schedules to be optimised to minimise downtime.

Co-operative robot systems, combining vision sensors with AI to see and interpret the local environment, can work safely alongside humans. These co-operative robots, or cobots, can be quickly trained to assist their human co-workers in a wide range of manual tasks from workholding to welding.

In future, a wide array of complex tasks could be performed by AI much more quickly, efficiently and effectively than by humans and could improve our quality of life.

Virtual reality (VR)

Virtual reality (VR) is an immersive technology that allows a user to experience a digitally generated scene as if they inhabited the virtual space. By wearing a head-mounted display and motion sensors, the user can move around and interact with the computer-simulated environment.

In the future, VR will become more realistic and absorbing. A full range of simulated training environments will be possible for complex tasks such as learning to drive, flying a plane or performing surgery. Meta (the parent company of Facebook) is already developing an entire virtual world it has called the Metaverse. Here, inhabitants can socialise, work and play together without ever having to leave their homes. The widespread use of VR will impact how we experience the world, build relationships and socialise, with online communities formed across geographical and social boundaries.

In manufacturing, VR is an important design tool that allows complex products and systems to be assembled and tested virtually before any physical parts are built. This is a key technology in aircraft design, pioneered by aerospace companies like Boeing and Airbus.

▲ Figure 2.6 An engineer exploring a product using virtual reality

Augmented reality (AR)

While VR describes a self-contained virtual world, **augmented reality (AR)** allows users to enhance their real-world experience with real-time, computer-generated augmentations or overlays.

> ### Key terms
>
> **Virtual reality (VR):** immersive technology that allows a user to experience a digitally generated scene as if they inhabited the virtual space.
>
> **Augmented reality (AR):** technology that allows users to enhance their real-world experience with real-time, computer-generated augmentations or overlays.

AR already has applications in medicine, allowing virtual anatomical models to be studied in the classroom. Its ability to provide real-time annotation of what the user can see could find applications in education, navigation and policing. It could also be used to guide field repairs on complex equipment or even assist in surgical procedures.

In manufacturing, Boeing uses AR to provide wiring diagram overlays for workers tasked with installing the thousands of electrical connections required in their aircraft.

> ### Test yourself
>
> What is the key difference between VR and AR?

Digitalisation

Digitalisation is the process of converting physical objects such as documents, paper-based systems or other artefacts into digital data that can be read, displayed, processed and distributed by computer. This will have an impact wherever instant access to current or real-time data is required.

For example:
- Digitalisation of medical records could mean better integration and communication between healthcare providers, leading to patients being treated in a more timely and effective way.
- Access to real-time domestic electricity consumption data will allow more efficient management of electricity generation and distribution.
- In manufacturing, assembly instructions and engineering drawings displayed on screens at individual workstations eliminate the need to manage paper documents.

Robotics

Robotics involves the design, development and manufacture of robots that are able to perform physical tasks that would previously have been carried out by humans.

Robots already have widespread applications in manufacturing. These include the assembling, welding and painting of car bodies, assembling electronic circuits, and the picking, packing and palletising of finished goods.

In the future, robots could be combined with AI to make entirely autonomous machines to perform complex or dangerous tasks, such as medical procedures, space exploration or mine clearance. When combined with suitable sensor technology, they could be used in exoskeletons to give independence to patients with paralysis or provide fully functioning prosthetic limbs.

Drones

Drones are unpiloted aerial vehicles. They vary in size and complexity, from small toys to full-sized aircraft.

> ### Key terms
>
> ***Digitalisation:*** process of converting physical objects such as documents, paper-based systems or other artefacts into digital data that can be read, displayed, processed and distributed by computer.
>
> ***Robotics:*** design, development and manufacture of robots that are able to perform physical tasks that would previously have been carried out by humans.
>
> ***Drones:*** unpiloted aerial vehicles.

Drones have already had a significant impact in military applications, where they are widely used for reconnaissance or in direct support of ground troops.

Drone cargo aircraft have also been developed. Royal Mail plans to use a fleet of these to deliver mail to dozens of remote islands around the UK.

In the future, drone technology is likely to find wider applications in law enforcement to conduct searches from the air. Autonomous drones could be used to monitor the condition of structures such as offshore wind-turbine installations, explore and map underground cave systems, or monitor the health and wellbeing of livestock.

In manufacturing, the use of drones is not yet commonplace. However, drones could be used to carry out stocktaking in large warehouses, deliver components or even help to monitor the condition of large machines and equipment.

> ### Case study
>
> In 2022, Royal Mail conducted trials to explore the use of autonomous drones to deliver letters and parcels to remote island communities. The goal is to establish permanent, regular delivery services by drone as soon as the technology is fully tested and UK regulations allow it. Drones are fast and reliable and have much lower emissions and running costs than larger, conventional aircraft.
>
> ### Questions
>
> 1. What are the potential dangers of using large autonomous drones in UK airspace?
> 2. What other businesses or organisations could benefit from using similar drone technology?

▲ Figure 2.7 Royal Mail plans to use drones to deliver mail to remote island communities

Autonomous systems

Autonomous systems integrate AI, vision systems, sensor technology and robotics to achieve independent and self-directed operation. This has applications in self-driving cars and autonomous robots and drones.

In manufacturing, autonomous robots can be used to automatically locate, collect and transport components to a manufacturing cell at exactly the right time.

Distributed energy systems

Distributed energy systems represent a move away from conventional centralised electricity generation. Instead of a small number of large power stations, a larger number of smaller and more localised generating schemes are used. This has been made possible by the emergence of reliable and efficient renewable energy technologies such as photovoltaic (PV) cells, wind turbines and battery storage.

These systems will reduce our reliance on fossil fuels and increase the resilience of our electricity network – factors that are key to reducing the impact and extent of any future climate change.

In manufacturing, the energy costs of running a factory are high. Consequently, the use of PV cells or wind turbines on industrial sites is becoming more common. Using its own renewable generating capacity reduces emissions, saves money and helps to isolate a business from any volatility in energy pricing.

Hybrid technology systems

Hybrid technology systems combine two or more different technologies in a single product or system to optimise utility and/or efficiency. For example, it is common for small-scale renewable electricity generation systems to use a combination of PV cells and wind turbines. The use of two distinct technologies helps ensure continuation of supply as generation continues whenever there is sufficient wind or daylight.

In manufacturing, it is becoming increasingly common for factories to generate a proportion of their electricity needs using PV and wind turbine hybrid systems.

Cyber-physical systems

Cyber-physical systems integrate computerisation, networking, and physical processes, equipment or machinery. Examples include:
- remote monitoring and control of wind-turbine installations
- precision farming, which directs resources such as water, fertiliser and herbicide exactly as needed to reduce costs and increase yield
- the integration of 3D modelling and manufacturing simulation software with physical machines and equipment.

Key terms

Autonomous systems: systems that integrate AI, vision systems, sensor technology and robotics to achieve independent and self-directed operation.

Distributed energy systems: small-scale, localised electricity generation systems that usually use renewable energy sources.

Hybrid technology systems: systems that combine two or more different technologies in a single product or system to optimise utility and/or efficiency.

Cyber-physical systems: systems that integrate computerisation, networking, and physical processes, equipment or machinery.

Internet of things (IOT)

The **internet of things (IOT)** describes a network of physical machines, equipment and devices that are embedded with computerised sensors and networking technology to make them capable of interconnection through the internet.

For example:
▶ Wearable technology that monitors health conditions can be accessed remotely by medical teams to improve patient outcomes.
▶ Monitoring and control of domestic heating equipment can reduce energy consumption by only heating rooms that are in use.
▶ Real time condition monitoring of machines and equipment using IOT enabled sensors can be used to optimise maintenance schedules.

Cloud computing

Cloud computing allows users to remotely access a wide range of computer services such as data storage, software and networking through the internet. It is often more cost-effective to pay a subscription to access these services instead of buying, operating and maintaining the physical equipment and software that would otherwise be required.

In manufacturing, real-time collaboration across multiple locations, access to software and shared resources, IOT and cyber-physical systems are all dependent on cloud-based data sharing.

> **Key terms**
>
> ***Internet of things (IOT):*** a network of physical machines, equipment and devices that are embedded with computerised sensors and networking technology to make them capable of interconnection through the internet.
>
> ***Cloud computing:*** where users remotely access computer services such as data storage, software and networking through the internet.

Sustainability

Sustainability requires us to fulfil our current needs without compromising the needs of future generations. Much of our industrial and economic development has been dependent on the unsustainable exploitation of non-renewable natural resources. As a consequence, we are now dealing with the effects of deforestation, pollution and, of course, climate change. As a society, it is important that we learn to live much more sustainably to minimise our impact on the environment.

> **Key term**
>
> ***Sustainability:*** ability to fulfil our current needs without compromising the needs of future generations.

Product life cycle

From design to eventual disposal, each step in the life cycle of a product should be sustainable:
▶ Design: decisions made during the design of a product have the greatest impact on its sustainability, as they determine how it functions and the materials used in its construction.
▶ Raw material extraction and processing: the use of new materials that require extraction and processing should be avoided if possible. Renewable or recycled materials are the most sustainable.
▶ Manufacturing: energy-efficient, low-waste and non-polluting processes are the most sustainable.
▶ Distribution: packaging materials should be lightweight and reusable or recyclable, and both shipping weight and volume should be minimised.
▶ Product use: low energy consumption and long service life maximise sustainability during the lifetime of a product.
▶ Disposal: materials should be easy to separate at the end of the product's life so that they can be reused or recycled.

Circular economy

The **circular economy** is a model for sustainable production that involves rethinking both the ownership and use of products. It involves sharing, leasing, reusing, repairing, refurbishing and recycling products to maximise their utilisation and useful service life. The principal aim of the circular economy is to make products last as long as possible, consume only renewable energy and create zero waste at the end of product life.

> **Key term**
>
> *Circular economy:* model for sustainable production that aims to make products last as long as possible, consume only renewable energy and create zero waste at the end of product life.

Exploring alternatives

Designers and engineers have a responsibility to explore alternative ways of achieving their goals that minimise their impact on the environment and maximise sustainability. This philosophy will impact future design as we migrate from traditional polluting or non-renewable technologies to more sustainable alternatives.

For example, in the search for a more sustainable replacement for petrol and diesel to power our cars, a wide range of alternatives is being explored:

▶ liquid petroleum gas (LPG), which is a by-product of the petrochemical industry
▶ biofuel derived from plant materials and not fossil fuels
▶ hydrogen derived from fossil fuels or by using electricity to split water into hydrogen and oxygen
▶ electricity stored in on-board batteries
▶ electricity generated by on-board hydrogen fuel cells.

In manufacturing, sustainable alternatives should be explored at every stage of the product life cycle, from design to disposal. This was explored earlier in the section on product life cycle.

Renewables

Use of renewables is essential in sustainable design. Unlike finite resources, renewables come from natural sources or processes that are constantly replenished and include:

▶ materials such as cotton, wool, straw and wood
▶ energy sources such wind, solar and tidal.

Waste and disposal

Sustainable practice will maximise reuse and recycling to keep the generation of waste materials to an absolute minimum. Where waste product cannot be avoided, it must be managed safely and in a way that avoids environmental contamination.

Some manufacturing processes, like machining, rely on material removal to create components. This generates waste metal as small chips or swarf that can be recycled into new raw materials.

> **Health and safety**
>
> Some waste products in the workplace are hazardous and pose a significant threat to health and/or the environment. These can include chemicals used in manufacturing, such as waste lubricants, solvents and paints. These must be handled and stored safely and be disposed of responsibly by approved specialist contractors, in accordance with the Control of Substances Hazardous to Health (COSHH) Regulations 2002 (amended 2004) (see page 188).

The least environmentally responsible approach to the disposal of general waste is landfill. This causes localised pollution and releases methane, which is a powerful greenhouse gas that contributes to climate change. Less harmful alternatives include incineration. There are numerous schemes that heat homes or generate electricity using the heat produced by incinerating waste.

Assessment practice

1. Processing equipment used to manufacture pharmaceuticals is designed and developed by businesses in which engineering sector?

2. Outline the activities performed by the logistics engineering sector.

3. Give an example of how the development of materials has impacted modern society.

4. Explain two economic impacts of the introduction of the internal combustion engine.

5. In the early part of the twentieth century, steam engines used to drive machinery in factories began to be replaced by electric motors. Give two advantages of using electric motors instead of steam engines.

6. Explain the importance of interchangeable parts when considering the maintenance and repair of complex machines.

7. Outline the three main technologies that have been used to display the picture on television screens from the 1950s to the present day.

8. Outline three ways in which widespread access to the internet has affected society.

9. Outline a manufacturing application that could make effective use of augmented reality technology.

10. Describe how cyber-physical systems could be used to increase crop yields in farming.

Project practice

You work as a graduate engineer for a manufacturer of high-technology farm equipment in the agricultural sector. One of your duties is to attend careers events to help promote your industry to students about to leave secondary school.

Design a flyer to promote the agricultural engineering sector that can be handed out at a careers fair.

Your flyer should include:
▶ an explanation of the main activities, products and services provided by the sector
▶ an explanation of why this sector is important
▶ examples of the latest technologies employed in the sector and how these might develop in the future.

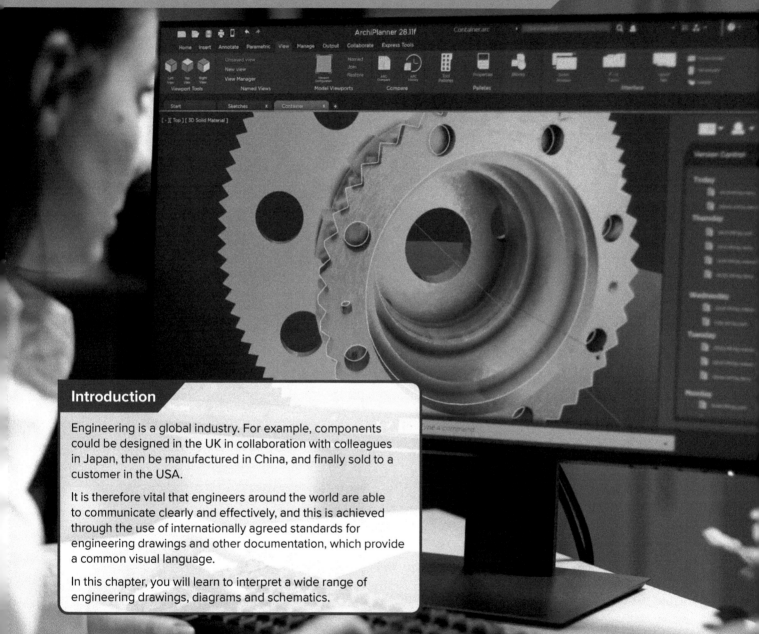

3 Engineering representations

Introduction

Engineering is a global industry. For example, components could be designed in the UK in collaboration with colleagues in Japan, then be manufactured in China, and finally sold to a customer in the USA.

It is therefore vital that engineers around the world are able to communicate clearly and effectively, and this is achieved through the use of internationally agreed standards for engineering drawings and other documentation, which provide a common visual language.

In this chapter, you will learn to interpret a wide range of engineering drawings, diagrams and schematics.

Learning outcomes

By the end of this chapter, you will understand:
1   the characteristics of, purposes of, and audience for different drawing types
2   the purpose and application of CAD systems and software
3   how to interpret and present information, symbols, conventions and annotations on engineering drawings, in accordance with the conventions of BS 8888 and BS 3939
4   how to interpret dimensions and related drawing symbols
5   how to calculate tolerances, limits and fits.

3.1 Drawings and information conveyed by drawings

Drawings

Drawings and diagrams are essential to communicate ideas and information in engineering. They range from quick, informal sketches that help to convey a basic concept, to formal engineering drawings containing sufficient information to manufacture a finished component.

Freehand sketches are an effective way of quickly recording your thoughts and are the basis of ideation techniques such as thumbnail sketching. There are no formal conventions for freehand pictorial sketches.

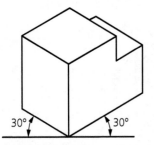

▲ Figure 3.1 Freehand sketch of a soft-faced hammer

Isometric drawings are used to achieve a basic three-dimensional (3D) appearance by using only vertical lines and lines receding at 30° from the horizontal.

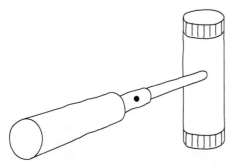

▲ Figure 3.2 Three-dimensional isometric drawing

Exploded views bring together a series of isometric component drawings, positioned on the page to illustrate how they fit together into an assembly. Individual components are labelled with a reference number that allows them to be identified in an accompanying table. This type of drawing is often included in manuals for machines and equipment, to allow maintenance engineers to identify replacement part numbers for worn or broken components.

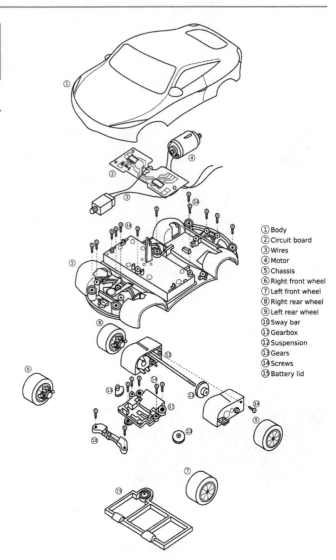

① Body
② Circuit board
③ Wires
④ Motor
⑤ Chassis
⑥ Right front wheel
⑦ Left front wheel
⑧ Right rear wheel
⑨ Left rear wheel
⑩ Sway bar
⑪ Gearbox
⑫ Suspension
⑬ Gears
⑭ Screws
⑮ Battery lid

▲ Figure 3.3 Exploded view of an engineering assembly

Key terms

Freehand sketches: drawings produced by hand without the use of drafting aids such as rulers, set squares or compasses.

Isometric drawings: drawings used to achieve a basic three-dimensional (3D) appearance by using only vertical lines and lines receding at 30° from the horizontal.

Exploded views: drawings that bring together a series of isometric component drawings, positioned on the page to illustrate how they fit together into an assembly.

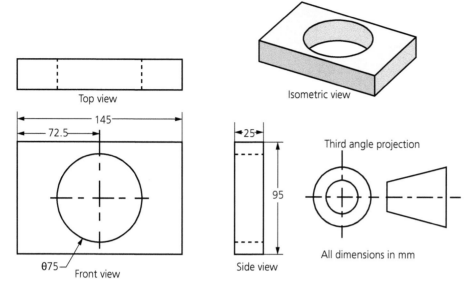

▲ Figure 3.4 Orthographic projections used in an engineering drawing

Orthographic projections include multiple views of a product or engineering component and are used in formal engineering drawings. They are discussed in greater detail later in this chapter.

In industry, only informal freehand sketches are drawn by hand. Instead, **computer-aided design (CAD)** software is used to create 3D models of engineering components that can be manipulated to produce isometric views, exploded views and orthographic projections. 3D models can also be rendered to make photorealistic representations of components or assemblies that can be viewed from any angle. This is a powerful way to illustrate to potential customers what a new product will look like, before a physical prototype is manufactured.

> **Test yourself**
>
> Identify the most appropriate type of drawing to use when:
> ▶ quickly recording a design idea in a notebook
> ▶ identifying the part number for a component in a larger assembly
> ▶ manufacturing a component part.

Diagrams

Complex engineering systems or processes are visually represented using diagrams.

Block diagrams

Block diagrams are used to simply represent complex systems and provide a basic overview of how their parts connect together (see Figure 3.5 on the next page for an example). Labelled boxes or blocks are used to represent all the sub-systems or major components in the system. The inputs and outputs from each block are labelled, and connections between blocks are indicated using arrowed lines.

Block diagrams are useful when designing complex systems and help end users, such as maintenance technicians, to understand how an installed system works.

> **Key terms**
>
> *Orthographic projections:* drawings that use multiple views of a product or engineering component and are used in formal engineering drawings.
>
> *Computer-aided design (CAD):* using computer software to create 3D models of engineering components that can be manipulated to create isometric views, exploded views and orthographic projections.
>
> *Block diagrams:* diagrams that use blocks and arrows to provide a top-down overview of a complex system in terms of its inputs, processes and outputs, and how they connect together.

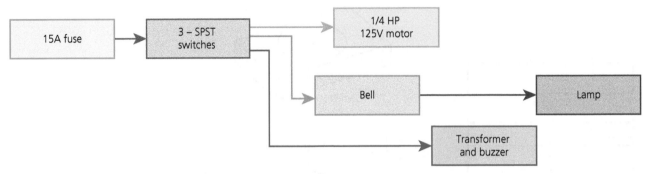

▲ Figure 3.5 Block diagram representing a simple system

Flowcharts

Flowcharts are used to visualise the steps in a process. They use a range of symbols connected by arrowed lines to represent each step, its type, how it links with other steps and the process flow direction.

Engineers use flowcharts to aid the design and development of processes and then to communicate how a final process works.

> ### Key term
>
> **Flowcharts:** diagrams used to visualise the steps in a process.

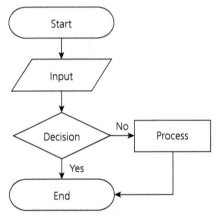

▲ Figure 3.6 A flowchart for a simple process

> ### Test yourself
>
> Draw a flowchart to illustrate the steps in a simple process like making a cup of tea.

Schematics

Schematics are simplified visual representations of the connections between components in an electrical, a pneumatic or a hydraulic system. Components are represented using standardised symbols to ensure they can be universally understood across engineering sectors and in different languages.

> ### Key term
>
> **Schematics:** simplified visual representations of the connections between components in an electrical, a pneumatic or a hydraulic system.

▼ Table 3.1 Basic symbols used in flowcharts

Symbol	Name	Purpose
(rounded rectangle)	Start/end	Indicates the start and end of the process
(parallelogram)	Input/output	Indicates inputs or outputs of the process
(rectangle)	Process	Indicates a processing function
(diamond)	Decision	Indicates a decision point between pathways in the process
(arrow)	Flow line	Connects symbols in a flowchart and indicates the direction of flow

Circuit diagrams

Circuit diagrams are drawn up by electronic design engineers to show how components (such as resistors, capacitors and integrated circuits) are connected in an electronic circuit. Each component is represented by a symbol and labelled with a reference number and component value.

Circuit diagrams are used by repair technicians, alongside component layout drawings, to help them understand how electronic devices work and to identify the physical location of components on a printed circuit board (PCB).

▲ Figure 3.7 A circuit diagram and PCB

> ### Key term
>
> **Circuit diagrams:** diagrams drawn up by electronic design engineers to show how components (such as resistors, capacitors and integrated circuits) are connected in an electronic circuit.

The symbols used in electrical and electronic circuit diagrams are defined by internationally agreed standards. In the UK, BS 3939 *Graphical symbols for electrical power, telecommunications and electronics diagrams* is still commonly referred to, although this was withdrawn some time ago. The content of BS 3939 was largely retained in the current international standard BS EN 60617 *Graphical Symbols for Diagrams*.

▼ Table 3.2 Basic graphical symbols used in electrical and electronic circuit diagrams and schematics

Symbol	Name
	Resistor
	Variable resistor
	Battery
	Capacitor
	Single-pole normally open switch
	Fuse

> ### Research
>
> Research the extensive range of components included in BS 3939 and BS EN 60617.
>
> Compile your own list of useful symbols for future reference.

Wiring schematics

Wiring schematics are drawn up by electrical design engineers to show how components (such as switches, relays, lamps and fuses) are wired together in an electrical system. They include information on the colour-coding of wires used to make each connection.

Wiring schematics are used during manufacturing to ensure that wires are routed appropriately and that connections between components are correct. It is essential that technicians refer to them when fault finding on complex electrical systems.

▲ Figure 3.8 An automotive wiring harness

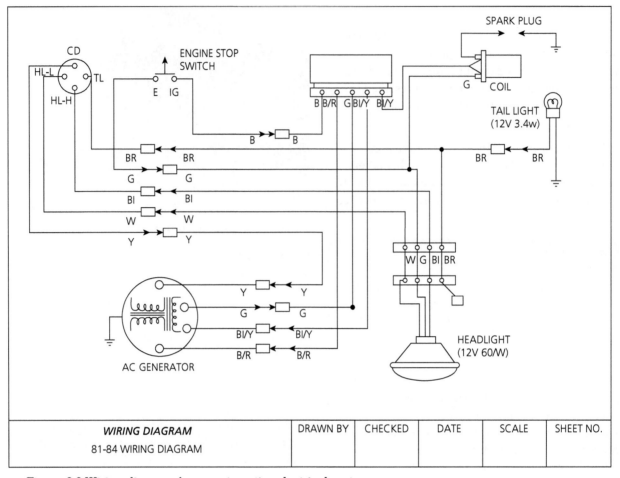

▲ Figure 3.9 Wiring diagram for an automotive electrical system

Hydraulic schematics

Hydraulic schematics are drawn up by fluid-power systems design engineers to show how hydraulic components (such as valves, actuators, regulators and gauges) are connected by pipework in a hydraulic system. Like wiring schematics, they are used during manufacture and when technicians are fault finding in complex hydraulic systems.

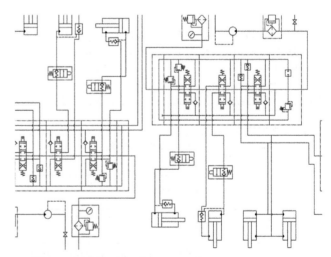

▲ Figure 3.10 Hydraulic circuit diagram

▲ Figure 3.11 Hydraulic system in an excavator

Pneumatic schematics

Pneumatic schematics are similar to their hydraulic equivalents and use similar component symbols. The difference is that pneumatic systems use air as the working fluid instead of incompressible hydraulic oil.

Computer-aided design (CAD)

Before the advent of computers, all drawings, diagrams and schematics were drawn by hand. This process was labour intensive, slow and repetitive. Making significant changes or updates to drawings often meant redrawing them completely, and the storage, management and distribution of paper drawings was problematic.

The development of computers and CAD software allowed vast improvements to be made over manual systems. CAD enables the creation, editing and copying of drawings to be performed much more quickly. Hundreds of thousands of drawings can be stored on a single computer hard drive, and facilities like email, file sharing and cloud storage mean that distribution is instant and secure.

▲ Figure 3.12 Prior to computers and CAD software, drawings were produced by hand

Test yourself

Outline the advantages of using CAD instead of manual drawing techniques.

Research

Research the following types of information commonly encountered in engineering contexts. Find out where they are used and what they are used for:
- graphical representations, for example engineering drawings, flowcharts and circuit diagrams
- technical information, for example routing sheets, purchasing specifications, inspection plans and manufacturing instructions
- reference information, for example *Machinery's Handbook*, *Presto Counsellor* and *Zeus Precision Data Charts and Reference Tables*.

Information conveyed by engineering drawings

Types of engineering drawing used during manufacturing include:
- single-piece drawings of individual component parts
- assembly drawings, which show how a number of components fit together in a sub-assembly
- general arrangement drawings, which show how sub-assemblies and other components fit together to make a complete finished product.

The format of engineering specifications, including engineering drawings, is specified in BS 8888 *Technical product documentation and specification*. This internationally recognised standard explains how engineering drawings are laid out, the symbols and conventions used, and the information that designers and engineers need to include.

Responsible department Engineering	Technical reference N. Makin		Document type Component drawing		Document status Released		
Legal owner C&G	**Created by** A. Buckenham		**Title** Mounting bracket		**Identification number** AB1211-4		
	Approved by S. Singh			**Rev.** A	**Date of issue** 2023-01-10	**Lang.** EN	**Sheet no.** 1/1

▲ Figure 3.13 A typical title block layout

Title block

Every engineering drawing must contain a **title block** that provides essential information about the drawing, what it shows and the identification number under which it is filed.

> **Key term**
>
> *Title block:* element of an engineering drawing that provides essential information about the drawing, what it shows and the identification number under which it is filed.

The exact contents and layout of title blocks can vary, but they must include the following mandatory elements:
- legal owner
- created by
- approved by
- document type
- title
- identification number
- date of issue
- sheet number.

Other information that is presented outside of, but adjacent to, the title block includes scale, projection symbol and general tolerances.

Scale

Using an appropriate scale is important if the information on a drawing is to be conveyed effectively:
- Many components can be conveniently drawn at their actual size using scale 1:1.
- Small components can be drawn larger than actual size using enlargement scales, for example 2:1, 5:1 or 10:1.
- Large components, assemblies or general arrangements can be drawn smaller than actual size using reduction scales, for example 1:2, 1:5 or 1:10.

It is worth noting that, by convention, scales are stated in the form *units on the drawing:units at full scale*. For example, a component drawn at 1:10 scale uses one unit on the drawing to represent ten units at actual size.

> **Improve your maths**
>
> Work out the length of the line used on a drawing to represent a feature that is 53 mm long when using the following scales:
> - 1:1
> - 1:5
> - 2:1.

Views

The views of a component shown on an engineering drawing are arranged as either first or third angle orthographic projections. It is vital that engineers are aware of the differences between these two arrangements so that drawings are not misinterpreted.

First angle orthographic projection

A drawing using first angle orthographic projection includes the graphical symbol shown in Figure 3.14.

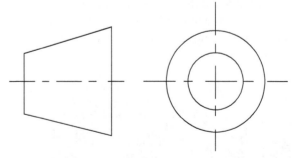

▲ Figure 3.14 Graphical symbol for first angle orthographic projection

The views or elevations of a component drawn in first angle orthographic projection are arranged as shown in Figure 3.15 on the next page.

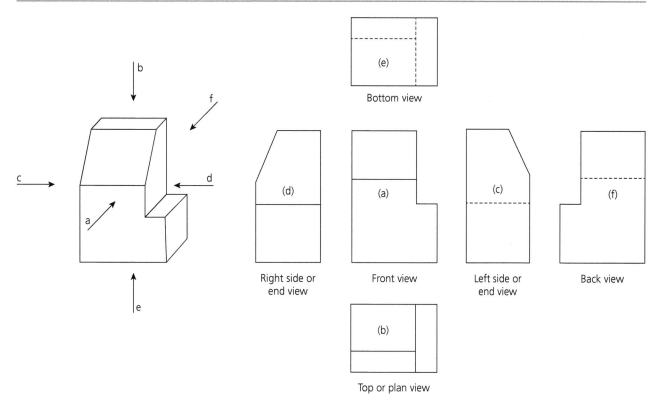

▲ Figure 3.15 Component drawn in first angle orthographic projection

Third angle orthographic projection

A drawing using third angle orthographic projection includes the graphical symbol shown in Figure 3.16.

The views or elevations of a component drawn in third angle orthographic projection are arranged as shown in Figure 3.17 on the next page.

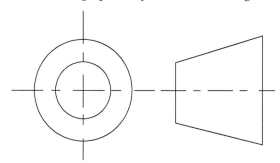

▲ Figure 3.16 Graphical symbol for third angle orthographic projection

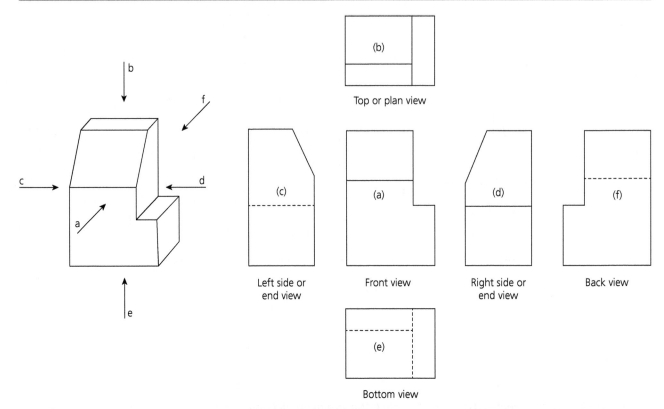

Top or plan view

Left side or
end view

Front view

Right side or
end view

Back view

Bottom view

▲ Figure 3.17 Component drawn in third angle orthographic projection

Auxiliary views

An **auxiliary view** of a component is used when a standard orthographic view is unable to show the required detail. This is commonly necessary where features such as holes are present on sloping or inclined surfaces.

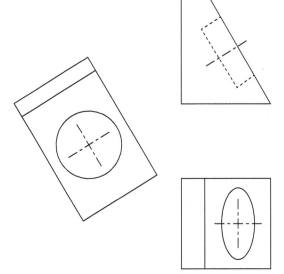

▲ Figure 3.18 First angle drawing with an auxiliary view showing the detail of the hole on the inclined surface

Section views

A **section view** is used to reveal a cross-section of a component when sliced along a specified cutting plane. The cutting plane in the example shown in Figure 3.19 is indicated by the line marked with an A at each end. The arrows indicate the direction in which the cut section is to be viewed.

The section view itself is shown on the right. It is labelled A–A to correspond with the associated cutting plane. Hatching indicates where solid material has been cut through.

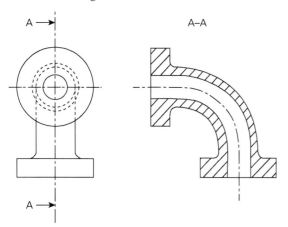

▲ Figure 3.19 Section view of a component

> **Key term**
>
> **Section view:** view used to reveal a cross-section of a component when sliced along a specified cutting plane.

Abbreviations

Although discouraged in drawing standards, the use of abbreviations is still commonplace in engineering drawings.

▼ Table 3.3 Abbreviations used in engineering drawings

Abbreviation	Meaning
AF	Across flats
CL	Centre line
DIA or Ø	Diameter
DWG	Drawing
MTL	Material
SQ	Square
CHAM	Chamfer
CSK	Countersink
HEX	Hexagon
R	Radius
THD	Thread
UCUT	Undercut
PCD	Pitch circle diameter

> **Test yourself**
>
> Write down the meaning of the following abbreviations:
> ► CHAM
> ► PCD
> ► MTL.

Line types

A number of line types are used in the construction of engineering drawings and each conveys a specific meaning.

▼ Table 3.4 Line types

Line	Description	Use
————————	Continuous, thick (0.7 mm)	Visible outlines and edges
————————	Continuous, thin (0.35 mm)	Projection, dimension and leader lines
– — – — – — — —	Dashed, thin (0.35 mm)	Hidden detail, outlines and edges
—— – —— – —— – —	Long-dashed, dotted, thin (0.35 mm)	Centre lines

Note: any construction lines used in the drawing process are removed prior to completion.

Surface finish

The roughness of a finished surface is defined using a numerical roughness average (Ra) and indicated on a drawing using the symbol shown in Figure 3.20.

▲ Figure 3.20 Symbol indicating surface finish requirement

Typical surface roughness values are as follows:
▶ highly polished surfaces 0.025–0.4μm
▶ turned roughing cuts on a lathe 6.3–25μm
▶ components manufactured using manual processes 1.6–6.3μm.

Manufacturing details

Engineering drawings often contain additional information relating to component manufacture, such as
▶ tapping drill sizes for threaded holes
▶ material grades and forms of supply
▶ heat-treatment requirements
▶ instructions to deburr sharp edges.

Standard features

To save time and to simplify the creation of engineering drawings, there are standard ways of representing common features. Some examples of these are shown in Table 3.5.

▼ Table 3.5 Standard features in engineering drawings

Feature	Standard representation
Internal screw thread	
External screw thread	
Nut	
Bolt	

Feature	Standard representation
Pin (split pin)	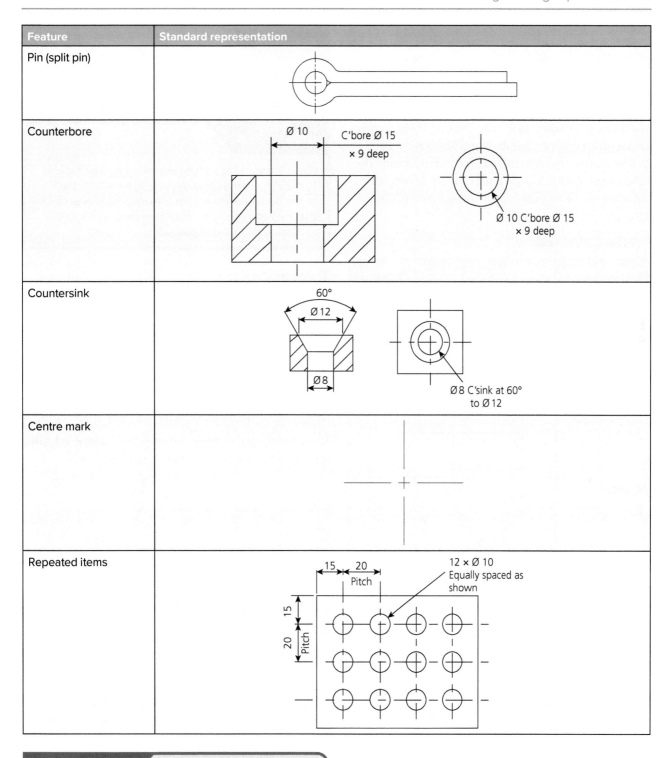
Counterbore	
Countersink	
Centre mark	
Repeated items	

Test yourself

Make quick sketches to show how the following standard features are represented in engineering drawings:
- ▶ internal screw thread
- ▶ external screw thread
- ▶ countersunk hole.

3.2 Dimensions and tolerancing on engineering drawings

Dimensions

Dimensions are used to specify the size of the features shown on an engineering drawing. There must be sufficient dimensions to fully define a component. The dimensions of each feature must only be shown once and must not be repeated on different views of the same component.

▼ Table 3.6 Dimensions

Dimension type	Example
Linear	(drawing showing dimensions 50, 60, 40)
Diameter	(circle with Ø50 shown twice)
Radius	(rounded rectangle with R10 and R30)
Angular	(shape with 143° angle)

Tolerances

A **tolerance** specifies the permissible variation in the size of a component feature. Tolerances are necessary to ensure that mating components fit together, and their use guarantees the interchangeability of standard parts.

Tolerances specific to individual features appear alongside, or are incorporated into, their dimensions.

There are several methods of showing **tolerance limits** on engineering drawings, as seen in Figure 3.21.

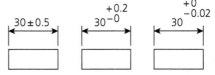

30 ± 0.5 $30 \, {}^{+0.2}_{-0}$ $30 \, {}^{+0}_{-0.02}$

▲ Figure 3.21 Three methods of expressing tolerance limits for a linear dimension

Statements specifying general tolerances are used to simplify dimensioning. Their use ensures that every dimension has an applicable tolerance, and nothing is left to the discretion of the reader.

Unless otherwise stated, general tolerances apply to all linear and angular dimensions and are specified by a note on the drawing that is in or near the title block.

```
GENERAL TOLERANCES
UNLESS OTHERWISE STATED
0 ± 0.25
0.0 ± 0.1
0.00 ± 0.05
ANGLES ± 2°
```

▲ Figure 3.22 General tolerances are specified in or near the title block

Limits and fits

In engineering, **fit** refers to the amount of clearance between two mating parts. It is chosen to allow the required amount of movement between parts for a given application. Each fit is defined by the tolerance limits applied to the mating parts.

Key term

Fit: amount of clearance between two mating parts.

Fit is most frequently specified when dealing with the clearance between holes and shafts.

There are three classes of fit (expressed here in terms of a shaft fitting into a hole):

▶ With a clearance fit, the hole is always larger than the shaft, allowing movement between the parts. Examples include loose-running fits in pivots and hinges, and location clearances used to accurately position stationary parts.
▶ With an interference fit, the hole is always smaller than the shaft, preventing movement and locking parts together. Examples include press fits, where force is used to cold press a shaft into the hole, and tighter driving fits that require heating to thermally expand the hole before the shaft can be pressed into place.
▶ With a transition fit, the hole is either slightly larger or slightly smaller than the shaft, presenting marginal clearance or interference. Transition fits can be located by hand or by using a soft-faced hammer.

Different classes of fit are achieved by using specific tolerance limits on both the hole and the shaft. These tolerance limits are standardised in BS EN ISO 286 *ISO code system for tolerances on linear sizes. Tables of standard tolerance classes and limit deviations for holes and shafts.*

ISO codes are used to identify the tolerance limits required for different classes of fit:

▶ Codes starting with a capital letter refer to hole tolerances, for example H11, H9 and H8.
▶ Codes starting with a lower-case letter refer to shaft tolerances, for example c11, d10 and e9.

In the UK, it is still common to use the original data tables first published in BS 4500 *ISO limits and fits. General, tolerances and deviations* (now withdrawn), as they include a helpful graphical interpretation of the tolerance limits. The tables cover:

▶ hole-basis fits, where the nominal size of the hole is known and the shaft is sized to produce the required class of fit
▶ shaft-basis fits, where the nominal size of the shaft is known and the hole is sized to produce the required class of fit.

For example, when sizing a shaft to run in a hole with a loose-running clearance fit, hole-basis fits of H11/c11 are used. For a 5 mm nominal hole size, the applicable tolerance limits can be found on the BS 4500A data sheet *Selected ISO fits – hole basis.*

From the data sheet, it can be seen that:

▶ for the hole, H11 corresponds to 5 +0.075/-0.000, giving a hole size between the tolerance limits 5.000 and 5.075
▶ for the shaft, c11 corresponds to 5 -0.070/-0.145, giving a shaft size between the tolerance limits 4.855 and 4.930.

Improve your maths

Using a copy of the BS 4500A data sheet *Selected ISO fits – hole basis,* look up the hole and shaft tolerance limits for an H7/g6 sliding clearance fit based on a nominal 35 mm diameter hole.

When starting a car, a small starter-motor pinion gear engages with a much larger ring gear fitted to the engine flywheel. There are no mechanical fixings holding the ring gear in place, and it relies entirely on a driving interference fit between the ring gear and the flywheel to stay in position.

▲ Figure 3.23 Engine flywheel complete with ring gear

Questions

1 What is the advantage of using an interference fit instead of mechanical fixings to secure the flywheel?
2 How would the ring gear be fitted to the flywheel?

Geometric dimensioning and tolerancing (GD&T)

Standard tolerancing used in engineering drawings specifies simple upper and lower dimensional limits. **Geometric dimensioning and tolerancing (GD&T)** allows tolerance limits to be applied to shape (or form), as well as orientation and location relative to a reference or datum surface, plane or axis.

Datums

A **datum** feature on a component defines the position of a reference point, surface, plane or axis. Datum features are identified in an engineering drawing using a filled (or unfilled) triangle linked by a leader line to a square box containing an identification letter. The same identification letter is used in geometric tolerances that refer to that datum.

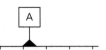

▲ Figure 3.24 Datum feature on an engineering drawing

Geometric dimensioning and tolerancing (GD&T): system that allows tolerance limits to be applied to shape (or form), as well as orientation and location relative to a reference or datum surface, plane or axis.

Datum: common reference point, surface, plane or axis from which dimensions are measured.

Geometric form tolerances

Geometric form tolerances control the shape of a feature. It should be noted that they do not require a datum.

Straightness

A straight line is defined as the shortest path between two points. Straightness defines the permissible variation from a straight line or axis.

For example, Figure 3.25 shows the allowable variation in the straightness of a line marked on the surface of a component. This is defined by a tolerance zone between two parallel lines a specified distance of 0.05 mm apart. Figure 3.26 shows how this is represented as a geometric form tolerance.

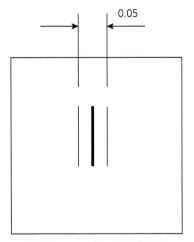

▲ Figure 3.25 Illustration of a straightness tolerance zone

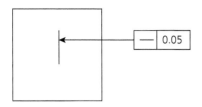

▲ Figure 3.26 Straightness represented as a geometric form tolerance

Geometric orientation tolerances

Geometric orientation tolerances control the orientation of a feature relative to a datum.

Parallelism

Parallel features maintain a constant distance between them. Parallelism defines the permissible variation in distance between a datum line, axis or surface and a parallel datum feature.

For example, Figure 3.27 shows the allowable variation in parallelism between a surface and a parallel datum surface X. This is defined by a tolerance zone between two parallel planes a specified distance of 0.5 mm apart. Figure 3.28 shows how this is represented as a geometric orientation tolerance.

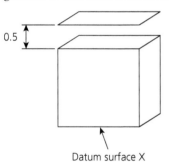

▲ Figure 3.27 Illustration of a parallelism tolerance zone and datum surface

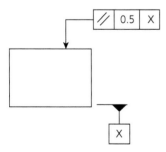

▲ Figure 3.28 Parallelism represented as a geometric orientation tolerance

Perpendicularity

Perpendicular features are at right angles to each other. Perpendicularity defines the permissible variation in alignment between a datum line, axis or plane and a component feature that is perpendicular to it.

For example, Figure 3.29 shows the allowable variation in perpendicularity between a datum surface X and a surface perpendicular to it. This is defined by a tolerance zone between two planes a specified distance of 0.5 mm apart. Figure 3.30 shows how this is represented as a geometric orientation tolerance.

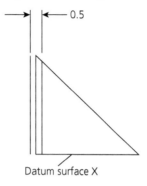

▲ Figure 3.29 Illustration of a perpendicularity tolerance zone and datum surface

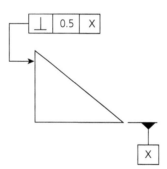

▲ Figure 3.30 Perpendicularity represented as a geometric orientation tolerance

Geometric location tolerances

Geometric location tolerances control the position of a feature relative to a datum.

Concentricity

Concentric circles (or cylinders) share the same centre point (or axis). Concentricity defines the permissible variation between the position of the centre (or axis) of a datum circle (or cylinder) and a feature concentric to it.

For example, Figure 3.31 shows that the allowable variation in concentricity between the centre of a datum circle X and the centre of a larger concentric circle is defined by a circular tolerance zone 0.05 mm in diameter. Figure 3.32 shows how this is represented as a geometric location tolerance.

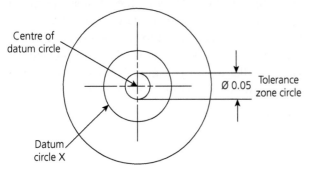

Centre of datum circle

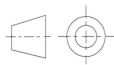
Datum circle X

▲ **Figure 3.31 Illustration of a concentricity tolerance zone and datum circle**

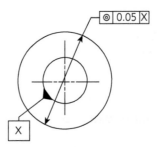

▲ **Figure 3.32 Concentricity represented as a geometric location tolerance**

Test yourself

Draw the geometric tolerance symbols for the following:
▶ straightness
▶ perpendicularity
▶ concentricity.

Assessment practice

1 Explain the purpose of a block diagram.
2 What electrical or electronic component is represented by the symbol shown?

3 What electrical or electronic component is represented by the symbol shown?

4 What does the following symbol represent on an orthographic engineering drawing?

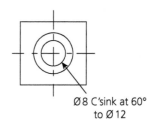

5 What is meant by the abbreviation UCUT on an engineering drawing?
6 Draw the line type used to represent a centre line on an engineering drawing.
7 Describe what is represented by the following engineering drawing.

8 Describe what is represented by the following engineering drawing.

Ø8 C'sink at 60° to Ø12

9 A ring gear (hole) is fitted to a flywheel (shaft) using an S7/h6 interference fit. The nominal diameter of the flywheel is 320 mm. Using suitable data tables, find the tolerance limits for the internal diameter of the ring gear and the external diameter of the shaft.
10 What does the following geometric dimensioning and tolerancing (GD&T) symbol mean?

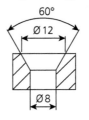

Project practice

You work as a machinist on the shop floor of a manufacturing company. You have been given an engineering drawing of a new component and asked to prepare to manufacture it. However, there are a couple of problems with the drawing.

▲ Identify what is missing from the concentricity geometric tolerance.

▲ Identify the dimension missing from a component.

▲ Explain the correct procedure for finding the missing information and adding it to the drawing.

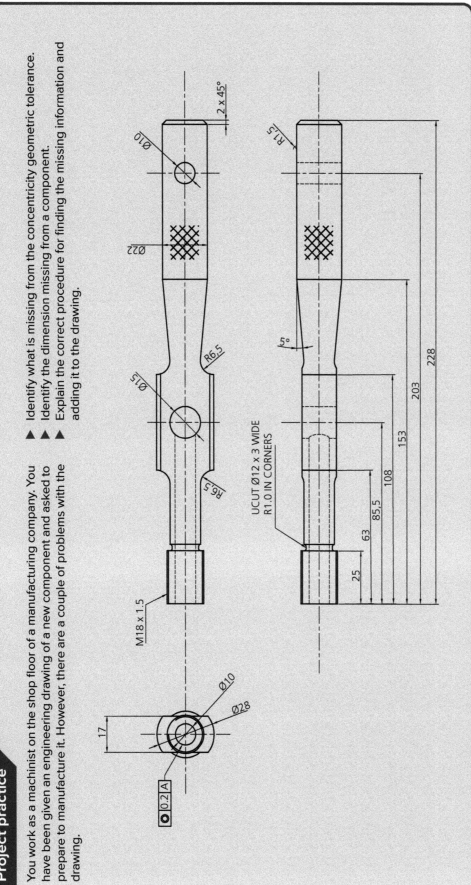

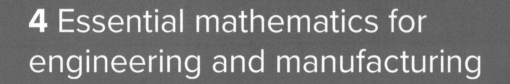

4 Essential mathematics for engineering and manufacturing

Introduction

Mathematical skills are an essential tool in engineering. For example, they can be used to:
- ▶ determine the properties required by products
- ▶ establish material requirements
- ▶ calculate the values of electrical components
- ▶ plot and identify trends in the performance of engineering processes.

This chapter identifies key mathematical skills and demonstrates how they are applied in engineering.

Learning outcomes

By the end of this chapter, you will understand how to:

1. perform arithmetic operations on integers, decimal numbers and numbers in standard form using rules of arithmetical preference (BIDMAS)
2. work to a specified number of decimal places or significant figures
3. carry out calculations using fractions, percentages, ratios and scale
4. simplify, factorise and manipulate equations to change the subject
5. solve simultaneous and quadratic equations
6. apply rules of indices
7. apply laws of logarithms (base 10 and natural), including solving problems involving growth and decay
8. determine numbers in a sequence using arithmetic progression, geometric progression and power series
9. calculate the area of 2D shapes (square, rectangle, triangle, circle) and the volume of 3D shapes (cube, cuboid, cylinder, cone)
10. interpret and express changes in an engineering system from a graph (straight line, trigonometrical and exponential relationships)
11. determine the equation of a straight line from a graph ($y = mx + c$)

12. determine standard differentials and integrals (basic arithmetic operations, powers/indices, trigonometric functions)
13. calculate maximum and minimum values in engineering contexts using differentiation
14. use Pythagoras' theorem and triangle measurement
15. apply circular measure, including conversion between radians and degrees
16. apply trigonometric functions (sin, cos, tan) and express their common values, rules and graphical representations
17. determine dimensions of a triangle using sine and cosine rules
18. use common trigonometric identities (sec, csc, cot)
19. use vectors, including addition, dot and cross product
20. add, subtract and multiply matrices in engineering contexts
21. calculate range, cumulative frequency, averages (mean, median and mode) and standard deviation for statistical data in an engineering context
22. determine probabilities in practical engineering situations
23. identify and convert between numbering systems
24. apply numbering systems typically used in engineering and manufacturing.

4.1 Applied mathematical theory in engineering applications

Standard arithmetic

Arithmetic deals with the properties and manipulation of numbers.

Ordering – rules of arithmetical preference

Most areas of engineering use formulae to calculate the values of required quantities or properties. These formulae often contain more than one mathematical operation.

When using formulae, there are some simple rules to ensure values are calculated correctly. These rules also apply when transposing or rearranging formulae.

There is a set order of priority for the sequence in which operations are carried out. This is known by the acronym **BIDMAS**, which stands for:

▶ **B**rackets
▶ **I**ndices (squaring and square roots)
▶ **D**ivision
▶ **M**ultiplication
▶ **A**ddition
▶ **S**ubtraction.

Example

$$12+(8÷4)×2-1=?$$

If this was calculated from left to right ignoring BIDMAS, the answer would be 9 (or 27 if the brackets were considered). Both of these answers would be incorrect.

Answer

Using BIDMAS:
▶ First, do the division in brackets: $8÷4=2$
▶ Then do the multiplication: $2×2=4$
▶ Then do the addition: $12+4=16$
▶ Finally, do the subtraction: $16-1=15$

Key term

BIDMAS: an acronym for remembering the set order of priority for the sequence in which operations are carried out: Brackets, Indices, Division, Multiplication, Addition, Subtraction.

Forms of numbers

Integers

These are whole numbers, such as 4, 17 and 645.

Fractions

A fraction represents a proportion of an object. This is typically shown as two integer values, the top number being divided by the bottom number:

▶ The top number is called the numerator.
▶ The bottom number is called the denominator.

In the fraction $\frac{3}{4}$, 3 is the numerator and 4 is the denominator.

If the numerator is larger than the denominator, the value is more than 1. This is called an improper fraction, for example $\frac{11}{4}=2\frac{3}{4}$.

Fractions should always be presented in their simplest form, for example $\frac{6}{12}$ should be written as $\frac{1}{2}$, and $\frac{4}{6}$ should be written as $\frac{2}{3}$.

To add, subtract, multiply or divide two fractions, they both need to be considered in terms of a common denominator. One way to do this is to multiply values by the denominators.

Addition of fractions

$$\frac{a}{b}+\frac{c}{d}=\frac{(a×d)+(c×b)}{b×d}$$

For example:

$$\frac{2}{3}+\frac{1}{4}=\frac{(2×4)+(1×3)}{3×4}=\frac{8+3}{12}=\frac{11}{12}$$

Subtraction of fractions

$$\frac{a}{b}-\frac{c}{d}=\frac{(a×d)-(c×b)}{b×d}$$

For example:

$$\frac{2}{3}-\frac{1}{4}=\frac{(2×4)-(1×3)}{3×4}=\frac{8-3}{12}=\frac{5}{12}$$

Multiplication of fractions

$$\frac{a}{b}×\frac{c}{d}=\frac{a×c}{b×d}$$

For example:

$$\frac{2}{3}×\frac{1}{4}=\frac{2×1}{3×4}=\frac{2}{12}=\frac{1}{6}$$

Division of fractions

$$\frac{a}{b} \div \frac{c}{d} = \frac{a \times d}{b \times c}$$

For example:

$$\frac{2}{3} \div \frac{1}{4} = \frac{2 \times 4}{3 \times 1} = \frac{8}{3} = 2\frac{2}{3}$$

Decimals

All fractions can be written as decimals by dividing the numerator by the denominator, for example:

$$\frac{1}{8} = 0.125$$

$$\frac{1}{4} = 0.25$$

$$\frac{1}{2} = 0.5$$

Some decimals have recurring digits. These are shown by a single dot above a single recurring digit, or by a dot above the first and last digit in a set of recurring digits. For example:

$$\frac{1}{3} = 0.3\dot{3}$$

$$\frac{1}{6} = 0.16\dot{6}$$

$$\frac{2}{11} = 0.\dot{1}\dot{8}$$

Standard form

Very large or very small numbers can be written in standard form, for example:
- 590 000 can be written as 5.9×10^5
- 0.0000024 can be written as 2.4×10^{-6} (the index is negative because the value is to the right of the decimal point).

Percentages

Percentages mean parts per hundred. To calculate a percentage, change the fraction to a decimal and multiply by 100, for example:

$$\frac{18}{60} = 0.3$$

$$0.3 \times 100 = 30\%$$

Ratios

Ratios can be used to show the relationship between two things, such as sizes, speeds, quantities or amounts. They are commonly used on engineering drawings to show the size of the drawing relative to the object that has been drawn, or for calculations involving gears or pulleys.

A fraction can be presented directly as a ratio by reducing it to its simplest form and separating the numerator and denominator with a colon, for example $\frac{7}{43}$ becomes 7:43.

To present a percentage as a ratio, it is first converted into a fraction, then reduced to its simplest form, for example:

$$55\% \rightarrow 0.55 \rightarrow \frac{55}{100} \rightarrow \frac{11}{20} \rightarrow 11:20$$

Significant figures

The application of **significant figures** is a method of managing the degree of approximation of a value. They specify the quantity of digits that are presented in the value. Zeros that are used to locate the decimal point are not significant.

> **Key term**
>
> *Significant figures:* the quantity of digits in a number that are required to present it to the correct approximation, not including zeros used to locate the decimal point.

To round a number to a given quantity of significant figures:

1. Start from the figure with the highest value and count the required number of figures.
2. Look at the next figure to the right of this and:
   - if the figure is 5 or more, round up
   - if the figure is less than 5, round down.
3. Add zeros, as necessary, to locate the decimal point and preserve the place value.

For example, the value 765 481 is:
- 765 000 to three significant figures
- 770 000 to two significant figures
- 800 000 to one significant figure.

When answering a problem using significant figures, the degree of approximation used should be stated, for example 'to three significant figures'.

Algebra

Algebra deals with formulae and equations, where letters are used to represent numbers and quantities.

Factorising and manipulating equations

Formulae and equations are used for almost every calculation in engineering, from establishing component values, to determining maximum structural loads, to converting measurements in different units. However, sometimes the value that needs to be determined is 'inside' the presented equation, rather than being the result ('outside' the equation).

For example, the formula for the mass of an object is $m = V\rho$, where V is the volume and ρ is the density of the material from which it is made. But what if we need to calculate the volume of material? If the mass and density are known, the equation can be rearranged to $V = \frac{m}{\rho}$.

The advantage of being able to rearrange formulae and equations is that it is not necessary to know all the possible different variations of equations using the same elements or variables. For any formula or equation, applying the rules shown in Table 4.1 allow it to be rearranged to make whichever element is required into the subject (or result).

▼ **Table 4.1 Rules for manipulating equations**

Rule	Example
If you add or subtract something on one side of an equation, you must do the same to the other side	If $x = y$, then: • $x + a = y + a$ • $x - a = y - a$
If you multiply or divide one side of an equation, you must do the same to the other side	If $x = y$, then: • $xa = ya$ • $\frac{x}{a} = \frac{y}{a}$
If you square or square root one side of an equation, you must do the same to the other side	If $x + y = z$, then $(x + y)^2 = z^2$ If $x^2 = y + z$, then $x = \sqrt{(y + z)}$

Table 4.2 lists some useful approaches when manipulating equations.

▼ **Table 4.2 Useful approaches when manipulating equations**

Approach	Example
To move or cancel something on one side of the equation, perform the opposite operation with it on both sides of the equation	$\frac{x}{y} = z$ can be rearranged by multiplying both sides by y, giving $x = yz$
Substitute any expression with another equal expression	If $x + y = z$ and $y = \frac{a}{b}$, then $x + \frac{a}{b} = z$
Expand out equations	$x(y + z) = 5$ is the same as $xy + xz = 5$
Factorise	$xy + xz = x(y + z)$

Example

Rearrange the equation $s = ut + \frac{1}{2}at^2$ to make a the subject.

Answer

▶ Subtract ut from both sides: $s - ut = \frac{1}{2}at^2$

▶ Multiply both sides by 2: $2(s - ut) = at^2$

▶ Divide both sides by t^2: $\frac{2(s - ut)}{t^2} = a$

Simultaneous equations

Simultaneous equations are a pair of equations that are to be solved at the same time by finding a pair of numbers that fit both equations.

The method for solving simultaneous equations is to eliminate one of the variables, allowing the other to be calculated. This value can then be substituted into the equations to allow the remaining variable to be calculated.

$2x + 6y = 30$ (equation 1)

$7x + 2y = 48$ (equation 2)

Answer

By multiplying the second equation by 3 and then subtracting the first equation from the result, the value of x can be determined. This value can then be inserted into the first equation and rearranged to calculate the value of y.

▶ Multiply the second equation by 3: $21x + 6y = 144$
▶ Subtract the first equation from this: $19x = 114$
▶ Rearrange to determine the value of x: $x = \frac{114}{19} = 6$
▶ Insert the value for x into the first equation: $12 + 6y = 30$
▶ Rearrange to determine the value of y: $6y = 18$

$y = \frac{18}{6} = 3$

Simultaneous equations may need to be rearranged to be in the required order, for example:

▶ $y = 4x - 6$ would need to be rearranged to $y - 4x = -6$ (or $4x - y = 6$)
▶ $5x - 2y - 7 = 0$ would need to be rearranged to $5x - 2y = 7$.

Example

The output force from a mechanical system is represented by the following equations:

$f = 3r - 5$

$3f - 5r - 1 = 0$

Determine the values of f and r.

Answer

▶ Rearrange the first equation: $3r - f = 5$
▶ Rearrange the second equation: $3f - 5r = 1$
▶ Multiply the first equation by 3: $9r - 3f = 15$
▶ Add this result to the second equation: $9r - 3f + 3f - 5r = 15 + 1$
▶ Resolve and determine the value of r: $4r = 16$

$r = \frac{16}{4} = 4$

▶ Insert the value for r into the first equation: $12 - f = 5$
▶ Rearrange to determine the value of f: $f = 12 - 5 = 7$

Solving quadratic equations

A quadratic equation is one that is in the form $ax^2 + bx + c$. When $ax^2 + bx + c = 0$, it can be solved by using the quadratic formula or by factorisation.

It should be noted that an equation may require rearranging into quadratic format. For example, $3x^2 = 5x + 17$ is in the form $ax^2 = bx + c$ and can be rearranged to $3x^2 - 5x - 17 = 0$.

Using the quadratic formula

The quadratic formula is:

$$x = \frac{-b \pm \sqrt{b^2 - 4ac}}{2a}$$

Example

Solve $3x^2 - 5x - 17 = 0$, giving your answers to three significant figures.

Answer

$$x = \frac{-b \pm \sqrt{b^2 - 4ac}}{2a} = \frac{-5 \pm \sqrt{(5)^2 - (4 \times 3 \times -17)}}{2 \times 3}$$

$x = -3.36$ and $x = 1.69$

Solving a quadratic equation by factorising

If quadratic equations have roots that are rational numbers, they can be solved by factorising. For the quadratic equation $ax^2 + bx + c = 0$, where $a = 1$, this requires finding a pair of numbers that satisfy the following:

▶ added together, the numbers must make the value b
▶ multiplied together, the numbers must make the value c.

For example, for $x^2 + 7x + 12 = 0$, the two numbers are 3 and 4:

$b = 7 = 3 + 4$

$c = 12 = 3 \times 4$

These two numbers can be used to factorise the equation into brackets:

$$x^2 + 7x + 12 = (x + 3)(x + 4)$$

Solving each bracket for zero gives the answers that satisfy this equation, $x = -3$ and $x = -4$.

Where $a \neq 1$, this requires finding a pair of numbers that satisfy the following:
▶ added together, the numbers must make the value b
▶ multiplied together, the numbers must make the value ac.

For example, for $4x^2 + 18x + 8 = 0$, $ac = 32$ and $b = 18$; the two numbers are 2 and 16. The equation can be rewritten as:

$$4x^2 + 2x + 16x + 8 = 0$$

The first two terms and the last two terms can be factorised:

$$4x^2 + 2x = 2x(2x + 1)$$

$$16x + 8 = 8(2x + 1)$$

Note that the two brackets are the same – this should be the case every time. Hence the equation becomes:

$$2x(2x + 1) + 8(2x + 1) = 0$$

Factorising again gives:

$$(2x + 8)(2x + 1) = 0$$

Solving each bracket for zero gives the final answers: $x = -4$ and $x = -0.5$.

Example

Solve $3x^2 - x - 10 = 0$ using factorising.

Answer

▶ Two numbers that satisfy $ac = -30$ and $b = -1$: 5 and -6
▶ Rewrite the equation: $3x^2 - 6x + 5x - 10 = 0$
▶ Factorise: $3x(x - 2) + 5(x - 2) = 0$
▶ Factorise again: $(3x + 5)(x - 2) = 0$
▶ Solve each bracket for zero to give the final answers: $x = -1.67$ (to three significant figures) and $x = 2$

Using indices

Indices show how many times a number or letter is multiplied by itself, for example:

$$3^4 = 3 \times 3 \times 3 \times 3$$

$$27^5 = 27 \times 27 \times 27 \times 27 \times 27$$

Table 4.3 outlines the rules for indices.

▼ Table 4.3 Rules for indices

Rule	Example(s)
Any number raised to the power of zero equals one	$3^0 = 1$ $27^0 = 1$
A negative power means the reciprocal of the term: $a^{-n} = \dfrac{1}{a^n}$	$6^{-3} = \dfrac{1}{6^3} = \dfrac{1}{216}$
If the power is a fraction with a number other than 1 as the numerator, then split the fraction up: $a^{\frac{m}{n}} = \left(\sqrt[n]{a}\right)^m$	$8^{\frac{2}{3}} = \left(\sqrt[3]{8}\right)^2 = (2)^2 = 4$
To multiply indices, add the powers	$2^2 \times 2^5 = 2^{(2+5)} = 2^7$
To divide indices, subtract the denominator from the numerator	$\dfrac{4^7}{4^3} = 4^{(7-3)} = 4^4$
To raise indices to a power, multiply the powers	$\left(7^3\right)^5 = 7^{(3\times5)} = 7^{15}$

Example

Solve $\left(\left(y^5 \times y^4\right) \div y^3\right)^2$.

Answer

$$\left(\left(y^5 \times y^4\right) \div y^3\right)^2 = \left(y^{5+4} \div y^3\right)^2 =$$

$$\left(y^9 \div y^3\right)^2 = \left(y^{9-3}\right)^2 = \left(y^6\right)^2 = y^{6\times2} = y^{12}$$

Using logarithms

Logarithms are the inverse functions to exponentials. They appear in many calculations in engineering, electronics and science, particularly in situations relating to growth or decay of variables.

Key terms

Indices: small superscript digits that appear after a number or letter to show many times it is multiplied by itself.

Logarithms: mathematical operations that determine how many times a base number is multiplied by itself to produce another number; they are the inverse functions to exponentials.

The logarithm of a number x is the exponent to which another number, the base b, must be raised to produce the number x, i.e. if $x = y^n$ then $\log_y x = n$. For example:

$$243 = 3^5 \rightarrow \log_3 243 = 5$$

Note that as $y^0 = 1$, therefore $\log_y 1 = 0$.

The two most common bases are base 10 and base e. $\text{Log}_{10} x$ is often just written as $\log x$. e is called the exponential constant and has a value of approximately 2.718.

Log_e is represented as ln and called a natural logarithm. Note that $\log_a x = \dfrac{\ln x}{\ln a}$.

The laws of logarithms are used when manipulating equations:

$$\log_a xy = \log_a x + \log_a y$$

$$\log_a x^m = m \log_a x$$

$$\log_a \frac{x}{y} = \log_a x - \log_a y$$

Example

Solve $9^{3x} = 3^{2x+2}$.

Answer

▶ Take logarithms of both sides:
$3x \log 9 = (2x + 2) \log 3$
▶ Resolve the brackets: $3x \log 9 = 2x \log 3 + 2 \log 3$
▶ Rearrange: $3x \log 9 - 2x \log 3 = 2 \log 3$
▶ Factorise: $x(3 \log 9 - 2 \log 3) = 2 \log 3$
▶ Rearrange: $x = \dfrac{2 \log 3}{3 \log 9 - 2 \log 3}$ $\quad x = 0.5$

Problems involving growth and decay

Logarithms are particularly useful for solving problems involving exponential growth (such as shown in Figure 4.1) or decay. This occurs when the change is proportional to the size of the population.

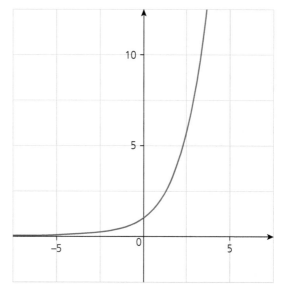

▲ Figure 4.1 Example of exponential growth: $y = 2^x$

If there is a requirement to determine the mathematical relationship between the two values, this could be achieved by directly plotting the data on a graph and manually finding the best fit, but this can be very time consuming and the accuracy may be limited. However, this is straightforward to do using logarithms.

An exponential relationship can be represented by the function $y = ax^b$, where a and b are constants.

Taking logarithms of both sides:

$$\log y = \log ax^b$$

Expanding this:

$$\log y = \log a + b \log x$$

When the log values are plotted, this gives a straight-line graph (see the section on calculus on page 62), as $\log y = \log a + b \log x$ is in the form $y = mx + c$ (in this case $m = b$, and $c = \log a$). By finding the intercept and gradient of a graph plotting the log values, the exact values of b and a can be determined.

Example

As a machine on a production line ages, it produces an increasing number of defective parts. The table below shows the number of defective parts produced per week by the machine over time. This can be modelled by the equation $D = at^b$, where D is the number of defects produced in the week, t is the number of weeks and a and b are constants.

Derive a suitable equation to find the predicted number of defects in week 50.

Week	1	2	3	5	7	11
Defective parts	14	17	19	23	25	29

Answer

Week	1	2	3	5	7	11
Log t	0	0.301	0.477	0.699	0.845	1.041
Defective parts	14	17	19	23	25	29
Log D	1.146	1.230	1.279	1.362	1.398	1.462

$Intercept = a = 14$

$$Gradient = b = \frac{1.462 - 1.146}{1.041 - 0} = 0.304$$

Hence the relationship is $D = 14t^{0.304}$

In week 50:

$$D = 14 \times 50^{0.304} = 45.98 = 46 \; defective \; parts$$

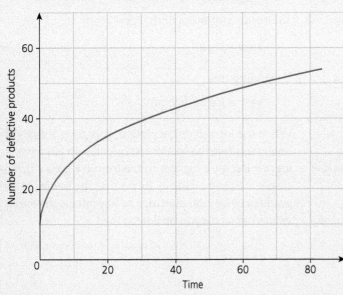

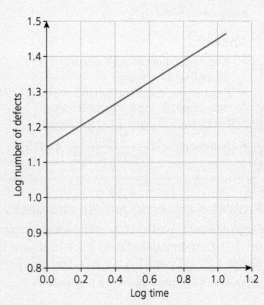

▲ Figure 4.2 Graphs of the number of defective products over time

Determining numbers in a sequence

The ability to understand how numbers progress in a sequence can be useful in activities such as calculating cumulative discounts on costs or determining the amount of time required to carry out certain tasks.

Types of sequence include **arithmetic progression**, **geometric progression** and **power series**.

> ### Key terms
>
> ***Arithmetic progression:*** sequence of numbers where the next term is found by adding or subtracting a common difference (d) to or from the previous term.
>
> ***Geometric progression:*** sequence of numbers where the next term is found by multiplying the previous term by a fixed (non-zero) number called the common ratio.
>
> ***Power series:*** infinite sequence of numbers which are power functions of variable x.

Arithmetic progression

An arithmetic progression is a sequence of numbers where the next term is found by adding or subtracting a common difference (d) to or from the previous term, for example 8, 10, 12, 14, 16 – each term has two added to the previous term.

The value of a specific term in the sequence can be calculated by:

$$u_n = a + (n-1)d$$

Where:
▶ a is the value of the first term in the sequence
▶ d is the common difference
▶ n is the term number in the sequence
▶ u_n is the nth term in the sequence.

For example, in the sequence 2, 5, 8, 11, 14, $a = 2$, $d = 3$. For 8, $n = 3$, as it is the third term in the sequence. The value of the twenty-third term in that sequence would be:

$$2 + \big((23-1) \times 3\big) = 68$$

The total sum of a sequence of n terms can be calculated by:

$$S_n = \frac{n}{2}(2a + (n-1)d)$$

For the sequence 2, 5, 8, 11, 14, if the sequence was 23 terms long the total sum would be:

$$S_n = \frac{23}{2}\big(2 \times 2 + ((23-1) \times 3)\big) = 805$$

Geometric progression

A geometric progression is a sequence of numbers where the next term is found by multiplying the previous term by a fixed (non-zero) number called the common ratio. For example, in the sequence 8, 12, 18, 27, each term is 1.5 (the common ratio) times the previous term.

The value of a specific term in the sequence can be calculated by:

$$u_n = a \times r^{n-1}$$

For example, in the sequence 8, 12, 18, 27, $a = 8$ and $r = 1.5$. The value of the twenty-third term in that sequence (to two decimal places) would be:

$$8 \times \left(1.5^{22}\right) = 59854.62$$

> ### Example
>
> A company is installing shelving in a warehouse. The first layer of shelving takes 40 minutes to assemble. Each subsequent layer takes 25% longer, as the shelving gets progressively higher.
>
> Determine how long it will take the company to install six layers of shelving.
>
> **Answer**
>
> | Shelf number | Time taken (minutes) | Cumulative time (minutes) |
> |---|---|---|
> | 1 | 40 | 40 |
> | 2 | 50 | 90 |
> | 3 | 62.5 | 152.5 |
> | 4 | 78.13 | 230.63 |
> | 5 | 97.66 | 328.29 |
> | 6 | 122.07 | 450.36 |

Power series

A power series is an infinite sequence of numbers which are power functions of variable x. It can be defined mathematically as:

$$\sum a_n (x-c)^n = a_0 + a_1 (x-c) + a_2 (x-c)^2 + \ldots + a_n (x-c)^n + \ldots$$

where a is a function independent of x and c is an arbitrary constant.

Power series are often used by calculators and computers to evaluate trigonometric, exponential and logarithm functions.

Standard matrices and determinants

A matrix is a rectangular array of algebraic or numerical elements (see Figure 4.3).

Matrices are useful as they can contain a lot of data in a form that makes computer calculations quick and efficient. Practical uses in engineering include the calculation of battery power outputs and solving alternating-current equations in electric circuits.

$A = \begin{bmatrix} 3 \\ 2 \end{bmatrix}$ 1 × 2 matrix

$A = \begin{bmatrix} 3 & 6 \\ 4 & 9 \end{bmatrix}$ 2 × 2 matrix

$A = \begin{bmatrix} 2 & 3 & 1 \\ 2 & 8 & 5 \end{bmatrix}$ 3 × 2 matrix

$A = \begin{bmatrix} 4 & 2 \\ 9 & 10 \\ -2 & 0 \end{bmatrix}$ 2 × 3 matrix

▲ **Figure 4.3 Examples of matrices**

The determinant is a special number that can be calculated from a square matrix, such as a 2×2, 3×3 or 4×4. It assists in finding the inverse of a matrix.

For the 2×2 matrix $\begin{bmatrix} a_{11} & a_{21} \\ a_{12} & a_{22} \end{bmatrix}$, the determinant is

$(a_{11} \times a_{22}) - (a_{12} \times a_{21})$.

Example

For matrix $A \begin{bmatrix} 3 & 4 \\ -2 & 6 \end{bmatrix}$, calculate the determinant.

Answer

$(3 \times 6) - (4 \times -2) = 26$

To find the determinant for the 3×3 matrix B, multiply each element by the determinant of the 2×2 matrix that is not in the corresponding row or column:

$$|B| = a_{11}(a_{22}\,a_{33} - a_{23}\,a_{32}) - a_{12}(a_{21}\,a_{33} - a_{23}\,a_{31})$$
$$+ a_{13}(a_{21}\,a_{33} - a_{22}\,a_{31})$$

Addition and subtraction of matrices

Matrices can only be added and subtracted if they are the same size. Each element is added to or subtracted from the element in the same position in the second matrix:

$$\begin{bmatrix} a_{11} & a_{12} \\ a_{21} & a_{22} \end{bmatrix} + \begin{bmatrix} b_{11} & b_{12} \\ b_{21} & b_{22} \end{bmatrix} = \begin{bmatrix} (a_{11}+b_{11}) & (a_{12}+b_{12}) \\ (a_{21}+b_{21}) & (a_{22}+b_{22}) \end{bmatrix}$$

For example:

$$\begin{bmatrix} 4 & -2 \\ 3 & 8 \end{bmatrix} + \begin{bmatrix} -3 & 7 \\ 0 & -4 \end{bmatrix} = \begin{bmatrix} 1 & 5 \\ 3 & 4 \end{bmatrix}$$

Multiplication of matrices

The number of rows in both matrices must be the same as the number of columns. Each row in the first matrix is multiplied by each column in the second matrix:

$$\begin{bmatrix} a_{11} & a_{12} \\ a_{21} & a_{22} \end{bmatrix} \times \begin{bmatrix} b_{11} & b_{12} \\ b_{21} & b_{22} \end{bmatrix} =$$

$$\begin{bmatrix} (a_{11} \times b_{11}) + (a_{12} \times b_{21}) & (a_{11} \times b_{12}) + (a_{12} \times b_{22}) \\ (a_{21} \times b_{11}) + (a_{22} \times b_{21}) & (a_{21} \times b_{12}) + (a_{22} \times b_{22}) \end{bmatrix}$$

For example:

$$\begin{bmatrix} 1 & 2 \\ 5 & -6 \end{bmatrix} \times \begin{bmatrix} -3 & -1 \\ 3 & 0 \end{bmatrix} =$$

$$\begin{bmatrix} (1 \times -3) + (2 \times 3) & (1 \times -1) + (2 \times 0) \\ (5 \times -3) + (-6 \times 3) & (5 \times -1) + (-6 \times 0) \end{bmatrix} = \begin{bmatrix} 3 & -1 \\ -33 & -5 \end{bmatrix}$$

Geometry

Geometry deals with the properties and relationships of two-dimensional and three-dimensional shapes.

Area of 2D shapes

When calculating the cost of material needed in a product or the stress in a component, it is typically necessary to calculate the area of simple shapes (see Figure 4.4 on the next page). The area of complicated shapes can be calculated by treating them as combinations of simple shapes (see Figure 4.5 on the next page).

Square

Rectangle

Width

Width

Length

Triangle

Circle

Height

Radius

Diameter

Base

▲ Figure 4.4 Common 2D shapes

▼ Table 4.4 Areas of 2D shapes

Shape	Formula for area	
Square	$width \times width$	$A = w^2$
Rectangle	$width \times length$	$A = wl$
Triangle	$\frac{1}{2} \times base \times height$	$A = \frac{1}{2}bh$
Circle	$\pi \times radius \times radius$	$A = \pi r^2$

Example

The shape in Figure 4.5 is made from a rectangle and a semicircle. What is its area?

Answer

Area of the rectangle:

$$wl = 0.2 \times 0.15 = 0.03\,\text{m}^2$$

Area of the semicircle:

$$\frac{1}{2}\pi r^2 = \frac{1}{2} \times 3.14 \times 0.1^2 = 0.0157\,\text{m}^2$$

To find the total area, add the area of the rectangle and the area of the semicircle:

$$0.03 + 0.0157 = 0.0457\,\text{m}^2 \; (4.57 \times 10^{-2}\,\text{m}^2)$$

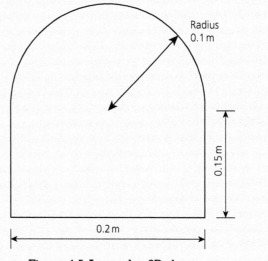

Radius 0.1 m

0.15 m

0.2 m

▲ Figure 4.5 A complex 2D shape

Volume of 3D shapes

When determining the mass of material needed in a product, it is necessary to calculate the volume of the shape (see Figure 4.6). Similar to area calculations, the volume of complicated shapes can be calculated by treating them as combinations of simple shapes.

▼ Table 4.5 Volumes of 3D shapes

Shape	Formula for volume	
Cube	$width \times width \times width$	$V = w^3$
Cuboid	$width \times length \times height$	$V = wlh$
Cylinder	$\pi \times radius \times radius \times length$	$V = \pi r^2 l$
Cone	$\frac{1}{3} \times \pi \times radius \times radius \times height$	$V = \frac{1}{3}\pi r^2 h$

Cuboid

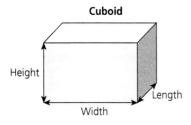

Height

Length

Width

Cylinder

Length

Cone

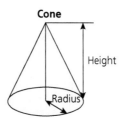

Height

Radius

▲ Figure 4.6 Common 3D shapes

Calculus

Calculus deals with the determination and analysis of change.

Graphs and charts relevant to engineering and manufacturing contexts

Graphs are often used to represent changes in engineering systems, such as the temperature of a machine over time or how the dimensions of different batches of products have varied. The trend shown by the graph indicates what is happening in the system.

The three main types of graphical relationship are shown in Figure 4.7. Being able to read values from each type of graph is an essential skill for many engineers.

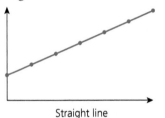
Straight line

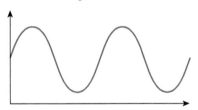
Trigonometrical (e.g. sine, cosine)

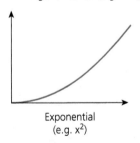
Exponential
(e.g. x²)

▲ Figure 4.7 Types of graphical relationship

Straight-line graphs

Straight-line graphs can be represented by the formula $y = mx + c$, where m is the gradient of the line and c is the value where the relationship crosses the y-axis, known as the intercept. The gradient can be found by dividing the rise of the graph by its step (see Figure 4.8). For example, if the increase in the y-axis value is 12 (the rise) for a step in the in the x-axis of 16, then the gradient is $\frac{12}{16} = 0.75$.

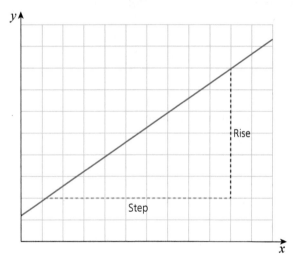

▲ Figure 4.8 Determining the gradient of a straight line

For example, Figure 4.9 shows the straight-line relationship between the resistance of a piece of copper and the amount of aluminium it contains. From the graph, $c = 8$ and $m = \dfrac{24 - 8}{0.2 - 0} = 80$.

Hence, the formula for the resistance:

$$R = (80 \times \% \, Al) + 8 \; ohms$$

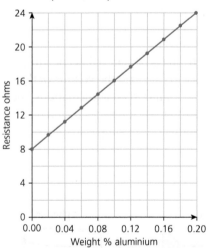

▲ Figure 4.9 Relationship between resistance and aluminium content in copper

Differentiation

In engineering, **differentiation** is often needed to determine the instantaneous rate of change of a function. For example:

▶ acceleration is the rate of change of velocity with respect to time

▶ velocity is the rate of change of position with respect to time.

> ### Key term
>
> **Differentiation:** process used to calculate the instantaneous rate of change of a function.

▼ Table 4.6 Standard derivatives

$Y = f(x)$	$\dfrac{dy}{dx} = f'(x)$
k (any constant)	0
x	1
x^n	nx^{n-1}
e^{kx}	ke^{kx}
$\sin x$	$\cos x$
$\sin kx$	$k \cos kx$
$\cos x$	$-\sin x$
$\cos kx$	$-k \sin kx$

For example:

▶ for the function $y = x^3$, the differential $\dfrac{dy}{dx} = 3x^2$

▶ for the function $y = 9x^2 + 2x - 3$, the differential $\dfrac{dy}{dx} = 18x + 2$

▶ for the function $y = \cos 3x$, the differential $\dfrac{dy}{dx} = -3\sin 3x$.

Calculating maximum and minimum values using differentiation

Differentiation can also be used to find the maximum and minimum values of a function. At the maxima or minima, there is a point of inflexion where the gradient and value of the differential will equal zero.

For example, the function $y = x^2 - 4x + 4$ has a single point of inflexion. The differential of this function is $2x - 4$. At the minima, $2x - 4 = 0$; rearranging $2x = 4$, therefore the point of inflexion occurs at $x = 2$. Using differentiation is much quicker than the alternative method of plotting a graph (see Figure 4.10).

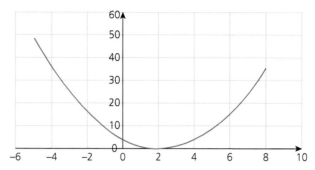

▲ Figure 4.10 Graph of $y = x^2 - 4x + 4$

For a more complex relationship, there may be both maximum and minimum values (see Figure 4.11). These can be determined by differentiating the function and then solving using the quadratic formula.

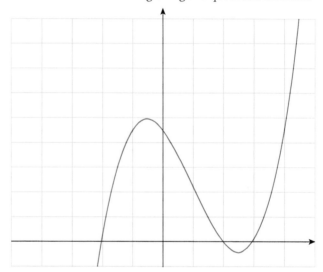

▲ Figure 4.11 Function with both a maximum and minimum point of inflection

For example, for the function $y = x^3 - 1.5x^2 - x + 1.5$, the differential $\dfrac{dy}{dx} = 3x^2 - 3x - 1$.

Equating this to zero and using the quadratic equation:

$$x = \frac{3 \pm \sqrt{9 - (4 \times 3 \times -1)}}{2 \times 3} = \frac{3 \pm \sqrt{21}}{2 \times 3}$$

Hence $x = 1.26$ and $x = -0.264$. To determine which of these values is a maximum and which is a minimum, the second differential can be used. The second differential of $\dfrac{dy}{dx} = 3x^2 - 3x - 1$ is $\dfrac{d^2y}{dx^2} = 6x - 3$.

When the previously substituted values of x are substituted into this, if $\dfrac{d^2y}{dx^2} > 0$ then it is a minimum, and if $\dfrac{d^2y}{dx^2} < 0$ then it is a maximum.

For $x = 1.26$, $\dfrac{d^2y}{dx^2} = 6x - 3 = 4.56$, therefore it is a minimum.

For $x = -0.264$, $\dfrac{d^2y}{dx^2} = 6x - 3 = -4.58$, therefore it is a maximum.

Integration

Integration is the opposite of differentiation. For example:
▶ the integral of a function for the acceleration of an object gives its velocity
▶ the integral of a function for the velocity of an object gives its change of position.

▼ Table 4.7 Standard integrals

$Y = f(x)$	$\int f(x)\, dx$
k (any constant)	$kx + c$
x	$\dfrac{x^2}{2} + c$
x^n	$\dfrac{x^{n+1}}{n+1} + c$
e^{kx}	$\dfrac{1}{k}e^{kx} + c$
$\dfrac{1}{x}$	$\ln x + c$
$\sin x$	$-\cos x + c$
$\sin kx$	$-\dfrac{1}{k}\cos kx + c$
$\cos x$	$\sin x + c$
$\cos kx$	$\dfrac{1}{k}\sin kx + c$

For example:
▶ for the function $y = x^3$, the integral
$$\int f(x)\, dx = \frac{1}{4}x^4 + c$$
▶ for the function $y = 9x^2 + 2x - 3$, the integral
$$\int f(x)\, dx = 3x^3 + x^2 - 3x + c$$
▶ for the function $y = \cos 3x$, the integral
$$\int f(x)\, dx = \frac{1}{3}\sin 3x + c.$$

When using standard integrals, it should be noted that a constant c is added to the value. If the integral is determining a change that occurs within a frame of two values, the constants c will cancel out.

Trigonometry

Trigonometry deals with the relationship between the sides and angles of a triangle – it is the study of triangle measurement. The internal angles in a triangle always add up to $180°$, irrespective of the relative proportions of the three sides.

Pythagoras' theorem

Pythagoras was a Greek mathematician who lived from 580 to 500BC. He discovered that for right-angled triangles, the square of the longest side (called the hypotenuse) was equal to the sum of the squares of the two shortest sides (see Figure 4.12): $a^2 + b^2 = c^2$.

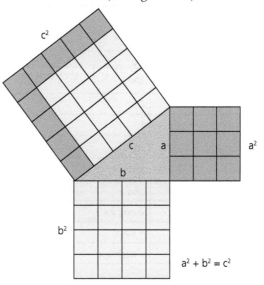

▲ Figure 4.12 Pythagoras' theorem

If two sides of a right-angled triangle are known, **Pythagoras' theorem** can be rearranged to find the third side. If the shorter sides measure 3 and 4 units long, then the hypotenuse will measure 5 units. Where right-angled triangles are multiples of the 3:4:5 relationship (for example 9:16:20 or 15:20:25), this can simplify the calculations required.

Key terms

Integration: the opposite of differentiation.

Pythagoras' theorem: for right-angled triangles, the square of the longest side (called the hypotenuse) is equal to the sum of the squares of the two shortest sides: $a^2 = b^2 + c^2$.

Example

Figure 4.13 shows a two-dimensional view of a taper. Calculate length A, giving your answer to three significant figures.

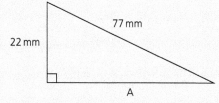

▲ Figure 4.13 Two-dimensional view of a taper

Answer

Rearranging Pythagoras' theorem:

$$A = \sqrt{\left(77^2 - 22^2\right)} = \sqrt{5445} = 73.79 = 73.8\,mm$$

(to three significant figures)

Trigonometric functions

If the length of only one side of a triangle is known, **trigonometric functions** can be used to calculate the other dimensions. The most commonly used trigonometric functions are sine, cosine and tangent, known by the abbreviations sin, cos and tan respectively. Determining which function to use depends on:

▶ which side has a known dimension
▶ the angle
▶ the side to be calculated.

Key term

Trigonometric functions: set of functions – sine (sin), cosine (cos) and tangent (tan) – that can be used to calculate the other dimensions of a triangle when only the length of one side is known.

For a right-angled triangle (see Figure 4.14):

▶ the hypotenuse opposite the right angle is the longest side
▶ the opposite is opposite the angle being used for calculations
▶ the adjacent is next to the angle being used for calculations.

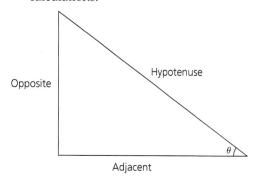

▲ Figure 4.14 Right-angled triangle

With knowledge of the angle θ and the length of one side, the dimensions of the other sides can be calculated (see Table 4.8). An easy way to remember and transpose these formulae is by using a formula triangle (see Figure 4.15), where you hold a finger over the unknown value to indicate the calculation. For example, to calculate the hypotenuse, it is $\frac{adjacent}{cosine\ \theta}$.

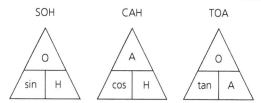

▲ Figure 4.15 Trigonometric formula triangle

▼ Table 4.8 Trigonometric formulae

$sin\ \theta = \dfrac{opposite}{hypotenuse}$	$opposite = sin\ \theta \times hypotenuse$	$hypotenuse = \dfrac{opposite}{sin\ \theta}$
$cos\ \theta = \dfrac{adjacent}{hypotenuse}$	$adjacent = cos\ \theta \times hypotenuse$	$hypotenuse = \dfrac{adjacent}{cos\ \theta}$
$tan\ \theta = \dfrac{opposite}{adjacent}$	$opposite = tan\ \theta \times adjacent$	$adjacent = \dfrac{opposite}{tan\ \theta}$

Example

The triangle below needs to be marked out for cutting.

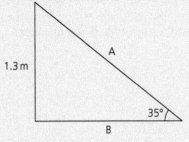

▲ Figure 4.16

Determine the lengths of sides A and B. Give your answers to three significant figures.

Answer

$$\text{Hypotenuse, } A = \frac{opposite}{\sin \theta} = \frac{1.3}{\sin 35°} = 2.27\,\text{m}$$

$$\text{Adjacent, } B = \frac{opposite}{\tan \theta} = \frac{1.3}{\tan 35°} = 1.86\,\text{m}$$

Common values of trigonometric functions

There are some common values of sin, cos and tan, as shown in Table 4.9. These can be worked out using special triangles (see Figure 4.17).

▼ Table 4.9 Exact common values of sin, cos and tan

	sin	cos	tan
0°	0	1	0
30°	$\frac{1}{2}$	$\frac{\sqrt{3}}{2}$	$\frac{1}{\sqrt{3}}$
45°	$\frac{1}{\sqrt{2}}$	$\frac{1}{\sqrt{2}}$	1
60°	$\frac{\sqrt{3}}{2}$	$\frac{1}{2}$	$\sqrt{3}$
90°	1	0	–

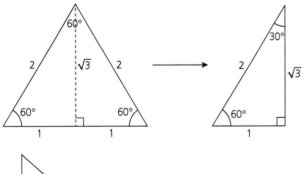

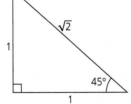

▲ Figure 4.17 Special triangles

Graphs of trigonometric functions

Relationships typical of trigonometric functions are often observed in graphs, for example of the output of electrical systems or the displacement of mechanical components. These graphs typically follow patterns that repeat at $360°$ intervals (or 2π if using radians). The characteristic graphs are different for each trigonometric function (see Figure 4.18a–c).

The time taken by the function to complete one cycle is called the period. The maximum height (or depth) of the function is called the amplitude.

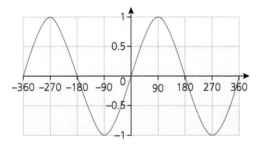

▲ Figure 4.18a Sine graph (degrees)

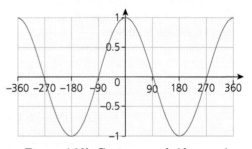

▲ Figure 4.18b Cosine graph (degrees)

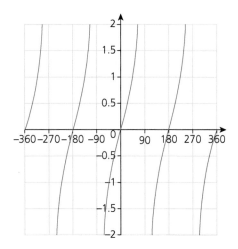

▲ Figure 4.18c Tangent graph (degrees)

Trigonometric identities

There are three trigonometric functions which represent the reciprocal functions of sin, cos and tan:

▶ $sec\theta = \dfrac{1}{cos\theta}$ ▶ $cosec\ \theta = \dfrac{1}{sin\theta}$ ▶ $cot\theta = \dfrac{1}{tan\theta}$

These are used to solve problems and represent mathematical relationships in civil, mechanical and electrical applications, for example when analysing alternating and direct currents.

Sine and cosine rules

Where a triangle does *not* contain two sides at a right angle to each other, the **sine rule** and **cosine rule** can be used to find unknown dimensions (see Figure 4.19):

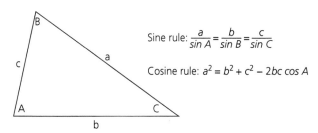

Sine rule: $\dfrac{a}{sin A} = \dfrac{b}{sin B} = \dfrac{c}{sin C}$

Cosine rule: $a^2 = b^2 + c^2 - 2bc\ cos A$

▲ Figure 4.19 Sine and cosine rules

> **Key terms**
>
> **Sine rule:** $\dfrac{a}{sin A} = \dfrac{b}{sin B} = \dfrac{c}{sin C}$
>
> **Cosine rule:** $a^2 = b^2 + c^2 - 2bc\ cos A$

> **Example**
>
> The part shown in Figure 4.20 needs to be marked out for cutting, to replace a damaged section on a metal casing. Determine the length x and the angles y and z.
>
>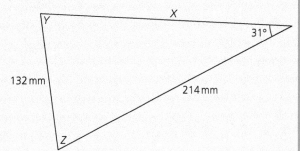
>
> ▲ Figure 4.20 Replacement part (not to scale)
>
> **Answer**
>
> ▶ Use the sine rule: $\dfrac{sin y}{214} = \dfrac{sin 31}{132}$
>
> ▶ Rearrange: $y = sin^{-1}\left(\dfrac{sin 31}{132} \times 214\right) = 56.6°$
>
> ▶ As the angles in a triangle add up to 180°: $z = 180 - 31 - 56.6 = 92.4°$
>
> ▶ Calculate: $x = \dfrac{132}{sin 31} \times sin 92.4 = 256\,mm$

Circular measure

It may seem odd to have a section on circular measure within trigonometry, however this is because circular measure is dependent upon angles. Some key terms referring to circles (see Figure 4.21 on the next page) are **diameter**, **circumference**, **arc**, **radius** and **sector**.

> **Key terms**
>
> **Diameter:** straight line passing from side to side through the centre of a circle.
>
> **Circumference:** outside border of a circle.
>
> **Arc:** partial length of the circumference of a circle.
>
> **Radius:** straight line from the centre of a circle to its circumference, equal to half the length of the diameter.
>
> **Sector:** pie-shaped part of a circle, comprising two radii at an angle of θ joined by an arc.

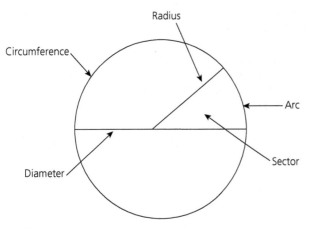

▲ **Figure 4.21 Characteristic features of a circle**

If the diameter of the circle was 1 unit, the radius would be 0.5 units and the circumference would be π units, i.e. *circumference* $= \pi d = 2\pi r$.

A full circle contains 360 degrees. However, there is an alternative measure for angles, called radians – this is based on the circumference of the circle. A full circle contains 2π radians. Values in radians are often left in terms of π and expressed as fractions.

To convert between degrees and radians:

▶ $radians = number\ of\ degrees \times \left(\dfrac{2\pi}{360}\right)$

▶ $degrees = radians \times \left(\dfrac{360}{2\pi}\right)$

For example:

$$20° = 20 \times \frac{2\pi}{360} = \frac{1}{9}\pi\ radians$$

$$\frac{\pi}{4}radians = \frac{\pi}{4} \times \left(\frac{360}{2\pi}\right) = 45°$$

Radians are particularly useful to simplify the calculation of the lengths of arcs or the area of sectors:
▶ The length of an arc when the angle is in degrees is $\dfrac{\theta}{360} \times 2\pi r$.
▶ As $360° = 2\pi$ radians, if the angular measure is in radians then the arc length is $\dfrac{\theta}{2\pi} \times 2\pi r$, where 2π cancels out, to give arc length $r\theta$.
▶ The area of a sector when the angle is in degrees is $\dfrac{\theta}{360} \times \pi r^2$.

Similar to converting to radians and cancelling out 2π, if the angular measure is in radians, then the area of a sector is $\dfrac{1}{2}r^2\theta$.

Example

Calculate the area of a sector where the radius is 3.2 m and the internal angle is 4 radians.

Answer

$$Area = \frac{1}{2}r^2\theta = \frac{1}{2}\ 3.2^2 \times 4 = 20.48\,m^2$$

Applications of vectors and coordinates

Vectors are commonly encountered in engineering, for example when programming the movement of a machine tool. A **vector** has both magnitude and direction and is usually denoted by an arrow (see Figure 4.22).

In contrast, a **scalar** has magnitude but no direction. Examples of scalar measures include density, volume, energy, mass and time.

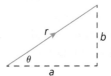

▲ **Figure 4.22 Characteristics of a vector**

Key terms

Vector: value with both magnitude and direction.

Scalar: value with magnitude but not direction.

Vectors can be defined in linear terms $r = a\boldsymbol{i} + b\boldsymbol{j}$ (note that the convention is to present the i and j in bold when printed) or as column vectors $\begin{pmatrix} a \\ b \end{pmatrix}$. The values a and b are the cartesian coordinates; the value r presented with the angle θ is the polar coordinate.

Magnitude and direction of a vector

The magnitude of a vector is represented by the vector being enclosed in parallel lines. It can be worked out using Pythagoras' theorem. For the vector in Figure 4.22:

$$|r| = \sqrt{a^2 + b^2}$$

The angle θ can be worked out using trigonometry. For the vector in Figure 4.22:

$$\tan\theta = \frac{b}{a}$$

Example

Find the magnitude and direction of the vector $r = 3i - 5j$.

Answer

$$|r| = \sqrt{a^2 + b^2} = \sqrt{3^2 + -5^2} = \sqrt{34} = 5.83$$

$$\theta = \tan^{-1}\frac{5}{3} = 59°$$

Addition of vectors

Vectors can be added together to make a resultant vector:

$$\begin{pmatrix} a \\ b \end{pmatrix} + \begin{pmatrix} c \\ d \end{pmatrix} = \begin{pmatrix} a + c \\ b + d \end{pmatrix}$$

For example:

$$\begin{pmatrix} 3 \\ -5 \end{pmatrix} + \begin{pmatrix} 2 \\ 2 \end{pmatrix} = \begin{pmatrix} 5 \\ -3 \end{pmatrix}$$

Alternatively:

$$(ai + bj) + (ci + dj) = (a + c)i + (b + d)j$$

For example:

$$(3i - 5j) + (2i + 2j) = 5i - 3j$$

Subtraction of vectors

Similarly, vectors can be subtracted:

$$\begin{pmatrix} a \\ b \end{pmatrix} - \begin{pmatrix} c \\ d \end{pmatrix} = \begin{pmatrix} a - c \\ b - d \end{pmatrix}$$

For example:

$$\begin{pmatrix} 3 \\ -5 \end{pmatrix} - \begin{pmatrix} 2 \\ 2 \end{pmatrix} = \begin{pmatrix} 1 \\ -7 \end{pmatrix}$$

Alternatively:

$$(ai + bj) - (ci + dj) = (a - c)i + (b - d)j$$

For example:

$$(3i - 5j) - (2i + 2j) = 1i - 7j$$

Dot product

The dot product is the sum of the products of the corresponding entries in two vectors. It is also known as the scalar product. Where vector $a = \begin{pmatrix} a \\ b \end{pmatrix}$ and $b = \begin{pmatrix} c \\ d \end{pmatrix}$, the dot product a.b is:

$$\begin{pmatrix} a \\ b \end{pmatrix} \times \begin{pmatrix} c \\ d \end{pmatrix} = (a \times c) + (b \times d)$$

For example:

$$\begin{pmatrix} 3 \\ -5 \end{pmatrix} \times \begin{pmatrix} 2 \\ 2 \end{pmatrix} = (3 \times 2) + (-5 \times 2) = -4$$

Alternatively:

$$(ai + bj) \times (ci + dj) = (ac) + (bd)$$

For example:

$$(3i - 5j) \times (2i + 2j) = (3 \times 2) + (-5 \times 2) = -4$$

Cross-product

The cross-product of two vectors a and b is a vector that is perpendicular to both a and b. One of the main uses of vector cross-products in engineering is to calculate the forces about a point or line. It is also known as the vector product:

$$a \times b = |a||b|\cos\theta$$

The cross-product can be rearranged to find the angle between two vectors.

Example

Using the cross-product, find the angle between the vectors $a = \begin{pmatrix} 3 \\ -5 \end{pmatrix}$ and $b = \begin{pmatrix} 2 \\ 2 \end{pmatrix}$.

Answer

$$\cos\theta = \frac{a.b}{|a||b|}$$

$$a.b = (3 \times 2) + (-5 \times 2) = -4$$

$$|a| = \sqrt{3^2 + (-5)^2} = \sqrt{34}$$

$$|b| = \sqrt{2^2 + 2^2} = \sqrt{8}$$

$$\theta = \cos^{-1}\frac{-4}{\sqrt{34}\sqrt{8}} = 104°$$

Statistical analysis

Statistics deals with the collection, organisation, analysis and interpretation of data.

Analysis of data

The ability to analyse data is used across all areas of engineering. Examples include monitoring the performance of a process or looking for trends in quality-control measurements.

When presented with a set of numbers, some of the key statistical values are the **range**, **cumulative frequency**, **mean average, $\bar{x}$, median, modal average** and **standard deviation**.

Standard deviation can be calculated for a population (full set of numbers) using the following formula:

$$\sigma = \sqrt{\dfrac{\sum f\left(x_i - \bar{x}\right)^2}{n}}$$

where f is the frequency with which the value x_i occurs.

For ungrouped data (i.e. individual values) $f = 1$. Grouped data is where there is more than one number of the same value or within the same range listed, with the frequency of occurrence.

When the standard deviation is calculated manually:
1. Calculate the mean average for the data set, $\bar{x}$.
2. For each value, subtract the mean and square the result.
3. Work out the mean of those squared differences.
4. Calculate the square root of the result.

In practice, this is often carried out by populating a table with these values. If the standard deviation is being calculated for a sample from a much larger set of numbers, then a slightly different formula is used:

$$\sigma = \sqrt{\dfrac{\sum f\left(x_i - \bar{x}\right)^2}{n-1}}$$

The mean, median and modal average measure the location of the data. The range, cumulative frequency and standard deviation measure the spread, which indicates how consistent the data is.

In a normal distribution, such as in Figure 4.23, approximately 68% of the data set falls within one standard deviation either side of the mean value; approximately 95% falls within two standard deviations and 99.7% within three standard deviations. This is important, as it is used to specify the action and control limits for statistical process-control activities.

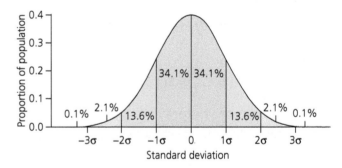

▲ Figure 4.23 Normal distribution of values

Answer

x_i	f	$x_i f$	Σf	$x_i - \bar{x}$	$(x_i - \bar{x})^2$	$f\,(x_i - \bar{x})^2$
49.5–50.5	3	150	3	-2.5	6.25	18.75
50.5–51.5	9	459	14	-1.5	2.25	20.25
51.5–52.5	19	988	33	-0.5	0.25	4.75
52.5–53.5	17	901	50	0.5	0.25	4.25
53.5–54.5	8	432	58	1.5	2.25	18
54.5–55.5	4	220	60	2.5	6.25	25
Totals	60	3150				91

Taking the mid-point of each range, to calculate x_i f, $\bar{x} = \dfrac{3150}{60} = 52.5\,mm$

Standard deviation:

$$\sigma = \sqrt{\frac{\sum f\left(x_i - \bar{x}\right)^2}{n}} = \sqrt{\frac{91}{60}} = 1.23\,mm$$

Probabilities

Probability is the measurement of the likelihood that an event will occur. It may be presented as a fraction (in its simplest form), a decimal or a percentage.

> **Key term**
>
> **Probability:** measurement of the likelihood that an event will occur, presented as a fraction (in its simplest form), a decimal or a percentage.

The probability of a single event can be presented directly as a value relative to the total number of events:

$$p\,(event) = \frac{number\ of\ events\ that\ occurred, x}{total\ number\ of\ events, n} = \frac{x}{n}$$

For example, if a batch of 30 otherwise identical components contains 17 with crack defects, the probability of picking a component at random with a crack defect is $p(crack) = \dfrac{17}{30}$.

Events that cannot occur simultaneously are called mutually exclusive events. For example, a switch can be either on or off, and a component can be either acceptable or faulty. Where events are mutually exclusive:

▶ The sum of the probabilities for a situation must equal 1 (or 100%). For example, if there are only two possible outcomes, A and B, to an event, then $p(A) + p(B) = 1$.

▶ In a situation with multiple outcomes, where the probability of outcome A occurring is $p(A)$, then the probability of outcome A not occurring is $p(A') = 1 - p(A)$.

▶ The probability of one out of two (or more) random events occurring can be found by adding the probabilities: $p(A\ or\ B) = p(A) + p(B)$.

▶ The probability of multiple random events occurring together can be found by multiplying the probabilities: $p(A\ and\ B) = p(A) \times p(B)$.

> **Example**
>
> A box contains 120 nuts that are identical apart from the threads: 60 nuts have fine threads, 40 nuts have medium threads and 20 nuts have coarse threads.
>
> Determine the probability that a nut selected at random has:
> ▶ a fine thread
> ▶ either a fine or a medium thread.
>
> **Answer**
>
> $$p\left(fine\right) = \frac{60}{120} = \frac{1}{2}$$
> $$p\left(fine\ or\ medium\right) = \frac{60 + 40}{120} = \frac{100}{120} = \frac{5}{6}$$

Estimation

Estimation is used in engineering to give a rough indication of a value or as a way of checking a value. It involves rounding any presented numbers so they can be quickly calculated using mental arithmetic.

For example, a company needs to produce 510 parts, each of which requires 0.76 kg of material. For an estimation, round the 510 to 500 and the 0.76 to 0.8; hence, the material required is approximately $500 \times 0.8 = 400$ kg.

It should be noted that this is different from the precise value ($510 \times 0.76 = 387.6$ kg). If an exam question states 'estimate' and the precise value is given instead, this would be marked as incorrect.

4.2 Number systems used in engineering and manufacturing

Numbering systems

Decimal notation

The most familiar numbering system in use is the decimal system. This works in base 10, which means that ten digits (0, 1, 2, 3, 4, 5, 6, 7, 8, 9) are used to create any required number. This is used for most practical daily calculations, such as totalling costs or calculating the strength of components.

Binary notation

Binary numbering is in base 2, using only the digits 0 and 1 (see Table 4.10). It is used by computers to store data, as they are based around electronic switches which have two states: open (also referred to as off or low) and closed (also referred to as on or high).

To convert a decimal value to a binary value, first find the highest binary value less than the decimal value. Then repeat this for each remainder until it is fully resolved; any binary value that is not exceeded by the remainder is zero.

For example, for the decimal value 22, the highest binary constituent would be 16, remainder 7; the first binary value would be 1 (remainder 7), the second 0 (remainder 7 is less than 8), the third 1 (remainder 3), the fourth 1 (remainder 0), the final 0. Hence the decimal value 22 in binary is 10110.

Similarly, by summing the individual digits, binary values can be converted to decimal values. For example, the binary value 101011 equals the decimal value 49.

Hexadecimal notation

Hexadecimal notation is in base 16 and uses the digits 0, 1, 2, 3, 4, 5, 6, 7, 8, 9, A, B, C, D, E, F (see Table 4.11). It is used when programming electronic devices, as each hex digit represents a 4-bit binary number. The practical implication of this is to reduce the need to input long strings of digits in binary values when programming.

For example, the binary value 11011010110101101 equals hexadecimal value 1B5AD (and the decimal value 112045). If there was an error of just one figure when entering a long binary value into a CNC machine, this could have disastrous implications for the item being manufactured.

Converting hexadecimal values to or from decimal or binary values uses the same methodology as for converting binary values to decimal. Some examples are shown in Table 4.12.

▼ Table 4.10 Binary and decimal values

Binary value	100000	10000	1000	101	100	11	10	1	0
Decimal value	32	16	8	5	4	3	2	1	0

▼ Table 4.11 Hexadecimal values

Hexadecimal value	F	E	D	C	B	A	9	8	7	6	5	4	3	2	1	0
Decimal value	15	14	13	12	11	10	9	8	7	6	5	4	3	2	1	0
Binary value	1111	1110	1101	1100	1011	1010	1001	1000	111	110	101	100	11	10	1	0

▼ Table 4.12 Values using different number systems

Decimal value	12	22	94	1494	46550	1750486
Binary value	1100	10110	1011110	10111010110	1011010111010110	110101011010111010110
Hexadecimal value	C	16	5E	5D6	B5D6	1AB5D6

Assessment practice

1 A container is filled with 1.8 litres of oil. A maintenance engineer pours half of this oil into a machine. He then uses a third of the remaining oil in a second machine. Calculate the amount of oil remaining in the container, giving your answer to one decimal place.

2 An engineering drawing of a component has been prepared using the scale 1:5. If the length of a critical dimension on the drawing is measured as 85 mm, calculate the actual length of the dimension on the component.

3 A solid metal cylinder of cross-sectional area 225 mm^2 is being used as part of a lifting rig. The maximum permitted tensile stress on the cylinder is 45 N/mm^{-2}. Rearrange the following formula to determine the maximum force that can be applied to the cylinder:

$$\sigma = \frac{F}{A}$$

where σ is stress, F is force applied and A is area.

4 The output from an electrical device has been recorded as:

$4x + 3y - 26 = 0$

$2x - 7y + 38 = 0$

Determine the values of x and y.

5 A trainee worker in a factory takes 30 minutes to assemble one product. With each hour of training, the trainee decreases the time taken to assemble one product by 10%, until they achieve the best possible time of 15 minutes.

How many hours of training are required for the trainee to be able to achieve 15 minutes? Give your answer as an integer.

6 The height of a cylinder in a machine is given by the relationship $s = 2t^2 - 3t + 4$, where s is in centimetres and t is in seconds. Determine the maximum height that the cylinder reaches in the machine.

7 The part shown below needs to be marked out for cutting. Determine length A and angles b and c.

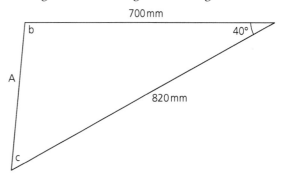

8 Determine the angle between the vectors $\begin{pmatrix} 6 \\ 8 \end{pmatrix}$ and $\begin{pmatrix} 4 \\ 0 \end{pmatrix}$.

9 During the manufacture of a batch of 20 products, the diameter of each product was measured in mm.

Find the mean, median, mode and standard deviation of the batch.

| 27.1 | 27.0 | 27.2 | 27.0 | 26.8 | 27.0 | 27.0 | 27.1 | 27.0 | 27.2 |
| 27.2 | 27.3 | 27.0 | 27.1 | 27.3 | 26.8 | 26.9 | 26.9 | 27.1 | 27.1 |

10 Convert the hexadecimal number 6A3 into binary form.

Project practice

1 Carry out the inspection of a key dimension for a batch of manufactured products. Determine the mean, median and modal averages of the dimension, along with the standard deviation. Investigate what these values mean for the quality-control activities used in subsequent batches.

2 Identify an application where CNC equipment is being used to manufacture a product. Determine the vectors required to programme the operation.

5 Essential science for engineering and manufacturing

Introduction

In this chapter, you will learn how scientific principles are used to solve problems in the real world. You will explore the measuring tools and systems used in the engineering sector, including how imperial units relate to SI units, as well as exploring various measuring techniques using a range of equipment, taking into consideration accuracy, precision and calibration. You will learn about types of measuring system, including vector and coordinate systems.

This chapter also covers the fundamentals of chemical understanding, such as atomic structure, chemical structure and chemical reactions. It describes how force and movement work and explains the characteristics of fluids and their applications, as well as including a thorough analysis of thermodynamic concepts and how they might be used in real-world situations.

Learning outcomes

By the end of this chapter, you will understand:
1. the difference between base and derived units
2. the units applicable to different properties
3. how to convert between SI units and comparable imperial units
4. how to convert between different multiples and submultiples
5. the definitions of, and differences between, scalar and vector coordinates
6. how to convert between Cartesian and polar coordinates where angles are in degrees
7. the concept of the scientific method (observation, questioning, making a hypothesis, prediction/simulation, testing, conclusion, iteration)
8. how to analyse, evaluate, synthesise and apply information, data, research findings, deliberation, and the processes, results and outcomes of testing, modelling and experimenting (accuracy, reliability, precision and replication)
9. what can be measured using different types of equipment
10. the techniques used to carry out measurements using different types of equipment
11. the accuracy and relative limitations and benefits of different measuring devices
12. how different principles and techniques are used in measuring and problem solving
13. the definitions of the terms atom, element, molecule, compound and mixture
14. the applications, characteristics, management and control of chemical interactions and reactions used in engineering (chemical etching, surface finishing, bonding, applications for oils and lubricants, high-risk operations)
15. the application of theory and calculations to solve practical engineering problems involving forces and motion
16. the application of theory and calculations to solve practical engineering problems involving fluids
17. the key differences between liquid flow and aerodynamics
18. the application of theory and calculations to solve practical engineering problems involving thermodynamics.

5.1 Units of measurement used in engineering

Measurement plays a key role in science, engineering and many other disciplines, as well as in day-to-day activities. Units of measurement have a vital role in describing the physical properties of an object or a component.

For instance, properties such as the dimensions or mass of an object may need to be defined by engineers. Each property can be described with a number followed by a unit of measurement.

The **International System of Units** (also referred to as the SI system of units) and the **imperial system of units** are the two major systems of measurement.

International System of Units (SI)

Base units

The SI system uses seven **base units**, which can be combined with each other further. These base units are shown in Table 5.1.

▼ **Table 5.1 SI base units**

Quantity	SI base unit
Length	metre (m)
Mass	kilogram (kg)
Time	second (s)
Electric current	ampere (A)
Thermodynamic temperature	kelvin (K)
Amount of substance	mole (mol)
Luminous intensity	candela (cd)

Test yourself

What are the seven SI base units?

Derived units

The SI system also includes a range of **derived units**, which are a combination of base units. Derived units either have special given names or are expressed as exponentials of base units.

Key terms

International System of Units: globally agreed set of base units and derived units used for standardised measurement in science and engineering.

Imperial system of units: traditional system of measurement introduced by the British Weights and Measures Act 1824 which includes units such as feet and inches, pints and gallons, and pounds and ounces.

Base units: fundamental units of the SI system that are defined arbitrarily and can be used to derive other units of measurement.

Derived units: units of the SI system that are derived by multiplying or dividing base units in specified combinations.

For example, force can be expressed as a combination of the base units kilograms, metres and seconds ($kg \times m \times s^{-2}$). The unit of force has a special name: the newton (N).

Similarly, pressure can also be expressed as a combination of the base units kilograms, metres and seconds ($kg \times m^{-1} \times s^{-2}$). The unit for pressure has the special name pascal (Pa).

However, the unit of volume is represented by an exponential of metres, m^3.

There are many derived SI units of measurement. Table 5.2 shows some of those relevant to engineering activities.

▼ **Table 5.2 SI derived units**

Quantity	SI derived unit
Acceleration	metres per second squared (ms^{-2})
Density	mass per unit volume (kgm^{-3})
Force	newton (N)
Pressure	pascal (Pa) or Nm^{-2}
Torque	newton metre (Nm)
Velocity	metres per second (ms^{-1})
Volume	cubic metre (m^3)
Area	square metre (m^2)

Imperial system of units

The imperial system of units was introduced in the UK by the British Weights and Measures Act 1824 and remained in official use until 1965, when the metric system was adopted. However, engineers in some countries, for example the USA, Myanmar and Liberia, still use the imperial system of units.

In the imperial system:
▶ length is measured in feet (ft), inches (in), yards (yd) and miles (mi)
▶ volume is measured in fluid ounces (fl oz), pints (pt) and gallons (gal)
▶ weight is measured in pounds (lb) and ounces (oz).

SI unit and imperial unit conversion

Engineers often need to convert units from one form to another, such as when a greater level of accuracy is needed for a certain measurement. For example, one minute is equivalent to 60 seconds, and one kilogram is equivalent to 1000 grams.

In addition, it is important to understand the relationships between SI units and imperial units, in order to be able to work with and interpret information from countries that may operate with a different system.

To convert imperial units to SI units, engineers can use the conversion factors shown in Table 5.3.

▼ Table 5.3 Conversion factors for imperial to SI units

Length	1 inch = 2.54 centimetres
	1 foot = 30.48 centimetres
	1 yard = 91.44 centimetres
	1 mile = 1.61 kilometres
Volume	1 fluid ounce = 28.41 millilitres
	1 pint = 0.57 litres
	1 gallon = 4.55 litres
Mass and weight	1 ounce = 28.35 grams
	1 pound = 0.45 kilograms

Examples

▶ To convert 140 pounds to kilograms:
$140 \times 0.45 = 63$ kilograms.
▶ To convert 36 inches to centimetres:
$36 \times 2.54 = 91.44$ centimetres.
▶ To convert 4 pints to litres: $4 \times 0.57 = 2.28$ litres.

Test yourself

Daweena buys a soil bag which has a mass of 13.5 kg. What is the mass of the soil bag in pounds?

Case study

Failure to take into account the correct unit conversions could have disastrous consequences.

For example, on 23 July 1983, an Air Canada Boeing 767 passenger aircraft ran out of fuel very early into its flight from Edmonton to Montreal. This caused the engines to shut down at 41000 feet in the air. Miraculously, the crew were able to land the plane at a nearby disused airbase with only minor injuries to a small number of passengers, but the outcome could have been catastrophic and the mistake put everyone on board at risk.

The reason that the aircraft ran out of fuel so soon was that the density ratio used for the fuel calculations was in pounds per litre rather than kilograms per litre. This resulted in a shortfall of over half of the fuel needed for the flight.

Questions

1 Discuss the consequences of using different units of measurement when fuelling the Boeing 767 Air Canada flight.
2 Find other examples where imperial/SI conversion errors have resulted in negative effects for organisations.

Converting between multiples and submultiples

It is important to use an appropriate size of SI unit for a measurement. For example, when measuring the distance between two towns, the metre unit would be too small and so a multiple is required – a kilometre. Multiples are simply the factors used to create larger forms.

Similarly, submultiples are used to make smaller forms of SI units; for example a centimetre is a submultiple of a metre.

▼ Table 5.4 Examples of multiples

Prefix	Symbol	Number	Factor
Kilo	k	1000	10^3
Mega	M	1000000	10^6
Giga	G	1000000000	10^9
Tera	T	1000000000000	10^{12}

▼ Table 5.5 Examples of submultiples

Prefix	Symbol	Number	Factor
Centi	c	0.01	10^{-2}
Milli	m	0.001	10^{-3}
Micro	μ	0.000001	10^{-6}
Nano	n	0.000000001	10^{-9}
Pico	p	0.000000000001	10^{-12}

Example

The distance between two points is measured as 0.01–m. Convert this into centimetres.

$0.01 \times 100 = 1 cm$

It is much easier to write and understand this measurement using the submultiple unit of cm.

Test yourself

1 Arrange the following measurements in order from smallest to largest:
   a 20 grams
   b 50 grams
   c 0.005 kilograms.
2 Arrange the following in order from least to greatest: terabyte, gigabyte, kilobyte megabyte.
3 The distance between the Sun and Venus is about 108 000 000 km. Convert this into terametres.

Research

Read the following articles that discuss the system of measurement used in the USA:

www.britannica.com/story/why-doesnt-the-us-use-the-metric-system

https://time.com/3633514/why-wont-america-go-metric

Discuss the advantages and disadvantages of the USA not adopting the International System of Units.

5.2 Vector and coordinate measuring systems

Scalar quantities

Scalar quantities only have magnitude (size) and are entirely described by a numerical value and a relevant unit. They do not have a direction. Examples include speed, distance, mass, volume, time, energy and power. Scalar quantities can be changed using standard arithmetic formulas.

Test yourself

Give two examples of scalar quantities.

Speed

Speed is a scalar quantity that describes the rate at which an object moves over a certain distance.

An object moving at a high speed will travel further than an object moving at a lower speed in the same length of time. Zero speed refers to an object at rest.

The formula for speed is:

$$s = \frac{d}{t}$$

where s represents speed, d represents distance and t represents time. The SI unit for speed is metres per second (ms⁻¹).

Distance

Distance is a scalar quantity that describes how far an object moves. The SI unit for distance is metres (m).

Key terms

Scalar: value with magnitude but not direction.

Speed: rate at which an object moves over a certain distance.

Distance: how far an object moves.

When moving in a straight line, the distance an object travels is the length of the line that connects the start point to the end point, as shown in Figure 5.1.

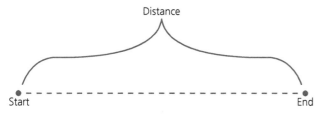

▲ Figure 5.1 Distance when moving in a straight line

Distance is the sum of an object's movements, regardless of direction. So, in Figure 5.2, distance can be calculated as $6\,m + 8\,m + 10\,m = 24\,m$.

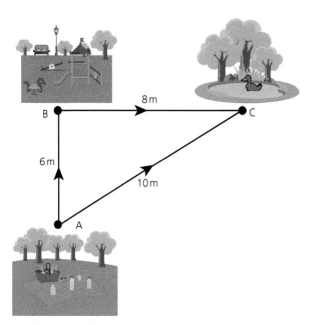

▲ Figure 5.2 Distance is the sum of an object's movements, regardless of direction

Adding and subtracting scalar quantities

Adding and subtracting scalar quantities simply involves the normal rules of addition and subtraction, assuming the quantities have the same units.

Examples

▶ Calculate the total mass of a 75 kg climber carrying a 15 kg backpack:

$75 + 15 = 90\,kg$

▶ Calculate the increase in temperature in a room that is heated from 12 °C to 21 °C using a heater:

$21 - 12 = 9°C$

Improve your maths

1 A person weighs 80 kg and carries a load of 28 kg. Calculate the total mass of the person and the load.

2 A car sets out on a journey of 220 km. After one hour, the car has travelled 100 km. Calculate the distance left to travel.

Vector quantities

Vector quantities have both magnitude and direction. Examples include displacement, velocity, acceleration, force, gravity, momentum and weight.

The direction of a vector can be given in a written description (for example 11 m east or $15\,ms^{-1}$ at 30 degrees to the horizontal) or it can be drawn as an arrow. The length of an arrow represents the magnitude of the quantity. Figure 5.3 shows three examples of vectors, drawn to different scales.

▲ Figure 5.3 Vector quantities have both magnitude and direction

---

Test yourself

Define the term vector quantity.

Displacement

Displacement is a vector quantity that describes the change in position of an object. Its length is the shortest distance between the object's start and end positions.

Key terms

Vector: value with both magnitude and direction.

Displacement: change in position of an object; its length is the shortest distance between the object's start and end positions.

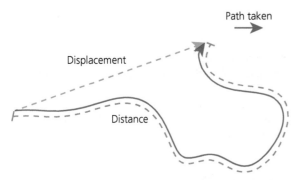

▲ Figure 5.4 Displacement measures the change in position of an object

Velocity

Velocity is a vector quantity that describes the speed at which an object moves in a particular direction.

Suppose two cars travel at 50 kilometres per hour on the same road. After two hours, the cars are 200 kilometres away from each other. How is this possible? Because the cars are moving in opposite directions, of course.

Speed describes distance and time but does not indicate direction. When the direction of an object's movement is included, you are describing its velocity.

The velocity of an object changes when either its speed or direction changes. If a downhill skier goes from 30 to 40 kilometres per hour, their velocity has changed. If they continue to move at 40 kilometres per hour but change direction, their velocity has changed again.

▲ Figure 5.5 The velocity of a downhill skier changes when either their speed or direction changes

Acceleration

Acceleration is a vector quantity that describes the rate at which velocity changes. It can result from a change in speed (increase or decrease), a change in direction

(back, forward, up, down, left, right), or changes in both. The unit for acceleration is metres per second squared (ms^{-2}).

In everyday usage, acceleration is usually thought of as a change in speed. However, it is important to remember that acceleration is a change in velocity. So, even if an object is moving at a constant speed, if it changes direction, the object is accelerating.

The formula for acceleration is:

$$a = \frac{\Delta V}{t} = \frac{\text{final velocity} - \text{initial velocity}}{\text{time}}$$

Cartesian coordinates

Cartesian coordinates are used to specify the distances of a point from the coordinate axes on a graph. They are shown as a pair of numbers or as three numbers, depending on whether they are being expressed in two or three dimensions. The points are therefore plotted as either (x, y) or (x, y, z).

> **Key terms**
>
> **Velocity:** speed at which an object moves in a particular direction.
>
> **Acceleration:** rate at which velocity changes.
>
> **Cartesian coordinates:** a pair of numbers or three numbers that specify the distances from the coordinate axes.

The three mutually perpendicular coordinate axes – the x-axis, the y-axis and the z-axis – that make up the Cartesian coordinate system in three dimensions are shown in Figure 5.6. The three axes come together at a point known as the origin (O).

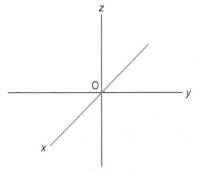

▲ Figure 5.6 The Cartesian coordinate system is made up of three mutually perpendicular coordinate axes

Let us use the example of a room to help visualise this:

▶ The point where the room's corner walls and the floor converge is the origin.

▶ The wall to the left connects to the floor along the horizontal x-axis.

▶ The wall to the right connects to the floor along the horizontal y-axis.

▶ Where the walls intersect vertically is the z-axis.

The parts of the lines you can see while you are in the room are the positive components of the axes.

Polar coordinates

A **polar coordinate system** is used to determine the location of a point on a plane in terms of the distance from a reference point (r) and the angle from a reference direction (θ). The points are therefore plotted as (r, θ). It is often used as a two-dimensional system but can also be extended into three dimensions.

> **Key term**
>
> **Polar coordinate system:** coordinate system used to determine the location of a point on a plane. Can be two- or three-dimensional.

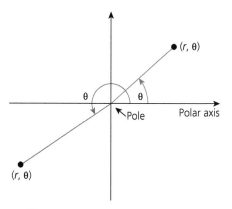

▲ Figure 5.7 The polar coordinate system

> **Test yourself**
>
> What is a polar coordinate?

Converting between Cartesian coordinates and polar coordinates

Trigonometry and Pythagoras' theorem can be used to convert between Cartesian coordinates (x, y) and polar coordinates (r, θ).

To convert Cartesian coordinates into polar coordinates:

$$r = \sqrt{(x^2 + y^2)}$$

$$\theta = \tan^{-1}\frac{y}{x}$$

> **Example**
>
> To convert the Cartesian coordinates (3, 4) into polar coordinates:
>
> $$r = \sqrt{3^2 + 4^2} = \sqrt{25} = 5$$
>
> $$\theta = \tan^{-1}\frac{4}{3} = \tan^{-1}1.33 = 53.1$$
>
> The polar coordinates are (5, 53.1°).

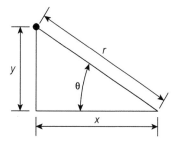

▲ Figure 5.8 Using trigonometry and Pythagoras' theorem to convert between polar and Cartesian coordinates

To convert polar coordinates into Cartesian coordinates:

$$x = r \times \cos\theta$$

$$y = r \times \sin\theta$$

> **Example**
>
> To convert the polar coordinates (5.05, 53.5°) into Cartesian coordinates:
>
> $x = r \times \cos\theta = 5.05 \times \cos53.5 = 3$
>
> $y = r \times \sin\theta = 5.05 \times \sin53.5 = 4.06$
>
> The Cartesian coordinates are (3, 4).

> **Test yourself**
>
> Convert the polar coordinates (8.41, 61.6°) into Cartesian coordinates.

5.3 Scientific methods and approaches to scientific inquiry and research

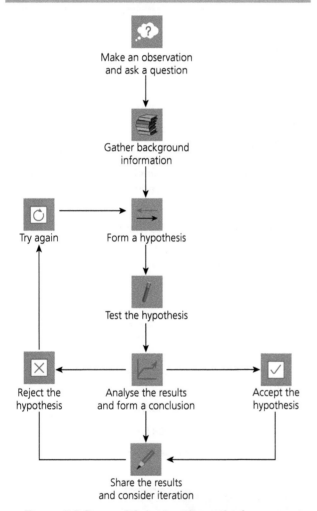

Make an observation and ask a question

Gather background information

Try again

Form a hypothesis

Test the hypothesis

Reject the hypothesis

Analyse the results and form a conclusion

Accept the hypothesis

Share the results and consider iteration

▲ Figure 5.9 Steps of the scientific method

The **scientific method** is a systematic and objective approach used to acquire knowledge. It follows standardised processes, including observation, questioning, formulation of hypotheses, testing and experimentation, data analysis and the formation of conclusions based on empirical evidence. The approach is applicable across all fields of science, as well as in technology and business.

Steps of the scientific method

Make an observation and ask a question

The first step is to ask a question about something that has been observed. For example, you have observed that visitors to an event do not read the instructions displayed, which state that they must register by completing a questionnaire. You question why this might be happening.

Gather background information

Once you have formulated your question, it is important to carry out research. This could include looking at previous studies on a similar topic, to make sure you understand the best techniques and equipment to use.

Form a hypothesis

The next step is to think about what has been observed and to consider a possible cause. A **hypothesis** is a testable statement of the expected outcome of a study following experimentation or any other verification process.

Key terms

Scientific method: systematic and objective approach used to acquire knowledge that includes observation, questioning, formulation of hypotheses, testing and experimentation, data analysis and the formation of conclusions based on empirical evidence.

Hypothesis: testable statement of the expected outcome of a study following experimentation or any other verification process.

Thinking about the example given above, your hypothesis might be: 'If the instructions are made bigger, more visitors will read them and register at the event.'

Test the hypothesis

Once you have formulated a hypothesis, it can be tested. This may involve gathering data from formal experiments. When deciding how to test, you need to consider the accuracy, reliability and precision of the data you gather: will it allow you to form reliable conclusions that support your hypothesis? Is it possible to carry out the experiment again and get the same or a similar result?

Returning to the earlier example, the hypothesis could be tested by creating two entry gates, A and B. Gate A has the current-sized instructions, while gate B has bigger-sized instructions. Half the visitors (group A) will enter via gate A, and the other half (group B) will enter via gate B. Such testing is developed to prove the hypothesis – is the size of instructions the only factor causing visitors not to register at the event?

Analyse the results and form a conclusion

The next stage is to analyse the results and draw conclusions from the data collected – do the results support the hypothesis?

For example, the results of the experiment regarding the size of instructions and number of registrants at an event could be observed after one month. If it was shown that the rate of registration rose from 15 to 28 per cent after increasing the size of the instructions, this would be reliable and accurate enough to prove the hypothesis and form a conclusion.

Share the results and consider iteration

The results of the experiment should be documented, so that they can be shared with others. If the hypothesis is not proven, it may lead to iteration – the creation of a new hypothesis for further testing to evaluate the cause of an action. Sometimes, an experiment may form other questions as well.

> **Research**
>
> Think about an engineering problem. Based on careful observation, consider a possible cause and develop a hypothesis. For example, why does one machine perform more efficiently than another of the same type? What are the possible reasons behind this and how can these be tested?

5.4 Measurement equipment, techniques and principles

Basic principles in measurement

The purpose of **measurement** is to obtain an accurate and precise value of a physical variable in accordance with national and international standards.

Measurement is important in the:
- design of a product or process
- assessment of the performance of a product
- verification of a response from input functions
- assessment of quality and effectiveness.

Accuracy and precision

Accuracy and precision are both important for manufacturing, when designing and producing a product and when evaluating its quality.

Accuracy can be defined as the degree of closeness between a measured value/dimension and its true value/dimension, often expressed as a percentage.

Precision refers to repetition in the measuring process. It is the degree of closeness between measurements taken repeatedly under the same conditions. As such, it defines the quality of a measurement.

However, it is important to note that the accuracy of measurement cannot be justified with precision. Accuracy gives the margin by which the measured dimension is away from the true value. It is possible to be accurate without being precise, and vice versa.

> **Key terms**
>
> **Measurement:** action performed to obtain an accurate and precise value of a physical variable in accordance with national and international standards.
>
> **Accuracy:** degree of closeness between a measured value/dimension and its true value/dimension, often expressed as a percentage.
>
> **Precision:** degree of closeness between measurements taken repeatedly under the same conditions.

Uncertainty

Uncertainty refers to the margin of doubt about a measurement, quantified by the width of the margin and a percentage confidence level.

For example, a tube of metal can be measured as 150 mm long, plus or minus 5 mm, with a 96% confidence level. This could be written as 150 mm ± 5 mm, with a confidence of 96%.

All measurements are affected by uncertainty and a corresponding statement should accompany measurement results. Uncertainty is important because it helps engineers to understand how likely measurements are to be accurate or within tolerance. This is vital to consider when checking the quality of manufactured products.

Resolution

Resolution is the smallest increment or decrement that can be displayed as a result of measurement. Accuracy and resolution are closely related: accuracy defines the proximity of the measured value and the true value, whereas resolution defines the smallest change detectable by the measuring equipment.

Calibration

The main purpose of **calibration** is to ensure the measurements taken by measuring equipment are accurate and repeatable.

Typically, the calibration procedure involves comparing the measurements obtained against a standard of known accuracy. In case of any deviation, adjustments will be carried out to achieve the required level of accuracy.

Key terms

Uncertainty: margin of doubt about a measurement, quantified by the width of the margin and a percentage confidence level.

Resolution: smallest increment or decrement that can be displayed as a result of measurement.

Calibration: process by which measuring equipment is checked to ensure measurements taken are accurate and repeatable.

Errors in measurements

As all measurements are associated with a certain level of uncertainty, they are not completely accurate. It is therefore important to understand the different types of errors: systematic and random.

Systematic errors

A systematic error involves a consistent amount of deviation, in the same direction, from the true measurement value. With proper analysis, such errors can be reduced, leading to increased accuracy. There are several reasons for the occurrence of systematic errors:

- ▷ calibration errors, for example when the measuring equipment has not been calibrated or set up correctly
- ▷ avoidable or human errors, such as improper use of measurement equipment or mistakes when reading values
- ▷ ambient and environmental conditions that could affect measurements taken, such as temperature, humidity or noise level.

Random errors

Random errors produce random deviations in measurement, and their cause cannot always be determined. Random errors sometimes go up and sometimes go down. Using a large sample size and taking an average of repeated measurements can help to reduce the effects of this type of error.

Industry tip

When working in any role that involves taking measurements, always ensure that equipment has been fully calibrated prior to its use. For example, doing this during quality checking of manufactured parts, products or components will help to ensure that measurements taken are accurate and will reduce the likelihood of errors. Failure to do this could result in faulty products passing inspections that they should have failed, causing customer dissatisfaction and/or potential safety issues.

Measurement equipment

The choice of measuring instrument depends on application and cost.

Rule

A rule is used to measure linear distance between a reference point and a measurement point. Care must be taken to align the points.

Steel rules are more popular than wood or plastic due to their durability and accuracy, although due to design limitations they can only measure to 1 mm, or sometimes to 0.5 mm if smaller graduations are marked.

Vernier calliper

French mathematician Pierre Vernier invented the Vernier calliper in 1631, which uses the Vernier scale to take accurate measurements to 0.01 mm.

Callipers can be used to measure the:
▶ external dimensions of an object, for example the external diameter of a metal bar
▶ internal dimensions of an object, for example the internal diameter of a tube
▶ depth of the inside of an object, for example the depth of a container.

They comprise a pair of jaws:
▶ an L-shaped fixed jaw with a main scale on it
▶ a moveable jaw with a Vernier scale plate that is able to slide over the main scale.

To read the measurement, add the values on the fixed scale and the Vernier scale. A clamping screw controls the movement of the Vernier scale, in order to improve the accuracy of the reading.

Digital and dial callipers tend to be used in preference to traditional callipers, as they are more accurate and easier to read.

Micrometer

The first micrometer was invented by William Gascoigne in the seventeenth century. It is a versatile instrument that measures dimensions using an anvil and a spindle face. It can provide a greater minimum count than the Vernier calliper, up to 0.001 mm.

There are three types of micrometer:
▶ An outside micrometer comprises a C-shaped frame with a fixed anvil and movable spindle, which can be positioned using a ground screw. The rotating thimble and stationary sleeve have a graduated scale engraved over them. A locknut is used to secure the spindle to facilitate easy readings.
▶ An inside micrometer is used for precision measurements, typically in metalworking and aircraft applications. It consists of a satin-chrome head with one or more rigid tubular-steel measuring rods.
▶ A depth micrometer has an interchangeable rod, a lapped measuring end, a ratchet stop and a measuring rod lock.

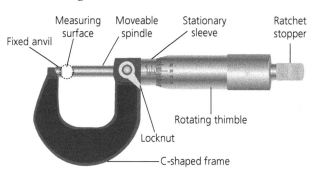

▲ Figure 5.11 An outside micrometer measures dimensions using an anvil and a spindle face

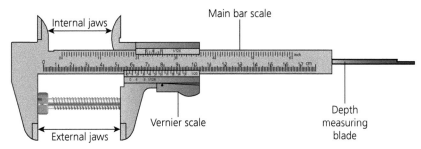

▲ Figure 5.10 Vernier calliper

Gauges

Gauges are used to measure thickness, diameter, gap distance and pressure. There are various types.

Slip gauge

Slip gauges are precision-machined blocks that are used to measure length. They can also be used to check the accuracy of other measuring instruments, such as micrometers and Vernier callipers. Their main advantages are their simplicity of use and high accuracy.

Angle gauge

Angle gauges provide a template for measuring angles between different lines or surfaces, for example when measuring the angle of steel beams fitted as part of a structure. They measure angles between 0 and 360 degrees.

Digital angle gauges show the results of measurements on a screen, hence reducing the likelihood of reading errors.

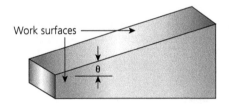

▲ Figure 5.12 An angle gauge measures the angle between two lines or surfaces

Plug gauges (go/no-go)

Plug gauges are cylindrical instruments used to check the precise dimensions (and therefore tolerance limits) of holes:

▶ Go plug gauges are used to check the lower limit of a hole.

▶ No-go plug gauges are used to check the upper limit of a hole.

They are convenient, highly accurate and economical if used correctly, as they eliminate the use of costly and complex instruments.

Dial test indicator (DTI)

Dial test indicators are used to measure small linear distances accurately, displaying the results on a dial. They are generally used for machine set up and measure the variation of the angle of a lever which is perpendicular to the indicator axis. They are also used to identify the difference in deflection of a beam under varying conditions.

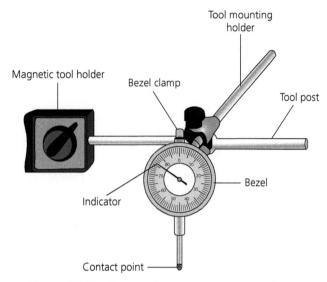

▲ Figure 5.13 Dial test indicators measure small linear distances accurately and display the results on a dial

Coordinate measuring machine (CMM)

A coordinate measuring machine (CMM) has three axes (x, y and z) which move perpendicular to each other, enabling it to measure point coordinates in three-dimensional (3D) space. It is therefore able to measure the physical geometrical characteristics of an object.

A CMM is able to perform fully automated measurement and is easily incorporated as part of a computer-integrated manufacturing (CIM) environment.

▲ Figure 5.14 A coordinate measuring machine (CMM) can measure point coordinates in three-dimensional space

Research

Write down a list of measurement equipment that you have studied or used in the engineering workshop environment. For each item, research what it is used to measure, how it works, and its benefits and limitations. You could record this as a table or produce a presentation of your findings to show to the class.

Look at the drawing of a flange housing plate. Suggest a measuring instrument for each of the dimensions.

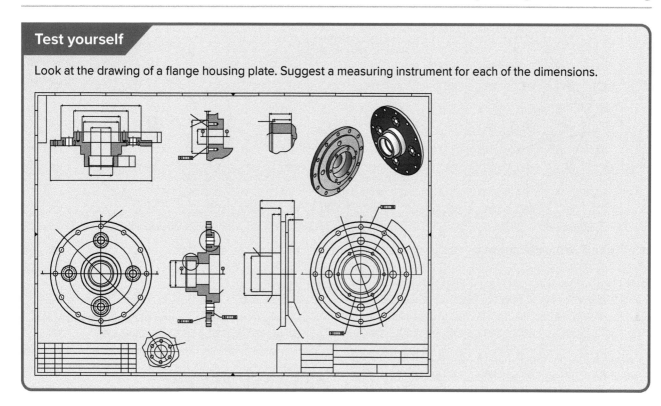

5.5 Chemical composition and behaviours

Chemical reactions and interactions are important in a range of engineering applications.

Atomic structure

An **atom** is the smallest unit of a chemical **element** that exhibits all the physical properties of that element. They are typically around 100 picometers in size.

Each atom is made up of three types of sub-atomic particle:
▶ **Protons** have a positive electric charge. The number of protons equals the atomic number.
▶ **Neutrons** have no electric charge and are the heaviest sub-atomic particles.
▶ **Electrons** have a negative electric charge and are the lightest sub-atomic particles.

The **nucleus** at the centre of an atom contains protons and neutrons, and its radius is measured in femtometres (quadrillionths of a metre). Electrons orbit the nucleus in a series of orbital layers or shells.

The electrons are bonded due to the attraction from the positively charged protons.

The mass of a proton and neutron are almost the same, and they are larger than electrons. The number of neutrons plus the number of protons forms the atomic weight.

Key terms

Atom: smallest unit of a chemical element that exhibits all the physical properties of that element.

Element: substance that cannot be broken down into smaller constituents and consists of only a single type of atom.

Protons: positively charged sub-atomic particles found in the nucleus of an atom.

Neutrons: sub-atomic particles without an electric charge found in the nucleus of an atom.

Electrons: negatively charged sub-atomic particles that orbit the nucleus of an atom.

Nucleus: centre of an atom made up of protons and neutrons.

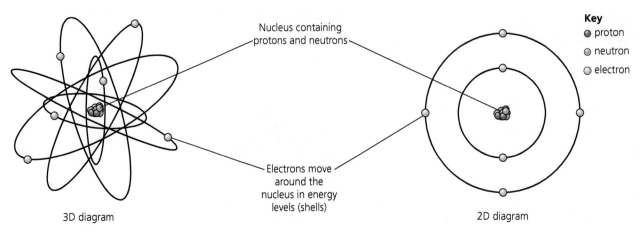

Nucleus containing protons and neutrons

Electrons move around the nucleus in energy levels (shells)

Key
- proton
- neutron
- electron

3D diagram

2D diagram

▲ Figure 5.15 Atomic structure

The chemical structure of an element varies according to **valence**, that is the number of bonds that its atoms can form with other atoms. The outermost electron shell of an atom is known as the valence shell.

A combination of more than one atom will form a **molecule**. They are bonded using a special type of chemical bond.

If the number of protons is equal to the number of electrons, the atom is electrically neutral. However, atoms can transfer electrons between them in order to form a more stable structure. If the number of electrons is greater than the number of protons, an element becomes negatively charged; if the number of protons is greater than the number of electrons, the element becomes positively charged. Elements that have an electric charge are known as **ions**.

Chemical bonding

Atoms are held together by bonds. There are four main types:
- ionic
- covalent
- metallic
- hydrogen.

> **Test yourself**
>
> 1 What is an atom?
> 2 What is the charge of an electron?
> 3 Which of the three sub-atomic particles is the lightest?

> **Research**
>
> Investigate the phenomenon of paramagnetism and how magnetic properties are achieved.

Chemical structure

Solutions are homogenous mixtures formed by the combination of two or more substances. In a solution, a solute is dissolved in a solvent.

Suspensions are heterogenous mixtures formed when solid particles are mixed in liquid without them dissolving.

Solubility is a measurement of the extent to which a chemical can dissolve in water.

> **Key terms**
>
> **Valence:** ability of an atom to combine with other atoms.
>
> **Molecule:** group of two or more atoms held together by chemical bonds.
>
> **Ions:** elements that have an electric charge due to losing or gaining electrons.
>
> **Solutions:** homogenous mixtures formed by the combination of two or more substances.
>
> **Suspensions:** heterogenous mixtures formed when solid particles are mixed in liquid without them dissolving.
>
> **Solubility:** measurement of the extent to which a chemical can dissolve in water.

A **compound** is a substance made from two or more elements, for example water (H_2O) or carbon dioxide (CO_2). The elements are bonded together and therefore difficult to separate.

Mixtures are formed by combining two or more chemical substances (elements, compounds, or both) which can be separated, as they are not bonded permanently. They can be separated using processes such as filtration, distillation and crystallisation.

Key terms

Compound: substance made from two or more elements where the elements are bonded together and difficult to separate.

Mixtures: substances formed by combining two or more chemicals which can be separated, as they are not bonded permanently.

Test yourself

What is the difference between a compound and a mixture?

Periodic table

The periodic table is a tabular arrangement of the chemical elements:

- The rows are known as periods and the elements are listed in order of increasing atomic number.
- The columns are known as groups, and elements with similar chemical and physical properties are grouped together.

The periodic table is useful to scientists and engineers as it allows them to quickly refer to key information about each of the different elements.

Group 1	Group 2											Group 3	Group 4	Group 5	Group 6	Group 7	Group 0
							1 **H** hydrogen 1										4 **He** helium 2
7 **Li** lithium 3	9 **Be** beryllium 4											11 **B** boron 5	12 **C** carbon 6	14 **N** nitrogen 7	16 **O** oxygen 8	19 **F** fluorine 9	20 **Ne** neon 10
23 **Na** sodium 11	24 **Mg** magnesium 12											27 **Al** aluminium 13	28 **Si** silicon 14	31 **P** phosphorus 15	32 **S** sulfur 16	35.5 **Cl** chlorine 17	40 **Ar** argon 18
39 **K** potassium 19	40 **Ca** calcium 20	45 **Sc** scandium 21	48 **Ti** titanium 22	51 **V** vanadium 23	52 **Cr** chromium 24	55 **Mn** manganese 25	56 **Fe** iron 26	59 **Co** cobalt 27	59 **Ni** nickel 28	63.5 **Cu** copper 29	65 **Zn** zinc 30	70 **Ga** gallium 31	73 **Ge** germanium 32	75 **As** arsenic 33	79 **Se** selenium 34	80 **Br** bromine 35	84 **Kr** krypton 36
85 **Rb** rubidium 37	88 **Sr** strontium 38	89 **Y** yttrium 39	91 **Zr** zirconium 40	93 **Nb** niobium 41	96 **Mo** molybdenum 42	[98] **Tc** technetium 43	101 **Ru** ruthenium 44	103 **Rh** rhodium 45	106 **Pd** palladium 46	108 **Ag** silver 47	112 **Cd** cadmium 48	115 **In** indium 49	119 **Sn** tin 50	122 **Sb** antimony 51	128 **Te** tellurium 52	127 **I** Iodine 53	131 **Xe** xenon 54
133 **Cs** caesium 55	137 **Ba** barium 56	139 **La*** lanthanum 57	178 **Hf** hafnium 72	181 **Ta** tantalum 73	184 **W** tungsten 74	186 **Re** rhenium 75	190 **Os** osmium 76	192 **Ir** iridium 77	195 **Pt** platinum 78	197 **Au** gold 79	201 **Hg** mercury 80	204 **Ti** thallium 81	207 **Pb** lead 82	209 **Bi** bismuth 83	[209] **Po** polonium 84	[210] **At** astatine 85	[222] **Rn** randon 86
[223] **Fr** francium 87	[226] **Ra** radium 88	[227] **Ac**** actinium 89	[261] **Rf** rutherfordium 104	[262] **Db** dubnium 105	[266] **Sg** seaborgium 106	[264] **Bh** bohrium 107	[277] **Hs** hassium 108	[268] **Mt** meitnerium 109	[271] **Ds** darmstadtium 110	[272] **Rg** roentgenium 111	[285] **Cn** copernicium 112	[284] **Uut** ununtrium 113	[289] **Fl** flerorium 114	[288] **Uup** ununpentium 115	[293] **Lv** livermorium 116	[294] **Uus** ununseptium 117	[294] **Uuo** ununoctium 118

Key
relative atomic mass
atomic symbol
name
atomic (proton) number

*	140 **Ce** cerium 58	141 **Pr** praseodymium 59	144 **Nd** neodymium 60	(145) **Pm** promethium 61	150 **Sm** samarium 62	152 **Eu** europium 63	157 **Gd** gadolinium 64	159 **Tb** terbium 65	162 **Dy** dysprosium 66	165 **Ho** holmium 67	167 **Er** erbium 68	169 **Tm** thulium 69	173 **Yb** ytterbium 70	175 **Lu** lutetium 71
**	232 **Th** thorium 90	231 **Pa** protactinium 91	238 **U** uranium 92	237 **Np** neptunium 93	(244) **Pu** plutonium 94	(243) **Am** americium 95	(247) **Cm** curium 96	(247) **Bk** berkelium 97	(251) **Cf** californium 98	(252) **Es** einsteinium 99	(257) **Fm** fermium 100	(258) **Md** mendelevium 101	(259) **No** nobelium 102	(260) **Lr** lawrencium 103

Metal   Non-metal   Difficult to classify

▲ Figure 5.16 Periodic table

Chemical behaviours

Chemicals in electricity

Chemical cells generate electrical current using chemical reactions. A **simple chemical cell** consists of two electrodes made from two different metals, which are placed in an **electrolyte** (a liquid that conducts electricity) and connected by a conductor such as a wire. Two or more cells can be connected together to form a battery.

Chemical cells can be primary or secondary:

▶ A **primary chemical cell** can only be used once and cannot be recharged.
▶ A **secondary chemical cell** can be recharged after discharge.

Key terms

Simple chemical cell: two electrodes placed in an electrolyte and connected by a conductor.

Electrolyte: liquid that conducts electricity.

Primary chemical cell: single-use chemical cell.

Secondary chemical cell: rechargeable chemical cell.

Cell capacity, power capacity and internal resistance

▶ Cell capacity refers to the amount of electricity generated by electrochemical reactions that is stored by a battery. It is defined by the area under the discharge curve (see Figure 5.18) and measured in watt-hours (Wh).
▶ Power capacity refers to the amount of electricity produced by a generator operating at full load.
▶ Internal resistance is opposition to the direction of flow of electric current produced in cells and batteries.

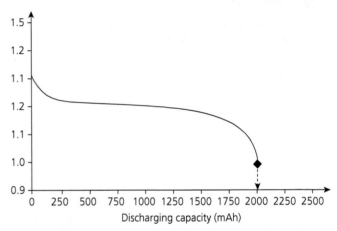

▲ Figure 5.18 Cell capacity is defined by the area under the discharge curve

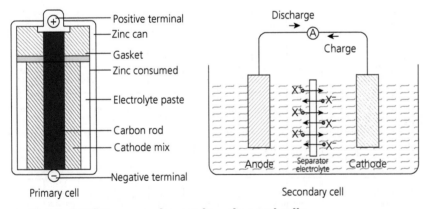

Primary cell   Secondary cell

▲ Figure 5.17 Primary and secondary chemical cells

Electrolysis

Electrolysis uses electricity to break down ionic compounds. For example, it is used to split water into hydrogen and oxygen when an electric current is passed through it. A positive electrode called an **anode** and a negative electrode called a **cathode** are connected to an electricity supply and placed in an electrolyte.

Negatively charged ions (**anions**) are attracted towards the positive anode; positively charged ions (**cations**) are attracted towards the negative cathode.

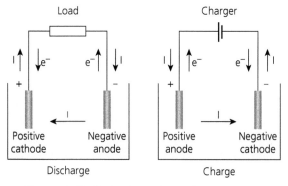

▲ Figure 5.19 Cathode and anode

Dissociation is the process of breaking up a compound into simpler components that could recombine under other conditions. Electrolytic dissociation occurs as a result of adding energy in the form of heat, resulting in the breaking up of molecules into ions.

Galvanic protection relies on an electrochemical reaction and involves protecting a base metal from corrosion by coating it with another more reactive metal. For example, steel is sometimes galvanised with a layer of zinc.

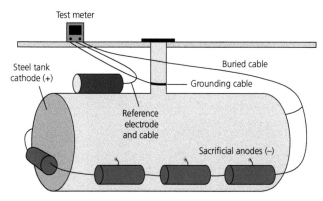

▲ Figure 5.20 Galvanic protection of a steel tank

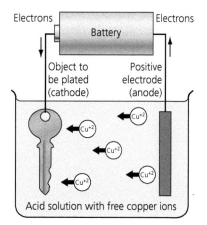

Key terms

Electrolysis: process that causes chemical decomposition through the use of electricity.

Anode: positive terminal of an electrolytic cell.

Cathode: negative terminal of an electrolytic cell.

Anions: negatively charged ions.

Cations: positively charged ions.

Dissociation: process of breaking up a compound into simpler components that could recombine under other conditions.

Galvanic protection: coating a base metal with another more reactive metal to protect it from corrosion.

Electroplating: metal finishing process that uses electrolysis to deposit a metal on the surface of a part or component.

Electroplating is a metal finishing process that uses electrolysis to deposit a metal (for example nickel, copper or zinc) on the surface of a part or component. The metal used for the coating is attached to the anode and the object to be plated is attached to the cathode, which are then placed in a bath containing an acid solution. As the electric current flows through the circuit, the metal coating is deposited on the component or part. This process is often used in the manufacture of vehicle body panels.

▲ Figure 5.21 Electroplating using electrolysis

Electroplating can improve the properties, as well as the aesthetics, of a part or component. It may be used to help protect the part from corrosion, abrasion and wear. It may also reduce friction and protect the part from impact or temperature. One application is the use of copper plating in circuit boards.

Test yourself

Explain the electrolysis process and give an example of its application in engineering.

Reactions of metals and alloys with weak and strong acids and alkalis

Acids react with most metals to produce salt and hydrogen gas:

Metal + Dilute acid → Salt + Hydrogen

Bases are substances that, in water, are capable of accepting hydrogen ions. They can be used to neutralise acids. Alkalis are chemicals that dissolve in water and can form salts when combined with acids. They make acids less acidic. Not all bases are alkalis, but all alkalis are bases. Bases also react with certain metals to produce salt and hydrogen gas.

Some metals (for example potassium and sodium) are very reactive and easily form new substances; others (for example platinum, copper, silver and gold) are less reactive.

Different metals can be combined to produce alloys. This is typically done to create a material with improved properties, such as greater resistance to corrosion.

Health and safety

It is important to be aware that metals react with acids and alkalis to produce hydrogen gas, as it is highly flammable.

Chemical etching is a process that uses the reaction of acids and alkalis to create markings on a metal surface. It is used to produce copper traces on a printed circuit board (PCB).

Key term

Chemical etching: process that uses the reaction of acids and alkalis to create markings on a metal surface.

Most reactive

<table>
<tr><th colspan="2">Metal</th><th>Reaction with oxygen</th><th>Reaction with water</th><th>Reaction with acids</th></tr>
<tr><td>Potassium</td><td>K</td><td rowspan="6">Burns to form oxide</td><td rowspan="5">Reacts and gives off $H_2(g)$</td><td rowspan="3">Reacts violently and gives off $H_2(g)$</td></tr>
<tr><td>Sodium</td><td>Na</td></tr>
<tr><td>Lithium</td><td>Li</td></tr>
<tr><td>Calcium</td><td>Ca</td><td rowspan="3">Reacts and gives off $H_2(g)$</td></tr>
<tr><td>Magnesium</td><td>Mg</td></tr>
<tr><td>Aluminium</td><td>Al</td><td></td></tr>
<tr><td>Carbon</td><td>C</td><td colspan="3"></td></tr>
<tr><td>Zinc</td><td>Zn</td><td rowspan="4">Forms oxide when heated (metal powder burns)</td><td rowspan="4">No reaction</td><td rowspan="4">Reacts slowly and gives off $H_2(g)$</td></tr>
<tr><td>Iron</td><td>Fe</td></tr>
<tr><td>Tin</td><td>Sn</td></tr>
<tr><td>Lead</td><td>Pb</td></tr>
<tr><td>Hydrogen</td><td>H</td><td colspan="3"></td></tr>
<tr><td>Copper</td><td>Cu</td><td>Forms oxide when heated</td><td rowspan="4"></td><td rowspan="4">No reaction</td></tr>
<tr><td>Silver</td><td>Ag</td><td rowspan="3">No reaction</td></tr>
<tr><td>Gold</td><td>Au</td></tr>
<tr><td>Platinum</td><td>Pt</td></tr>
</table>

Least reactive

▲ Figure 5.22 Reactivity series of common metals

Applications for oils and lubricants

Oils and lubricants are used to reduce friction when two moving surfaces are in contact with each other, such as meshing gears. They are often used as coolants, as reducing friction also reduces heat. The viscosity of a lubricant is very important because it determines how efficient it is at doing its job.

5.6 Forces and motion in engineering

Force can be defined as a push or pull on an object that can cause it to change its:

▶ speed (either accelerating or slowing down)
▶ direction
▶ shape.

We encounter forces every day; for example all bodies or objects are drawn to the surface of the Earth by a force of attraction known as gravity.

For a fixed mass, force (F) is defined as **mass** (m) multiplied by acceleration (a):

$$F = ma$$

The SI unit for force is the newton (N).

Motion refers to the change in position of an object with respect to time:

▶ Motion along a line or a curve is known as translation.
▶ Motion that changes an object's orientation is known as rotation.

In both types of motion, velocity and acceleration are the same for every part of the object.

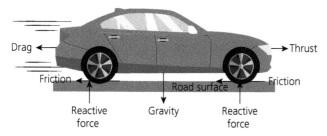

▲ Figure 5.23 The forces acting upon a car

Pressure

Pressure is defined as the force applied perpendicular to the surface of an object per unit area over which that force is distributed. It is measured in pascals (Pa); one pascal is equal to one newton per square metre.

Key terms

Force: push or pull on an object that causes it to change its speed, direction or shape.

Mass: quantitative measure of an object's inertia.

Motion: change in position of an object with respect to time.

Pressure: force applied perpendicular to the surface of an object per unit area over which that force is distributed.

The formula for pressure is:

$$p = \frac{F}{A}$$

where p is pressure, F is force and A is area.

Example

A force of 500 N is applied to an area of 100 cm^2. Calculate the pressure.

$$p = \frac{F}{A} = \frac{500}{100} = 5\,\text{Ncm}^{-2}$$

$$5\,\text{Ncm}^{-2} \times 10\,000 = 50\,000\,\text{Pa} = 5 \times 10^4\,\text{Pa}$$

Test yourself

A force of 15 N is acting on an area of 6 m^2. Calculate the pressure.

Vector representation of forces

Forces have both magnitude and direction and can therefore be considered vector quantities.

Vectors are shown as directed line segments. The size or magnitude of the vector is shown by the length and the direction is shown by the arrow.

▲ Figure 5.24 A vector as a directed line segment

Balanced forces

Balanced forces are equal in size but act in opposite directions; in other words, they cancel each other out (see Figure 5.25). When the forces exerted on an object are balanced, there is no change in its motion or velocity – if stationary it will remain stationary, or if moving it will keep travelling at the same speed in the same direction (uniform motion).

Unbalanced forces

Forces are said to be unbalanced when the force acting in one direction is greater than the force acting in the other direction (see Figure 5.26). **Unbalanced forces** result in a change in speed and/or direction.

▲ Figure 5.25 Balanced forces are equal in size and act in opposite directions

▲ Figure 5.26 Forces are unbalanced when the force acting in one direction is greater than the force acting in the other direction

Moments about a force

The turning effect of a force is called the **moment**. Moments can be clockwise or anticlockwise.

The size of a moment (M) can be calculated by multiplying the force (F) by the perpendicular distance from the pivot to the line of action of the force (d):

$$M = F \times d$$

The SI unit of a moment is newton metres (Nm).

Moments are used in a wide range of engineering applications, for example when lifting loads and in levers and gears.

> **Key terms**
>
> **Balanced forces:** forces that are equal in size and act in opposite directions.
>
> **Unbalanced forces:** when the force acting in one direction is greater than the force acting in the other direction.
>
> **Moment:** turning effect of a force.

A person is standing at one end of a seesaw. The distance between the person and the seesaw pivot is 1.5 m. The person has a weight of 600 N. The moment is clockwise and can be calculated:

$M = F \times d$

$M = 600 \times 1.5 = 900 \text{ Nm}$

A force of 12 N is applied to turn a handle. The force is applied 100 mm from the pivot of the handle. Calculate the moment.

Torque

A force that causes an object to rotate around an axis or another point in an angular direction is known as **torque**. In a similar way to how force causes acceleration in linear kinematics, torque causes angular acceleration. The SI unit of torque is newton metres (Nm).

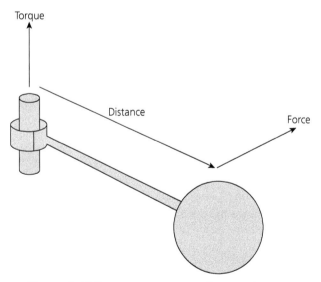

Torque

Distance

Force

▲ Figure 5.27 Torque causes angular acceleration

A force of 25 N is applied to a 5 m long radial vector at an angle of 60°. Calculate the torque, giving your answer to three significant figures.

$Torque = r \times F \times \sin\theta$

$Torque = 5 \times 25 \times \sin 60 = 108 \text{ Nm to 3 s.f.}$

Conditions for equilibrium

There are two conditions for **equilibrium**:
1. The sum of the forces acting on an object is zero ($\Sigma F = 0$).
2. The sum of the torque acting on an object is zero ($\Sigma \tau = 0$).

In other words, the magnitude and direction of the forces acting on an object are balanced.

A rigid object can be considered as being in equilibrium when the linear and angular accelerations are both zero with respect to an inertial frame of reference. Consequently, an object can move while in equilibrium, but its linear and angular velocities must be constant.

Static equilibrium occurs when an object is at rest in a specified coordinate system.

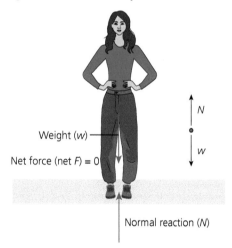

Weight (w)

Net force (net F) = 0

N

w

Normal reaction (N)

▲ Figure 5.28 A stationary person is in static equilibrium

Coplaner forces

In a coplaner force system, all forces act in just one plane. This is different from concurrent forces, where the forces act at the same point.

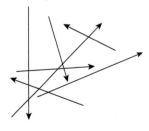

Concurrent coplaner forces   Parallel coplaner forces   Non-concurrent and non-parallel coplaner forces

▲ Figure 5.29 In a coplaner force system, all forces act in just one plane

Test yourself

Define the term 'coplanar force system'.

Research

According to the National Oceanic and Atmospheric Administration (NOAA), around 1000 tornadoes are reported in the USA every year. More tornadoes are formed in mid-western USA than anywhere else in the world.

Investigate the forces, motion and **momentum** which cause damage from tornadoes.

Key term

Momentum: product of the mass and velocity of an object.

Types of motion

There are four basic types of motion: rotary, linear, reciprocating and oscillating.

Rotary motion

In rotary motion, all parts of a rigid object move in circular orbits around a common axis with the same angular velocity. The axis can pass through the object. Examples of rotary motion include the Earth rotating on its axis or the wheels turning on a car.

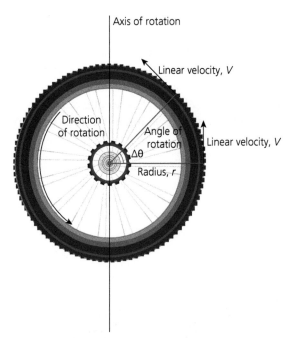

▲ Figure 5.30 In rotary motion, all parts of a rigid object move in circular orbits around a common axis with the same angular velocity

Linear motion

Linear motion (also referred to as uniform or rectilinear motion) is motion in one spatial dimension.

According to Newton's first law of motion (often known as the principle of inertia), a body with no net force acting on it would, in accordance with its original condition of motion, either remain at rest or continue moving with uniform speed in a straight line.

▲ Figure 5.31 Bowling is an example of linear motion

Reciprocating motion

Reciprocating motion is motion that moves back and forth in a straight line, for example up and down, side to side, or in and out. This principle is what allows most internal combustion engines to function. It is also used in water pumps and different types of machinery.

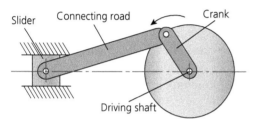

▲ Figure 5.32 Reciprocating motion is repetitive motion in a straight line

Oscillating motion

Oscillating motion is motion that moves back and forth about an equilibrium position. This occurs due to torque acting upon an object. A swinging pendulum is a classic example of oscillating motion.

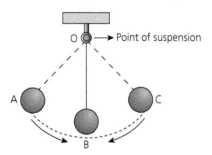

▲ Figure 5.33 Oscillating motion repeats itself after a specific interval of time

Test yourself

Explain the difference between reciprocating and oscillating motion.

5.7 Fluid dynamics in engineering

Fluid dynamics is a branch of fluid mechanics that is concerned with the movement of liquids (hydrodynamics) and gases (aerodynamics). It is of particular importance in the aerospace, automotive, process and emerging-technology industries.

Hydraulic pressure on an immersed plane surface

The pressure exerted by water or liquid on a surface is known as **hydraulic pressure**. It is measured as force per unit area, but area is calculated as the product of density and liquid depth. Hydraulic pressure varies according to the motion of the water or liquid.

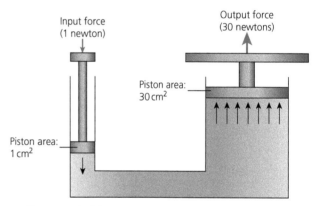

▲ Figure 5.34 Hydraulic pressure (input pressure = output pressure)

Key terms

Hydraulic pressure: pressure exerted by water or liquid on a surface.

Hydrostatic pressure: pressure applied at a given point by fluid at equilibrium due to gravity.

Hydrostatic pressure is the pressure applied at a given point by fluid at equilibrium due to gravity. It increases with depth measured from a reference surface due to the increased fluid weight causing a downward force. This phenomenon can be explained by the following equation:

$$P = \rho g h$$

where P is pressure, ρ is the density of the fluid, g is acceleration due to gravity and h is the height of the fluid. So, this static fluid pressure is independent of total mass, surface area and container geometry.

Research

A hydraulic crane is a powerful tool for lifting objects. The most common system uses a jib, which lifts using a piston assembly.

Investigate the working principles of a hydraulic crane and the role fluid power plays.

Hydrostatic thrust on an immersed plane surface

Hydrostatic thrust is a force acting on a surface that is submerged in a liquid.

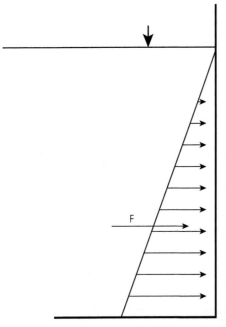

▲ Figure 5.35 Hydrostatic thrust on an immersed plane surface

To calculate the hydrostatic thrust acting on an immersed plane surface:

$$F = p_a A = \rho g h_a A$$

where:

F = thrust force (N)

$p_a = \rho g h_a$ = average pressure on the surface (Pa)

A = area of submerged surface (m^2)

ρ = density (kgm^{-3}) (water $1000\,kgm^{-3}$)

g = acceleration due to gravity ($9.81\,ms^{-2}$)

h_a = average depth (m).

Example

A container is submerged in water. The bottom of the container is 1.5 m wide, 2 m long and 1.2 m deep.

Calculate the hydrostatic thrust acting on the bottom of the container.

$$F = 1000 \times 9.81 \times 1.2 \times (1.5 \times 2)$$

$$F = 35316\,N$$

$$F = 35.3\,kN$$

Centre of pressure

The centre of pressure is the point at which the total amount of pressure acts on a body. This results in a force acting through that point. For example, on a sail boat, the position where the aerodynamic force is concentrated is represented using this concept. Another example is on an aircraft wing, or aerofoil, where the centre of pressure moves with the angle of attack, as shown in Figure 5.36.

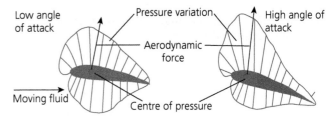

Centre of pressure is the average location of the pressure. Pressure varies around the surface of an object. p = p(x)

▲ Figure. 5.36 Centre of pressure

Viscosity

Viscosity measures a fluid's resistance to flow, and is sometimes defined as the friction between fluid particles. It is measured in centipoise or Nsm^{-2}. Highly viscous fluids, such as honey, flow at a slower rate than fluids with a low viscosity, such as water.

Key term

Viscosity: measurement of a fluid's resistance to flow in centipoise or Nsm^{-2}.

Water
Low viscosity:
1 centipoise

Honey
High viscosity:
2000–10 000 centipoise

▲ **Figure 5.37 Water has a much lower viscosity than honey**

Bernoulli's principle

Bernoulli's principle describes the energy per unit volume balance before and after the flow (i.e. at two points) and derives from the law of conservation of energy. It states that an increase in the speed of a fluid decreases static pressure, and vice versa.

This is represented by the equation:

$$P_1 + \frac{1}{2}\rho v_1{}^2 + \rho g h_1 = P_2 + \frac{1}{2}\rho v_2{}^2 + \rho g h_2$$

Where:
▶ P_1 is the pressure at elevation 1.
▶ v_1 is the velocity at elevation 1.
▶ h_1 is the height at elevation 1.
▶ P_2 is the pressure at elevation 2.
▶ v_2 is the velocity at elevation 2.
▶ h_2 is the height at elevation 2.
▶ ρ is the fluid density.
▶ g is the acceleration as a result of gravity.

This principle explains how aircraft wings work. When the air flowing past the top surface of the wing is moving more quickly than the air moving past the bottom surface, then the pressure below the wing will be higher. This creates upwards lift.

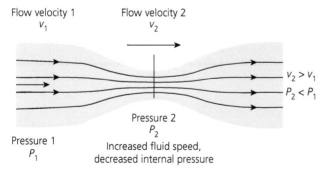

Flow velocity 1
v_1

Flow velocity 2
v_2

$v_2 > v_1$
$P_2 < P_1$

Pressure 2
P_2
Increased fluid speed,
decreased internal pressure

Pressure 1
P_1

▲ **Figure 5.38 Bernoulli's principle states that an increase in the speed of a fluid decreases static pressure, and vice versa**

Immersion of a body

When a body's weight and the **buoyancy** of a liquid are equal, the body will be suspended at that depth in the liquid when fully submerged. In this case, the weight of the body and the weight of the liquid being displaced are equal. When a body's weight is less than the buoyancy of a liquid, the body will float when partially submerged. The body's weight is less than that of the water that has been displaced in this situation. See Figure 5.39 for a visual representation of this.

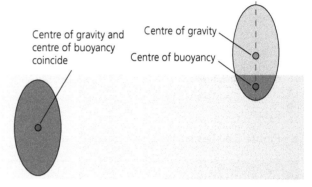

Centre of gravity and centre of buoyancy coincide

Centre of gravity

Centre of buoyancy

Fully submerged   Partially submerged

▲ **Figure 5.39 Immersion of a body. The centre of buoyancy is the centre of gravity for the volume of water that has been displaced. It is the point of equilibrium for buoyancy**

Flow characteristics around a two-dimensional shape

When fluid flows around a two-dimensional shape, at every point the flow velocity remains parallel to a fixed plane.

There are several flow characteristics you need to understand:
▶ **Laminar flow** – this is when fluid particles move smoothly in parallel layers with little or no mixing. Laminar flow can be observed when fluid flows in a pipe or air flows over an aircraft wing.

> ### Key terms
>
> **Buoyancy:** an upward force exerted by a fluid that opposes the weight of an immersed object.
>
> **Laminar flow:** flow pattern where fluid particles move smoothly in parallel layers with little or no mixing.

▶ **Turbulent flow** – this is when fluid particles move in such a way that they mix in a zig-zag pattern. It usually occurs when fluid flows with a large velocity or in large-diameter pipes. Eddies arise because of the zig-zag motion, which results in significant loss of energy.

▶ **Vortices** – these are regions of a fluid where the flow revolves about an axis line. They may appear straight or curved, for example whirlpools in the wakes of boats and winds in cyclones.

▶ **Separation point** – as a fluid flows around an object it forms what is known as a boundary layer. The separation point is the point where the boundary layer detaches from the surface and forms a wake, or turbulent flow.

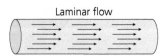 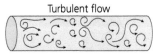

Laminar flow   Turbulent flow

▲ Figure 5.40 Laminar and turbulent flows

Test yourself

Explain the difference between laminar and turbulent flows.

Principles of aerodynamics

Aerodynamics is a branch of fluid dynamics that is concerned with how air moves around objects. There are four main forces to consider:

▶ **Drag** acts in opposition to the movement of an object through air – it is also referred to as air resistance. It causes objects to slow down.

▶ **Thrust** is the force that makes an object move forward by overcoming drag. For example, for a small aircraft, a propeller may provide enough thrust, while a large aircraft may require a jet engine.

▶ **Lift** is the upward force generated in response to an object moving through the air that acts against gravity. In order for an object to fly, lift must be greater than weight.

▶ **Weight** is the force that opposes lift. It acts through the object's centre of gravity to pull it downwards.

Key terms

Turbulent flow: flow pattern where fluid particles move in such a way that they mix in a zig-zag pattern.

Vortices: regions of a fluid where the flow revolves about an axis line.

Separation point: the point where the boundary layer detaches from the surface and forms a wake, or turbulent flow.

Aerodynamics: branch of fluid dynamics that is concerned with how air moves around objects.

Drag: force that acts in opposition to the movement of an object through air; also referred to as air resistance.

Thrust: force that makes an object move forward by overcoming drag.

Lift: upward force generated in response to an object moving through the air that acts against gravity.

Weight: downwards force that opposes lift.

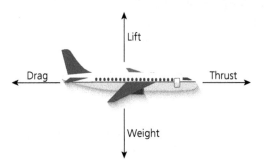

▲ Figure 5.41 Aerodynamic principles

Test yourself

What is the main function of thrust in an aircraft?

5.8 Thermodynamics in engineering

Thermodynamics is the branch of physics that is concerned with the relationship between heat, energy and work. It is governed by numerous laws, including Boyle's law, Charles' law and the general gas equation.

Heat-transfer mechanisms

There are three main types of heat-transfer mechanism:

▶ **Conduction** is the transfer of heat through direct contact between atoms or molecules. It most readily occurs in solids (where the particles are closer together), although it can also happen in liquids and gases. An example of conduction is a hand getting warm when touching a hot seatbelt in a car.

▶ **Convection** is the transfer of heat through the movement of heated particles in liquids or gases. Warmer liquids and gases have faster-moving particles, making them less dense, so they rise, while cooler particles sink. A refrigerator uses the principle of convection to circulate cold air around its contents.

▶ **Radiation** is the transfer of heat via electromagnetic waves. An example of radiation is using an open fire to heat a room.

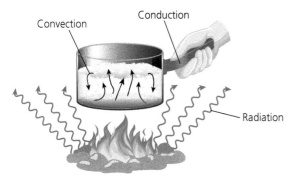

▲ Figure 5.42 Heat-transfer mechanisms

Test yourself

1 Why are basements usually cooler than the rest of a building?
2 What happens to the temperature of soup when you turn off the stove's element? Explain why this happens.

Thermodynamic systems

Thermodynamic systems can be classified as one of three main types:

▶ In an **open system**, both mass and energy transfer take place. An example is a car engine, which exchanges matter and heat with its surroundings.

▶ In a **closed system**, only energy transfer takes place and mass remains constant. An example would be a saucepan with a fitted lid, where the lid prevents matter from leaving or entering the saucepan.

▶ In an **isolated system**, neither energy nor mass is transferred. An example would be a vacuum flask.

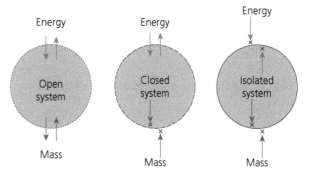

▲ Figure 5.43 Thermodynamic systems

Key terms

Thermodynamics: branch of physics that is concerned with the relationship between heat, energy and work.

Conduction (thermal): transfer of heat through direct contact between atoms or molecules.

Convection: transfer of heat through the movement of heated particles in liquids and gases.

Radiation: transfer of heat via electromagnetic waves.

Open system: thermodynamic system where both mass and energy transfer take place.

Closed system: thermodynamic system where only energy transfer takes place and mass remains constant.

Isolated system: thermodynamic system where neither energy nor mass is transferred.

Typical parameters in thermodynamic systems are:
▶ pressure – the force applied perpendicular to the surface of an object per unit area over which that force is distributed
▶ **volume** – the amount of three-dimensional space occupied by something
▶ **temperature** – the amount of kinetic energy produced by particles (it can also be defined as the rate of internal energy that changes with **entropy**).

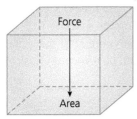

Pressure = Force / Area

▲ **Figure 5.44 Pressure is equal to force divided by area**

▲ **Figure 5.45 Volume is the amount of three-dimensional space occupied by something**

▲ **Figure 5.46 Temperature is the amount of kinetic energy produced by particles**

Different types of heat

Sensible heat

Sensible heat is heat added to or removed from a substance that:
▶ causes a change in the temperature of the substance
▶ does not cause a change in the phase (state) of the substance.

It is literally heat that can be felt.

The equation for sensible heat is:

$$Q = mc\Delta t$$

where Q is the amount of heat energy, m is mass, c is heat capacity and Δt describes the temperature change.

Latent heat

Latent heat is heat added to or removed from a substance that:
▶ causes a change in the phase of the substance
▶ does not cause a change in the temperature of the substance.

> ### Key terms
>
> **Volume:** amount of three-dimensional space occupied by something.
>
> **Temperature:** amount of kinetic energy produced by particles.
>
> **Entropy:** the measure of a system's thermal energy per unit temperature.
>
> **Sensible heat:** heat added to or removed from a substance that results in a change in its temperature without causing a change of phase.
>
> **Latent heat:** heat added to or removed from a substance that results in a change of phase without causing a change in temperature.

The equation for latent heat is:

$$Q = mh$$

where Q is the amount of heat energy, m is mass and h is specific latent heat.

Latent heat is used for phase changes:
▶ Latent heat of fusion is the heat used to form a liquid state from a solid state and vice versa. For example, the heat required to convert ice to water and water to ice.

▶ Latent heat of evaporation is the heat used to form a gaseous state from a liquid state and vice versa. For example, the heat required to convert water to water vapour and water vapour to water.

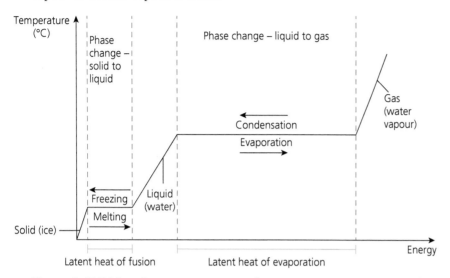

▲ Figure 5.47 Adding heat to water can either increase its temperature (sensible heat) or change its phase (latent heat)

Expansivity and the coefficient of heat transfer

When materials are heated, they expand. **Expansivity** is the amount by which a material expands or contracts due to a change in temperature of one degree.

The **coefficient of heat transfer** (α) indicates how well heat will be transferred through a series of resistant mediums.

The equation for thermal expansion is:

$$\Delta L = \alpha L \Delta t$$

where ΔL is the change in length, α is the linear coefficient of thermal expansion, L is the original length and Δt is the change in temperature.

Equations

Absolute temperature

Absolute temperature is the temperature of an object measured on the Kelvin scale, where zero is absolute zero.

The equation for absolute temperature is:

$$T = t + 273.15$$

where T is absolute temperature and t is the temperature in °C.

Absolute pressure

Absolute pressure is zero-referenced against a perfect vacuum; it is the sum of gauge pressure and atmospheric pressure.

The equation for absolute pressure is:

$$P_{abs} = P_g + P_{atm}$$

where P_{abs} is absolute pressure, P_g is gauge pressure and P_{atm} is atmospheric pressure.

Example

A gauge has read a pressure of 1400 kPa on an aircraft tyre while it is on the ground. If the atmospheric pressure is 101 kPa, then the absolute pressure would be:

$$1400\,kPa + 101\,kPa = 1501\,kPa$$

Volume, mass and density

Volume is the measure of three-dimensional space, given in m³. Mass is the amount of matter present in an object, measured in kg. Density is the mass per unit volume of an object, measured in kgm⁻³.

These three quantities are all therefore related via the formula:

$$density\,(\rho) = \frac{mass\,(m)}{volume\,(v)}$$

This formula can be rearranged to find any one of the quantities when the other two are known.

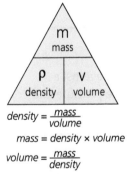

$density = \frac{mass}{volume}$

$mass = density \times volume$

$volume = \frac{mass}{density}$

▲ Figure 5.48 The relationship between volume, mass and density

Example

A manufactured object has a mass of 100 kg and volume of 2 m³. To calculate the density of the object:

$$\rho = \frac{100}{2} = 50 kgm^3$$

Boyle's law

Boyle's law describes the relationship between pressure and temperature for an **ideal gas**. It states that for a fixed amount of gas at a constant temperature, pressure and volume have an inverse relationship:

$$P_1 V_1 = P_2 V_2$$

where P_1 is the initial pressure, V_1 is the initial volume, P_2 is the final pressure and V_2 is the final volume.

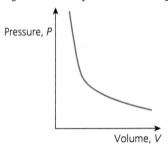

▲ Figure 5.49 Boyle's law

Example

A balloon has a volume of 3l (V_1) and is on the ground at around sea level, where there is a pressure of 101 kPa (P_1). The balloon is allowed to climb to an altitude where the pressure is reduced to 60 kPa (P_2). Assuming a fixed temperature, the final volume (V_2) can be calculated:

First, rearrange the formula:

$$P_1 V_1 = P_2 V_2$$

$$V_2 = \frac{P_1 V_1}{P_2}$$

$$V_2 = \frac{101\,000 \times 3}{60\,000}$$

$$V_2 = \frac{303\,000}{60\,000}$$

$$= 5.05l$$

Key terms

Boyle's law: law stating that for a fixed amount of gas at a constant temperature, pressure and volume have an inverse relationship.

Ideal gas: a gas composed of molecules that do not attract or repel each other.

Charles' law

Charles' law states that for a constant amount of gas and pressure, volume (V) and temperature (T) have a direct relationship:

$$\frac{V}{T} = \text{constant}$$

$$V_1 T_1 = V_2 T_2$$

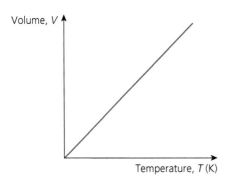

▲ Figure 5.50 Charles' law

Example

An example of Charles' law is the spark ignition internal combustion engine. Gases from the combustion process are subjected to very high temperatures. The increase in temperature results in the volume of the gases increasing. This in turn increases the force against the cylinder and piston head, resulting in crankshaft rotation.

General gas equation

The **general gas equation** shows that the product of pressure (p) and volume (V) is directly proportional to temperature (T) for a given amount of gas:

$$\frac{pV}{T} = \text{constant}$$

Characteristic gas equation

The **characteristic gas equation** shows that the product of pressure (p) and volume (V) is directly proportional to the temperature (T) and the amount of gas (m):

$$pV = mRT$$

where R is a characteristic gas constant.

Key terms

Charles' law: law stating that for a constant amount of gas and pressure, volume and temperature have a direct relationship.

General gas equation: equation showing that the product of pressure and volume is directly proportional to temperature for a given amount of gas.

Characteristic gas equation: equation showing that the product of pressure and volume is directly proportional to the temperature and the amount of gas.

Research

As thermodynamics deals with the production, storage, transfer and conversion of energy, its principles have great importance in power plants.

Investigate the thermodynamic principles used in a thermal power plant, including the fundamental operating cycle for the generation of mechanical and electrical energy.

Assessment practice

1 Describe the difference between base units and derived units.

2 State three examples of scalar quantities.

3 Explain why velocity is a vector quantity.

4 Define the term 'hypothesis'.

5 State which piece of measuring equipment could be used to check that the diameter of a hole is 5 ± 0.1 mm.

6 A quality control engineer is measuring the depth of holes drilled in a workpiece by a machinist. Another part is to be inserted into the holes at the next stage of the manufacturing process.

Explain the importance of the quality control engineer using calibrated measurement equipment when measuring the hole depths.

7 A company uses zinc to electroplate some machinery parts. Describe how the electroplating process is carried out.

8 A spanner is being used to tighten a nut. A force of 10 N is applied at the end of the spanner to achieve this. The spanner is 150 mm long. Determine the moment of the force in terms of both its magnitude and direction.

9 An oil storage tank is 10 m deep. The oil has a density of 700 kgm^{-3}. Calculate the hydrostatic pressure at the bottom of the tank. Give your answer in kPa to three significant figures.

10 Explain the difference between conduction and convection.

Project practice

You work for an organisation that manufactures closed conduits such as pipe and tube. You have been asked to deliver a presentation to a group of new apprentices that outlines Bernoulli's principle and types of flow in pipes.

Use appropriate digital software to produce the presentation and then deliver it to the students in your class. It should be no longer than 20 minutes.

6 Materials and their properties

Introduction

It is vital that engineers understand the range of materials available and their properties, to ensure that the materials chosen when designing and planning the manufacture of mechanical components behave as anticipated, are safe and are cost effective.

This chapter explores a wide range of engineering materials, their properties and how they are affected by different manufacturing processes.

Learning outcomes

By the end of this chapter, you will understand:
1 the difference between physical and mechanical properties
2 the definitions of physical and mechanical properties
3 how to calculate density
4 the common forms of supply, relative properties, applications and methods of disposal of various materials
5 the differences between: pure metals and alloys, ferrous and non-ferrous metals, thermoplastic and thermosetting polymers, composites and alloys
6 the definition, characteristics and typical applications of smart materials
7 the relationship between the structure of a material and its properties
8 the difference between crystalline and non-crystalline materials
9 how processes affect the structure and physical and mechanical properties of materials
10 how heat-treatment and surface-treatment processes affect the structure and properties of materials
11 common applications of each heat treatment and surface treatment
12 how materials fail due to corrosion as a result of material consumption, chemical composition and attack, reduction in thickness and perforation
13 the factors that contribute to fatigue failure and the three stages of creep
14 the different methods of preventing corrosion and their relative benefits and limitations
15 the advantages and limitations of different materials testing methods
16 the steps involved in the materials testing methods and how these determine the material properties
17 how to interpret load-extension graphs.

6.1 Physical and mechanical properties of materials

Physical properties

The **physical properties** of a material include a wide range of physical, thermal and electrical characteristics that can be observed or measured without changing the nature of the material. Examples are given in Table 6.1.

> **Key term**
>
> **Physical properties:** material characteristics that can be observed or measured without changing the nature of the material.

▼ Table 6.1 Physical properties of materials

Property	Description
Density (ρ)	The mass (m) of material contained per unit volume (V) where $\rho = \dfrac{m}{V}$ Measured in kgm^{-3}
Melting point	The temperature at which a material changes state from a solid to a liquid
Thermal conductivity (K)	The ability of a material to conduct heat Measured in $Wm^{-1}K^{-1}$
Electrical conductivity (s)	The ability of a material to conduct electricity Measured in Sm^{-1}
Electrical resistivity (ρ)	The ability of a material to resist the flow of electricity (resistivity is the inverse of conductivity) Measured in Ωm
Coefficient of thermal expansion (α)	The rate at which a material expands or contracts when subject to a change in temperature Measured in $°C^{-1}$
Corrosion resistance	Qualitative indication of the ability of a material to resist the chemical, electrochemical and other processes that cause corrosion
Specific heat capacity (c)	The quantity of heat energy required to raise the temperature of 1kg of a material by $1°C$ Measured in $Jkg^{-1°}C^{-1}$
Hardenability	Qualitative indication of whether heat treatment is able to alter the crystalline structure of a metal alloy to increase its hardness
Weldability	Qualitative indication of the ability of a material to be welded without cracking or adversely affecting the material's mechanical properties around the joint
Magnetic permeability (μ)	The degree to which a material becomes magnetised in the presence of an external magnetic field; the ratio between the applied magnetic field strength and the magnetic flux density present in the material Measured in Hm^{-1}
Electrical permittivity (ε)	The degree to which a material becomes polarised by an electric field; the ratio between the applied electric field strength and the electric flux density present in the material Measured in Fm^{-1}
Recyclability	Qualitative indication of the degree to which a material is suitable for recycling and subsequent reuse

> **Improve your maths**
>
> A rectangular block of mild steel measures $30\,mm \times 40\,mm \times 80\,mm$ and has a mass of $0.756\,kg$.
>
> Calculate the density of the steel in kgm^{-3}.

Mechanical properties

Mechanical properties are characteristics that define the behaviour of a material in response to an applied stress. Stress is force per unit area and is independent of the size of the component on which the force is applied. This means it can be applied in general terms as a property of the material itself.

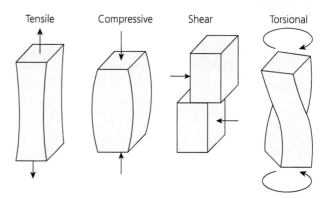

▲ Figure 6.1 Types of externally applied stresses

> ### Key term
>
> **Mechanical properties:** characteristics that define the behaviour of a material in response to an applied stress.

▼ Table 6.2 Mechanical properties of materials

Property	Description
Tensile strength (or ultimate tensile stress)	The maximum pulling or tensile stress that a material can withstand before failure Measured in Nm^{-1} or Pa
Compressive strength (or ultimate compressive stress)	The maximum squeezing or compressive stress that a material can withstand before failure Measured in Nm^{-1} or Pa
Shear strength (or ultimate shear stress)	The maximum shear stress that a material can withstand before failure Measured in Nm^{-1} or Pa
Torsional strength (or ultimate torsional stress)	The maximum twisting force or torque that a material can withstand before failure Measured in Nm^{-1} or Pa
Hardness	How well a material resists surface indentation, scratching or abrasion Units of hardness depend on the test methodology used and include the Rockwell hardness number
Toughness	The amount of impact energy a material can absorb up to the point when it fractures Measured in Jm^{-3}
Brittleness	Qualitative description of low toughness in materials that are not impact resistant and shatter easily, or those that undergo little or no plastic elongation before failure when subject to a tensile stress
Ductility	The ability of a material to undergo plastic deformation when subject to a tensile stress without failure Can be measured as either the: • percentage elongation of a test specimen at fracture • percentage reduction in surface area in a test specimen at fracture
Elasticity	Qualitative description of the ability of a deformed material under load to recover its original shape once the load causing the deformation is removed
Plasticity	Qualitative description of the ability of a deformed material under load to retain that deformation permanently when the load is removed
Malleability	The ability of a material to undergo permanent plastic deformation when subject to a compressive stress without failure Usually expressed in terms of the compressive strength of the material

> ### Test yourself
>
> Describe the difference between physical properties and mechanical properties.

6.2 Types of material and their structures

Metals

Pure metals

Pure metals contain a single type of metal atom. The atoms are packed together in a regular and repeating crystal lattice structure. For example, iron is a metallic chemical element and so is considered a pure metal.

Pure metals tend to be low strength, ductile and malleable. As a result of their relatively poor mechanical properties, very few pure metals find applications in engineering.

Alloys

Alloys contain two or more metals in solid solution with one another. Atoms of the different alloying metals are different sizes, changing the way they pack together in the crystal lattice structure. This has an impact on the mechanical properties of the alloy. Carefully chosen alloying metals, added in the right quantities, can significantly improve both material strength and hardness.

For example, bronze is a solid solution of copper and tin and so is an alloy. Both constituent metals are soft and ductile in their pure form. However, bronze has much higher strength and hardness than either parent metal.

Test yourself

Explain the difference between a pure metal and an alloy.

Ferrous metals

Ferrous metals are a small but extremely important group of metal alloys that contain iron. Pure iron has low tensile strength but is tough, ductile and malleable. As wrought iron, it has been hot forged by blacksmiths for centuries, into everything from gates to horseshoes.

Key terms

Pure metals: metals that contain a single type of metal atom and are not mixed or alloyed with other elements.

Alloys: materials comprising two or more metals in solid solution with one another.

Ferrous metals: iron and alloys containing iron.

When iron is alloyed with carbon in the correct proportions to form steel, it becomes much harder, and its strength increases significantly. The ability to manufacture inexpensive steel on a large scale has led to its widespread use in a vast range of applications, in everything from nails to skyscrapers.

Types of ferrous metal alloys are described in Table 6.3.

▼ Table 6.3 Ferrous metal alloys

Ferrous alloy	Composition	Properties	Applications
Low-carbon steel	Iron with 0.1–0.3% carbon	High ductility Low corrosion resistance, tensile strength and hardness Cannot be hardened by heat treatment (although can be hardened by cold working)	General purpose Steel beams Car body panels Nails
Medium-carbon steel	Iron with 0.3–0.6% carbon	Low corrosion resistance Medium ductility, tensile strength and hardness Can be hardened by heat treatment	Railway tracks Crankshafts Gears

Ferrous alloy	Composition	Properties	Applications
High-carbon steel	Iron with 0.6–1.0% carbon	Low ductility and corrosion resistance High tensile strength and hardness Can be hardened by heat treatment	Cutting tools Springs
Stainless steel	Iron with: • >11.5% chromium • <1.0% carbon	Low ductility High corrosion resistance Medium tensile strength and hardness	Cutlery Kitchen sinks Cookware Medical equipment
Cast iron	Iron with 2–4% carbon	Low ductility, corrosion resistance, tensile strength and hardness	Engine blocks Machine beds

Test yourself

Name two common uses of high-carbon steel.

Key term

Non-ferrous metals: metals that do not contain iron.

Non-ferrous metals

Non-ferrous metals and their alloys do not contain iron and are used in a wide range of important engineering applications.

Types of non-ferrous pure metals are described in Table 6.4.

▼ Table 6.4 Non-ferrous pure metals

Name	Properties	Applications
Aluminium	High ductility and electrical conductivity Moderate corrosion resistance Very low tensile strength and hardness	Electrical cables
Copper	High ductility Very high electrical conductivity High corrosion resistance Low tensile strength and hardness	Electrical cables
Nickel	High ductility, electrical conductivity and corrosion resistance Moderate tensile strength and hardness	Plating Batteries
Zinc	Low ductility (at room temperature) High electrical conductivity and corrosion resistance Extremely low tensile strength and hardness	Galvanising Sacrificial anodes Batteries

Types of non-ferrous metal alloys are described in Table 6.5.

▼ Table 6.5 Non-ferrous metal alloys

Name	Composition	Properties	Applications
Aluminium alloy (general purpose)	Aluminium with: • 1.2% manganese • 0.12% copper	High ductility, electrical conductivity and corrosion resistance Low tensile strength and hardness	General purpose Kitchenware Sheet
Brass (cartridge brass)	Copper with 30% zinc	High ductility, electrical conductivity and corrosion resistance Moderate tensile strength and hardness	Ammunition cartridges Plumbing fittings Hinges
Bronze (phosphor bronze)	Copper with: • 5% tin • <0.4% phosphorous	Moderate ductility High electrical conductivity and corrosion resistance Moderate/high tensile strength and hardness	Valve components Springs Fasteners

Forms of supply – metals

Metals used in engineering come in a wide range of standard forms, including plate, sheet, round bar (or rod), square bar, flat bar, round tube, square tube and channel.

To ease manufacturing, the form of supply used is that with the nearest net shape to the component being made.

Polymers

Thermoplastic polymers

When heated, **thermoplastic polymers** soften and can flow, which makes them suitable for injection moulding, vacuum forming and other high-volume manufacturing processes.

> **Key term**
>
> **Thermoplastic polymers:** polymers that soften when heated and can be moulded before being allowed to cool and reharden.

Types of thermoplastic polymer are described in Table 6.6.

▼ Table 6.6 Thermoplastic polymers

Name	Chemical name	Properties	Applications
ABS	Acrylonitrile butadiene styrene	Moderate strength and stiffness	Plumbing waste pipes Computer keyboards TV remote controls
HIPS	High-impact polystyrene	Moderate strength High impact resistance	Packaging Food trays
PLA	Polylactic acid	Low strength and thermal resistance	Single-use packaging (film, bottles etc.) 3D printing

Name	Chemical name	Properties	Applications
PS	Polystyrene	Moderate strength and toughness	DVD cases Disposable cutlery
EPS	Expanded polystyrene	Very low-density aerated foam able to absorb impact energy	Packing peanuts Egg boxes
PC	Polycarbonate	High strength, stiffness, toughness and transparency	Safety glasses Machine guards
PP	Polypropylene	Moderate strength and stiffness High toughness and fatigue resistance	Toys Document wallets Cutting boards Bottles
PMMA or acrylic	Polymethyl methacrylate	High strength and transparency Moderate toughness	Lighting components Lenses

Forms of supply – thermoplastic polymers

The form of supply for thermoplastic polymers depends on the manufacturing process being used:

▶ Pellets are required for moulding processes such as injection moulding, extrusion or rotational moulding.
▶ Sheets are required for thermo-forming processes such as vacuum forming.

Thermosetting polymers

Thermosetting polymers undergo a chemical curing process when components are manufactured that permanently sets their shape. They are generally much harder and more rigid than thermoplastic polymers.

> **Key term**
>
> ***Thermosetting polymers:*** polymers that set permanently during manufacture and do not soften when heated.

Types of thermosetting polymer are described in Table 6.7.

▼ Table 6.7 Thermosetting polymers

Name	Chemical name	Properties	Applications
UF	Urea formaldehyde resin	High strength, hardness, stiffness and thermal stability Good adhesive properties	Electrical fittings Adhesives
MF	Melamine formaldehyde resin	High strength, hardness, stiffness and thermal stability	Worktops Tableware
PF	Phenol formaldehyde resin	Moderate/high strength, hardness, stiffness and thermal stability	Pan handles Knobs
EP	Epoxy resin	High strength and toughness Excellent resistance to chemicals and adhesive properties	Adhesives Encapsulation Composites
PE	Polyester resin	Moderate strength and toughness Excellent adhesive properties	Coatings Composites

▼ Table 6.8 Elastomers

Common name	Chemical name	Properties	Applications
Natural rubber	Polyisoprene	High abrasion resistance and tear strength High flexibility at low temperatures Low resistance to chemicals, heat and UV light	Tyres Gaskets Latex gloves
Neoprene	Polychloroprene	High abrasion resistance and flexibility High resistance to chemicals, heat and UV light	Hoses Wetsuits Gaskets O-rings

Elastomers

Elastomers are a type of flexible polymer and consist of tangled long-chain molecules. Large elastic deformations of these materials are possible as these tangled chains are pulled straight.

Types of elastomers are described in Table 6.8.

> **Test yourself**
>
> To which family of materials does polychloroprene belong?

Forms of supply – thermosetting polymers and elastomers

Thermosetting polymers and elastomers are unusual in that the polymer itself does not exist as a raw material. Instead, it is created by a chemical reaction as the component is being made.

Thermosets are supplied in one of two forms:
▶ a powdered mix of precursor chemicals that require heat and pressure to cure into the solid polymer
▶ a two-part liquid consisting of a resin and an activator that when mixed will cure into the solid polymer.

Composites

Composites are materials made up of a mixture or combination of distinctly different materials that work together to provide improved mechanical properties.

Unlike metal alloys, the constituent materials in composites remain physically distinct from one another and are referred to as the matrix and the reinforcement. The matrix material, often a thermosetting polymer resin, is used to hold together or suspend the fibres or particles of the reinforcement material.

> **Key term**
>
> **Composites:** materials made up of a mixture or combination of distinctly different materials that work together to provide improved mechanical properties.

Types of composites are described in Table 6.9.

▼ Table 6.9 Composites

Name	Matrix	Reinforcement	Properties	Applications
Glass-reinforced polymer (GRP)	Polyester resin	Glass-fibre strands or woven matting	High strength and strength-to-weight ratio	Boat hulls Printed circuit boards Canoes
Carbon-reinforced polymer (CRP)	Epoxy resin	Carbon-fibre strands or woven matting	Very high strength, strength-to-weight ratio and stiffness	Aircraft structural components Bicycle frames
Medium-density fibreboard (MDF)	UF	Wood fibres	Low strength High density	Furniture

Forms of supply – composites

Fibre-based composites are usually supplied as separate matrix and reinforcement materials. For example, to make a component from carbon-reinforced polymer (CRP), you would need:

▶ a two-part thermoset epoxy resin
▶ carbon-fibre woven matting.

Test yourself

Name the matrix and reinforcement materials used in GRP composite.

Engineering ceramics

Engineering ceramics are non-metal, non-organic compounds characterised by their high melting points and extreme hardness.

Types of engineering ceramics are described in Table 6.10.

▼ Table 6.10 Engineering ceramics

Name	Properties	Applications
Silicon carbide	Crystalline structure Very high hardness, strength and melting point Very low ductility (brittle) Chemically unreactive	Abrasives: wet and dry paper, grinding wheels, cutting discs
Glass (borosilicate glass)	Amorphous structure High hardness and strength Very high melting point Very low ductility (brittle) Chemically unreactive	Cookware Laboratory glassware

Forms of supply – engineering ceramics

Crystalline engineering ceramics are usually supplied as powders. These can be used as shot-blasting media, in abrasives, or combined with a suitable matrix to make ceramic-particle composites.

Timber

Natural woods are strongly **anisotropic**, with high tensile strength in the direction of the grain and much lower strength across the grain. Engineered woods are composites designed to be **isotropic** by reinforcing a resin matrix with wood veneers, chips or fibres orientated to provide the same strength in all directions.

Key terms

Anisotropic: having different mechanical properties when measured in different directions.

Isotropic: having the same mechanical properties when measured in any direction.

Types of timber are described in Table 6.11.

▼ Table 6.11 Timber

Name	Properties	Applications
Soft wood (European redwood)	Moderate strength and durability Anisotropic	General woodwork Floorboards Shelves Roofs
Hard wood (oak)	High strength, toughness and durability Anisotropic	Furniture Doors Window frames
Engineered wood (plywood)	High strength in all orientations Manufactured in large sheets Isotropic	Floors Furniture

Forms of supply – timber

Engineered woods like plywood, block board, chip board and oriented strand board (OSB) are supplied in sheets of various thicknesses.

Natural timber is supplied as boards, strips, dowels or shaped mouldings.

Smart materials

Smart materials are a class of advanced modern materials characterised by their ability to react in response to an external stimulus.

Key term

Smart materials: advanced modern materials characterised by their ability to react in response to an external stimulus.

Types of smart material are described in Table 6.12.

▼ **Table 6.12 Smart materials**

Name	Properties	Applications
Shape memory alloy (SMA)	Changes shape in response to a change in temperature Heat treatment sets the shape memory of an SMA; the material always returns to this shape when heated above its critical temperature	Orthodontic braces Robotic actuators
Quantum tunnelling composite (QTC)	Changes electrical resistance in response to an externally applied force	Pressure sensors Touch pads
Thermochromic pigment	Changes colour in response to a change in temperature	Strip thermometers Cold-chain indicator labels
Photochromic material	Changes colour in response to a change in light	Reactive sunglasses
Piezoelectric crystal	Generates an electrical potential or voltage in response to an externally applied force and conversely generates a force in response to an applied voltage	Microphones Sounders

Test yourself

Which type of smart material is used in reactive sunglasses?

Methods of disposal and recycling

All products eventually come to the end of their working lives, at which point they will be recycled or disposed of in landfill. Many materials can be recycled but only if they can be successfully separated from other materials and contaminants.

Methods of disposal and recycling are described in Table 6.13.

▼ **Table 6.13 Methods of disposal and recycling**

Material	Disposal
Metals and metal alloys	All metals and metal alloys can be recycled without deterioration of their physical or mechanical properties
Thermoplastic polymers	Thermoplastic polymers can be reheated and combined with virgin polymer to make new products. However, after being recycled two or three times, the long-chain polymer molecules begin to break down and the properties of the material deteriorate
Thermosetting polymers and elastomers	Recycling of thermosetting polymers and elastomers is extremely difficult, as their polymer chains are permanently crosslinked and are not readily reshaped. These are generally disposed of in landfill
Ceramics	Most glass materials can be recycled without deterioration. Crystalline ceramics cannot be recycled and are generally disposed of in landfill
Composites	Recycling of composites (including engineered woods) is extremely difficult. The matrix and reinforcement materials in composites cannot be readily separated. Even if this were possible, many composites use thermoset resins which are themselves extremely difficult to recycle, even in ideal circumstances. Most composite materials are disposed of in landfill or incinerated
Natural timber	Natural timber is repurposed and used again where possible. It can also be shredded for use as animal bedding, chippings, mulch or biofuel

Structures

Atomic structure

Atoms

An atom is the smallest possible quantity of an element that exhibits all the physical properties of that element.

Atoms are composed of three types of sub-atomic particle. The nucleus at the centre of an atom contains neutrons and protons, while electrons orbit the nucleus in a series of orbital layers or shells. Protons are positively charged and electrons are negatively charged.

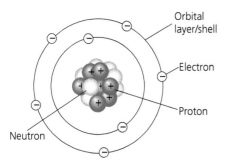

▲ Figure 6.2 Bohr's atomic model of the structure of an atom

Compounds

Compounds are formed when atoms of two or more elements bond together into molecules. A molecule is the smallest possible quantity of a compound.

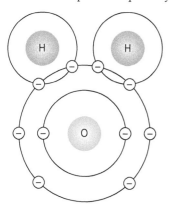

▲ Figure 6.3 H_2O is a compound of hydrogen (H) and oxygen (O)

Bonding mechanisms in solids

Metallic bonding

Metal atoms join into large structures in which their outer electrons are free to move. These structures are held together by metallic bonding caused by the electrostatic forces between positive metal ions and their free-moving, delocalised electrons.

Metals form large crystalline lattice structures. They tend to be solids at room temperature, with high melting and boiling points. The delocalised electrons present in metallic bonding means they have very high electrical conductivity.

Ionic bonding

Ionic bonding happens between a metal and a non-metal and is due to the strong electrostatic forces between positively and negatively charged ions. Ionic compounds are solid at room temperature and have very high melting and boiling points. As solids, they do not conduct electricity. However, when melted into a liquid or dissolved in a solution, they have high electrical conductivity.

Covalent bonding

Covalent bonding happens between non-metal atoms that share outer-shell electrons. All covalent materials have very low electrical conductivity. Covalent atomic bonding is very high strength, and large crystalline lattice structures make materials that are extremely hard with very high melting points.

Covalent molecular solids have strong covalent bonds between the atoms in individual molecules but only weak intermolecular forces (called Van der Waals forces) between the molecules themselves. This means that they have low hardness and low melting and boiling points.

> ### Test yourself
>
> A material does not conduct electricity as a solid but becomes a good conductor when melted into a liquid or dissolved in a solution. What kind of chemical bonding is present in the material?

Microstructure and the formation of grains in metals

As molten metal cools to form a solid, small groups of atoms come together and start to arrange themselves into regular lattice structures, forming numerous seed crystals throughout the material. As the metal cools, more atoms join each seed crystal and the structures become larger and form distinct grains. Individual grains continue to grow until they meet other grains that were grown from adjacent seed crystals. The orientation of the crystal lattice in each grain is different, so they do not combine but remain separated by grain boundaries.

Very large grains are visible to the naked eye, but the microstructure of metals is best seen under a microscope (above ×25 magnification).

Lattice structure in metals

Solid metals are **crystalline**. That means that when they solidify, they spontaneously form into large, well-ordered lattice structures.

> ### Key term
>
> **Crystalline:** where atoms, ions or molecules come together to form solids with well-ordered, repeating lattice structures.

The elastic behaviour of metals is a consequence of the elastic deformation of these structures under load. When the load is removed, the crystal structure regains it shape.

Plastic behaviour of metals happens when the force acting on the lattice causes permanent shifts in the crystal structure.

When crystal structures form, they do so imperfectly. Line defects or dislocations are inconsistencies or gaps in the lattice structure, and these play a key role in plastic deformation.

When load is applied to the structure, very little work is required to break and reform individual bonds around a dislocation. As this happens, the dislocation starts to move along the line of atoms in the lattice, causing permanent realignment and permanent deformation of the structure. The movement of the dislocation is only stopped when it reaches a grain boundary, or when it is pinned by some other inconsistency in the structure. Pinning is caused by the presence of an atom of an alloying element or another dislocation.

Grains and the lattice structure in steel

Steel is an alloy of iron and carbon, and its mechanical properties rely on the crystal structures formed as the molten steel solidifies. In liquids, atoms or molecules are free to move, and in molten iron, carbon is completely soluble. However, as the iron cools, iron atoms begin to spontaneously arrange themselves into crystal lattice structures as they find the most energy-efficient way to pack together. Carbon is much less soluble in these structures.

Above the critical temperature (around $800^{\circ}C$ in steel), the crystal lattice structure formed by iron atoms is called austenite. Austenite can dissolve a maximum of about two per cent carbon by weight.

As the steel cools below the critical temperature, the most energy-efficient way to pack the iron atoms together changes and so the structure of the crystal lattice spontaneously reorders itself. The new structure formed is called ferrite. Ferrite can only dissolve 0.02 per cent carbon by weight.

If the cooling process is gradual, any excess carbon is forced out of the saturated ferrite and bonds with iron atoms to form a hard and brittle intermetallic compound called cementite (Fe_3C).

However, if the cooling is rapid, the material is forced to form an alternative and extremely hard non-equilibrium structure. If the dissolved carbon in the austenite does not have time to move out of solution as the ferrite forms, then it becomes trapped in the crystal lattice. This results a highly strained, hard structure called martensite.

Key terms

Non-equilibrium structure: structure that is not arranged in the most efficient way of packing atoms together into a crystal lattice.

Non-crystalline or amorphous: having randomly arranged molecules without a regular lattice structure.

Test yourself

The movement of dislocations in the lattice structure of metals plays a key role in what type of material deformation process?

Crosslinking of polymers

Polymers consist of large molecular chains that fold together into irregular patterns and do not form regular crystal lattices. In thermoplastics, these chain molecules can slide over one another when heated, allowing them to be reformed into different shapes.

However, in thermosetting polymers, the structure of the materials is set permanently by the formation of additional chemical crosslinking between the individual long-chain molecules.

Ceramic structures

Engineering ceramics are non-metal, non-organic (not based on compounds of carbon and hydrogen) compounds held together by ionic or covalent atomic bonding that usually form crystalline solids.

Non-crystalline or amorphous ceramics are known as glasses. These have much of the rigidity and strength of a regularly arranged crystal lattice structure but contain molecules that are arranged randomly. This means that there are no grains or grain boundaries and no lattice dislocations. Amorphous solids have high hardness but low impact toughness. Unlike crystalline solids, amorphous materials do not have a specific melting point but instead soften over a temperature range.

Composites

As we learned earlier, composites use a combination of at least two discrete material components to create a new material that has enhanced mechanical properties. These components are referred to as the matrix and the reinforcement.

Particle-based composites

Particle-based composites consist of a resin or metal matrix that bonds together particles of the reinforcement. For example, tungsten carbide cutting tools are made from tungsten carbide particles suspended in a cobalt metal matrix.

Fibre-based composites

Fibre-based composites consist of a resin matrix that bonds together fibres of the reinforcement. For example, glass-reinforced polymer (GRP) components are made from glass fibres held in a polyester resin matrix.

Laminated or sandwich composites

Laminated or sandwich composites use a resin matrix to bond together thin layers or sheets of reinforcement material. For example, plywood is made from thin wooden sheets or veneers held together by a thermosetting resin matrix of urea formaldehyde.

6.3 The effects of processing techniques on materials

Effects of processing on metals

Casting

Casting involves pouring molten metal into a prepared mould and allowing it to cool rapidly and solidify into the desired shape. Rapid, uncontrolled cooling in large castings can result in:
▶ gas porosity caused by trapped gas bubbles
▶ shrinkage porosity, cracks and distortion caused by thermal shrinkage during rapid cooling
▶ areas of segregation where alloying elements are concentrated
▶ the formation of large, coarse, irregular grains
▶ areas with **inclusions**, where impurities or foreign matter were trapped during casting.

These factors mean that cast metals in general have lower strength and hardness but higher ductility than **worked or wrought metals**, but these can be inconsistent through the casting.

> ### Key terms
>
> **Inclusions:** foreign matter or impurities trapped in the structure of a metal.
>
> **Worked or wrought metals:** metals that have undergone plastic deformation processes.

> ### Test yourself
>
> What term is used to describe metals that have undergone plastic deformation processes?

Recrystallisation temperature

When metals undergo plastic deformation, they **work harden** and store some of the deformation energy as additional dislocations and grain distortions in the metal microstructure. When heated above the recrystallisation temperature, the dislocations and distortions in the structure become the nucleation points for the growth of new grains. These grow and replace the existing structure, relieving any internal stresses or work hardening as they do so.

The effect of deformation processes on metals depends on whether they are carried out above or below the recrystallisation temperature.

> ### Key term
>
> **Work harden:** increase in hardness and reduction in ductility due to a build-up of dislocations and grain distortions caused by plastic deformation.

Hot working

Material processing carried out above the recrystallisation temperature of the metal is called hot working. During hot working, the effects of deformation cause simultaneous recrystallisation, preventing the build-up of dislocations that cause work hardening and allowing very large deformations to be achieved without the risk of fracturing the material.

Impurities and intermetallic compounds present at the initial grain boundaries of the material do not recrystallise during hot working. Instead, these flow along in the direction of grain distortion and are redistributed throughout the material. This leaves hot-worked components with a visible internal flow structure that provides enhanced mechanical properties in the direction of flow and can also inhibit **crack propagation** through the material.

> ### Key term
>
> ***Crack propagation:*** process of crack enlargement and movement through a material that will eventually lead to material failure.

Hot-working processes result in poor surface finish due to oxidation of the metal at high temperatures where it contacts atmospheric oxygen. The hard oxide layer formed on the surface is called mill scale.

In addition, dimensional accuracy of hot-worked components is poor due to thermal expansion and contraction during processing.

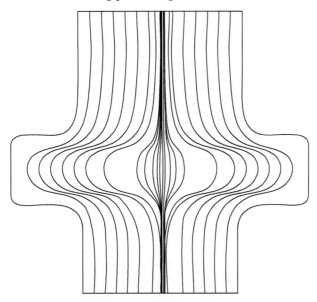

▲ Figure 6.4 Grain flow visible in a hot-worked forging

When cast metal **ingots** are hot worked:
- the coarse, irregular grain structure is refined into regular, small, spheroidal grains
- voids and other internal structural flaws are eliminated
- any segregation of alloying elements is reduced, and the solid solution becomes homogenous
- any concentrations of inclusions or impurities are broken up and distributed in the direction of the grain flow
- strength and hardness are increased
- ductility decreases.

> ### Key term
>
> ***Ingots:*** rough rectangular castings used as the raw material for worked or wrought metals.

When previously cold-worked or work-hardened metals are hot worked:
- any internal stresses are relieved
- distorted, elongated grains are refined into regular, small, spheroidal grains
- surface finish is poor due to surface oxidation at high temperatures
- dimensional accuracy is poor due to thermal expansion and contraction during processing
- strength and hardness are decreased
- ductility is increased.

> ### Test yourself
>
> What term is used to describe plastic deformation processes carried out above the recrystallisation temperature?

Hot rolling

Hot rolling between heavy rolls is used to reduce the thickness of plates or sheets of metal. The full depth of the metal undergoes recrystallisation during processing.

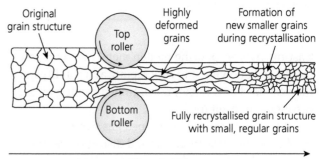

▲ Figure 6.5 Changes to grain size and shape that occur during recrystallisation in hot rolling

Drop forging

Drop forging uses an impact between dies to shape a billet of heated metal. The forging blow causes plastic deformation mainly in the outer layer of the metal. As recrystallisation is initiated by plastic deformation, this only takes place in the deformed outer layer. So, in drop forging, recrystallisation does not occur all the way through the forged component.

Hot press forging

Press forging squeezes the material between dies at very high pressure. This tends to cause deformation and therefore recrystallisation through the full thickness of the metal.

Cold working

Material processing carried out below the recrystallisation temperature of the material is called cold working. During cold working, there is no recrystallisation and the build-up of dislocations and grain distortions work hardens the material, reducing its ductility. Once work hardened, the material becomes much more difficult to process. Large deformations can result in cracking or a complete fracture of the material.

Cold working gives an excellent surface finish, results in dimensional accuracy, and increases both strength and hardness by work hardening the material.

When previously hot-worked metals are cold worked:
▶ distortion and elongation of grains causes work hardening
▶ strength and hardness are increased
▶ ductility is decreased.

Cold rolling

Cold rolling between heavy rolls is used to reduce the thickness of previously hot-rolled metal plates or sheets to their final, finished dimensions.

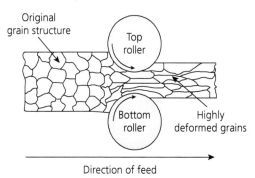

▲ Figure 6.6 Cold rolling of previously hot-rolled plates

Test yourself

Which type of plastic deformation processes cause work hardening?

Welding

Welding is a process of permanently joining components made from similar metals and is most commonly carried out on steel, stainless steel and aluminium. It involves localised heating of the parent metal until both edges to be joined start to melt and the introduction of additional metal from a filler rod to form a molten weld pool. When the weld cools, the material at the joint is fused together.

At the joint itself, similar issues are encountered to those seen in casting, as molten metal cools rapidly. These issues reduce the strength and increase the ductility of the weld metal and include:
▶ gas and shrinkage porosity
▶ shrinkage cracks
▶ formation of large and irregular grains
▶ inclusions.

In the area around the joint where the parent metal did not actually melt, it is still affected by the heating process. Metal in this zone undergoes varying degrees of extreme heating and rapid cooling, in what amounts to a poorly executed and highly variable heat treatment. This area is called the heat-affected zone (HAZ).

When thin sections, sheets or thin-walled tubes are welded, they are prone to distortion caused by different rates of thermal expansion experienced in the HAZ.

There are also unpredictable changes to the microstructure of the material in the HAZ. The exact nature and effects of these changes depend very much on the parent metal being welded and its processing history. Effects include recrystallisation and grain growth, like those seen in normalising (see Section 6.4) that increase ductility and lower strength, or structural phase changes like those seen in quench hardening (see Section 6.4) that increase hardness and decrease ductility.

Test yourself

In welding, what does HAZ stand for?

Brazing

Brazing is a process used to join a range of metals, including steel, stainless steel, brass, bronze and cast iron. Unlike welding, brazing does not melt and fuse together the component parent metal. Instead, it relies on the surface adhesion of a brass or bronze filler material to make the join. The melting point of the brass is well below the melting point of the component parent metal. This avoids the problems of thermal distortion and the unwanted effects caused by the high temperatures encountered in welding.

Sintering

Sintering is the key process in powder metallurgy, which is used to fuse together fine metal powders into near full density metal parts.

In sintering, fine metal powders held in a resin binder are moulded into the shape of the required component. This so-called green component is then heated to just below the melting point of the metal. This initially burns off the binder and then fuses together the metal particles so that a solid shape is formed.

The amount of compaction and the duration and temperature of the sintering process controls the final density of the part, which can range from around 75 per cent to near full density.

Coating

Coating metal parts in paint or powder coating significantly increases corrosion resistance by forming a physical barrier that excludes oxygen, water or corrosive chemicals.

Effects of processing

Processing thermoplastic polymers

Injection moulding is a widely used manufacturing process for thermoplastic polymer components.

Two important process parameters in injection moulding are material temperature and injection pressure. A number of problems are encountered when these process parameters are not set correctly:

▶ Low injection pressure and temperature can cause short shots. This is where the molten material does not fill the mould cavity and leaves voids in the component.
▶ High injection pressure can cause jetting. This is when the initial jet of molten material entering the mould solidifies before the mould cavity fills. This can cause substantial weakness in the material.
▶ High temperature and injection pressure can cause discolouration and burn marks where trapped air has been compressed and overheated as the mould cavity fills.

Processing thermosetting polymers

Thermosetting polymers supplied as powders must be compacted into moulds and then heated under pressure to start the chemical curing process. They must be given sufficient time to fully cure, as the chemical crosslinking between polymer chains takes time to fully develop. Failure to do so will result in parts with low hardness and low tensile strength.

Two-part liquid polymers must be thoroughly mixed and are often degassed before use to eliminate air bubbles that might affect the strength of the final polymer. These can take 24 hours to cure at room temperature and even longer to attain their full tensile strength.

Processing ceramics

Sintering has long been an important manufacturing process for industrial ceramics that have both high hardness and high melting points. The basics of the process of sintering ceramic powders are the same as used in powder metallurgy, the key process parameters being pressing force or compaction and firing temperature.

High compaction and high firing temperatures will give high-density and extremely high-strength ceramic components.

Processing composites

Fibre-based composites can be manufactured to have anisotropic properties by aligning more of the reinforcement fibres in the direction requiring most strength. Areas of additional fibre can also be built up to reinforce local areas of the component around joints and fixings.

By increasing the ratio of reinforcement to resin matrix, the tensile strength of fibre-based-composite components can be increased.

6.4 Heat treatments and surface treatments

Heat treatments

Heat treatments are used to manipulate the crystal structure of metals in order to change their mechanical properties.

Case hardening

Mild steels (<0.3 per cent carbon) do not contain sufficient carbon to create martensite and so **quenching** from above the critical temperature has no effect on the hardness of the steel. For case hardening to work, additional carbon needs to be absorbed into the surface of the steel so that it becomes hardenable. This is achieved using high-temperature diffusion of carbon atoms into the surface layer of the mild steel by heating it above its critical temperature in a carbon-rich atmosphere.

> ### Key term
>
> **Quenching:** rapid cooling of hot metals, usually achieved by submerging them in oil or water.

Once sufficient carbon has been dissolved into the surface layer, quenching the carbon-rich austenite will cause hard and brittle martensite to form. Case hardening allows mild-steel components to retain the toughness of a mild-steel core with an outer-surface layer up to 1 mm in depth of much harder steel, which provides enhanced wear resistance.

Quench hardening

When medium- and high-carbon steels (0.3–1 per cent carbon) are quench hardened, the structure of the steel is affected all the way through the component.

Through hardening involves heating the steel to above its critical temperature so that the crystal structure changes from a mixture of ferrite and cementite into a more homogenous mix of austenite and dissolved carbon. When this is rapidly cooled by quenching in water or oil, carbon is trapped inside the highly strained lattice structure, forming martensite. This increases both strength and hardness but reduces ductility, making the material brittle.

> ### Test yourself
>
> Name the high-strength, high-hardness but brittle lattice structure formed when high-carbon steels are quench hardened.

Tempering

Quench-hardened carbon steel consists of extremely hard and brittle martensite. Although the material hardness and increased strength are both desirable characteristics of martensite, its brittleness is problematic. This low toughness can make quenched carbon steel unsuitable in many practical applications.

Tempering is a heat-treatment process that eliminates brittleness. Tempering is carried out well below the critical temperature of the steel, anywhere between 200 and 600 $^\circ$C, depending on the desired outcome.

At these temperatures, some of the martensite begins to dissociate and release its trapped carbon atoms to form the stable equilibrium structures of ferrite and cementite. The presence of soft and malleable ferrite in the grain structure reduces hardness but also gives a significant increase in toughness.

The temperature and duration of the tempering cycle depends on the required characteristics of the final structure. The more martensite that dissociates back into ferrite and cementite, the lower the overall hardness but higher the toughness of the material.

Normalising

Normalising is a heat-treatment process used to refine the grain structure in steels and other metals.

In steel, the material is heated to just above critical temperature and held there until a fully uniform mix of austenite and dissolved carbon forms. This is then allowed to cool in air. This high cooling rate promotes the formation of seed crystals throughout the material which grow into numerous small grains of uniform size and shape.

> ### Test yourself
>
> Name the main effects of normalising on the grain structure of steel.

Annealing

Annealing is a similar process to normalising but after heating above critical temperature, the cooling process is much slower and more controlled. The gradual cooling allows larger grains to form and eliminates any non-equilibrium structures or internal stresses.

Once fully annealed, a material is as ductile and soft as possible.

Precipitation hardening

Precipitation or age hardening is a heat-treatment process typically used to strengthen non-ferrous alloys.

For example, aluminium is commonly alloyed with copper to increase its hardness and overall strength. Molten copper and aluminium are mixed freely together; when left to cool slowly, the alloy solidifies and the crystal structure of aluminium begins to form. The copper mixes in solid solution with the aluminium until it becomes saturated. At this point, no additional copper can be dissolved in the aluminium and excess copper is forced out of solution and precipitates to form an intermetallic compound $CuAl_2$.

Precipitation hardening is carried out in three stages:

1 Solution treatment – solution treatment involves heating the material until all the precipitates are dissolved back into the aluminium crystal lattice. This is possible because the solubility of copper in aluminium increases with temperature.

2 Quenching – slow cooling would give time for excess copper to precipitate back out of the solution. However, rapid quenching prevents this from happening and traps the dissolved copper in the aluminium crystal lattice, forming a supersaturated solid solution.

3 Aging – over time, the non-equilibrium structures in the supersaturated solid will dissociate back into saturated aluminium and precipitates of intermetallic compound $CuAl_2$. These precipitates are dispersed throughout the crystal lattice and at grain boundaries. They cause pinning and prevent the movement of dislocations, improving the hardness and strength of the aluminium.

Surface treatments

Surface treatments are designed to enhance the aesthetics of a component and protect its surface from direct contact with water, oxygen or other chemicals in the environment that might otherwise cause corrosion.

Painting

Paint is a mixture of coloured pigments and binders dissolved in a solvent. Once applied to the surface of a component, the solvent evaporates leaving a solid protective film on the material.

Plastic coating

As an alternative to wet paint, powder coating is a common method of applying a coat of thermoplastic polymer to the surface of metal components.

In powder coating, an electrostatic charge is applied to fine particles of thermoplastic polymer and sprayed onto a metal component with the opposite charge. The charged particles are attracted to the component and form a thin, even layer over its surface.

The component is then heated in an oven to melt and fuse together the particles of polymer into a continuous protective layer.

Galvanising

Steel components that are meant for use outdoors are often coated in zinc or galvanised. The coating is applied by dipping the component in a tank of molten zinc. This ensures that the entire surface area of the component, inside and out, is coated.

Galvanising provides superior corrosion protection to paint or powder coating. It achieves this in two ways:
▶ by forming a protective barrier to prevent the steel from making contact with oxygen or water, which are the two main drivers of corrosion in steel
▶ by providing electrolytic (galvanic) protection from corrosion.

6.5 Causes of material failure and their prevention

Causes of material failure

Corrosion

Corrosion is caused by the action of chemical processes that attack and consume the material or degrade its mechanical properties. Failure due to corrosion is usually a consequence of thinning and perforation of the material to an extent where there is insufficient material remaining to withstand working loads.

Oxidation of metals including rusting of ferrous metals

The most common form of corrosion found in engineering is the formation of rust on iron and steel. Rust is a form of iron oxide that does not adhere to the surface of the steel but instead flakes off, and in doing so causes material loss and exposes more steel to the conditions required for continued corrosion.

When steel is exposed to oxygen in the presence of water, a series of weak electrochemical cells are established between different areas of the material surface.

Areas that become slightly anodic lose mass as iron atoms are lost from the surface and transported away as positive ions in the water which forms the electrolyte in the cell. This is the first step in a series of complex electrochemical reactions that ultimately end in the formation of familiar red, flaky hydrated iron (III) oxide, or rust.

Rusting accelerates in the presence of common salt. This dissolves in the water, making a much more effective electrolyte, and increases the rate of corrosion.

Stress corrosion

Stress corrosion is a form of electrochemical or galvanic corrosion that occurs between stressed and unstressed regions of the same material. It is often found in cold-worked materials where residual stresses are present in work-hardened areas that are next to unworked areas with little or no residual stress.

Chemical attack

Many working environments expose materials to reactive chemicals. These can chemically attack and eat away material from exposed surfaces or affect their physical and mechanical properties.

Even in low concentrations, corrosive chemicals acting over long periods of time can cause significant degradation, weakening and material loss.

For example, in a hydrogen-rich working environment, steel components can suffer hydrogen embrittlement as hydrogen atoms diffuse in the microstructure of the material. This can lead to the formation of microcracks in the surface of the material which can initiate fatigue cracks.

Aging

Aging is a combination of gradual degradation processes that only become significant when their effects are combined after long periods of time.

A wide range of factors combine to limit the service life of materials, which include the effects of:
- exposure to ultraviolet light (sunlight)
- exposure to atmospheric pollutants
- weathering
- humidity
- chemical attack
- accumulated wear
- accumulated impact damage
- thermal cycling.

> **Test yourself**
>
> Name three factors that contribute to aging.

Physical causes of material failure

Overloading

When a ductile material is loaded beyond its yield point, it will undergo plastic deformation and eventually fracture. Regular visual inspection can identify signs of plastic deformation where a component has stretched or bent. This gives engineers a warning that the component is overloaded, and time to act in order to rectify the problem.

When brittle materials are overloaded, there is little or no plastic deformation and so no warning signs before the material fractures.

Fatigue

Fatigue is a phenomenon that affects materials subject to cyclic loading, that is, any application where loading and unloading is repeated over an extended period.

> **Key term**
>
> *Fatigue:* material degradation process that can cause components to fail when they undergo repeated cycles of loading and unloading.

> **Industry tip**
>
> Component failure caused by fatigue cracking often comes without warning, occurs at normal working loads, and is immediate and catastrophic. Failure can occur after hundreds of loading cycles or millions, so it is a crucial factor to be considered in component design and material selection.

Fatigue begins with the formation of microcracks in the surface of the material around stress concentrations. Stress raisers on or near the surface of a component can include voids or inclusions, surface damage or tooling marks. The geometry of the component itself can also cause stress concentrations on the inside of sharp corners or at sudden changes in cross-sections.

Once a crack initiates, the tip of the crack acts as its own stress raiser, ensuring that with every loading cycle the crack propagates a little further through the material. This cyclic propagation causes the fracture to grow and leaves the fracture surface with characteristic beach lines. Once the surface area of intact material becomes insufficient to support the working load on the component, the material becomes overloaded and fails.

Creep

Creep is a process where materials subject to constant loading well below their yield point experience gradual elongation due to plastic deformation and structural degradation.

Key term

Creep: process where materials subject to constant loading well below their yield point experience gradual elongation due to plastic deformation and structural degradation.

Creep occurs in three distinct stages:

1 Primary creep is rapid and occurs over a brief period after the material is put under load for the first time. It is caused by the movement of dislocations within the crystal lattice and it stops when the material work hardens.

2 Secondary creep occurs over a much longer period and continues to degrade the internal structure of the material. This is caused by the continuation of dislocation effects inside individual grains but mostly by the movement between grains at grain boundaries.

3 Tertiary creep is the final stage of the process, in which the structure of the material has become dangerously weakened and normal loading conditions become enough to cause rapid elongation and material failure.

The effects of creep are usually insignificant in most normal working environments. However, at elevated temperatures, all forms of creep are accelerated. For example, turbine blades in jet engines are exposed to high loads at temperatures at or near their melting point. In this situation, creep is a critical consideration in component design and material selection.

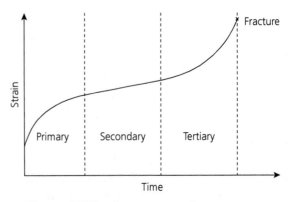

▲ Figure 6.7 The three stages of creep

Case study

Turbine blades in jet engines operate at temperatures close to their melting point and, as a consequence of their high rotational speeds, have to withstand huge loads for extended periods of time. These conditions are ideal for the development of creep, and early jet engines with forged steel alloy blades suffered greatly from its effects.

Most of the elongation in secondary creep is caused by slip at grain boundaries. Knowing this, engineers at Rolls Royce came up with a manufacturing technique to cast blades in a nickel-based superalloy as a single crystal. These blades have no grain boundaries at all and so virtually eliminate the effects of secondary creep.

Questions

1 Why is the failure of jet engine turbine blades such an important issue?
2 Name two factors that accelerate the effects of creep.

Erosion

Erosion is the physical wear of a surface caused by the flow of liquids or gases over it. This can be due to the impact of liquid particles in gases or solids suspended in a flowing liquid. The rate of surface erosion is increased with increasing temperature, pressure or flow rate.

Another mechanism of surface erosion is cavitation in pipework. Cavitation is the formation of tiny bubbles in flowing liquids caused by localised pressure variation. When these reach areas of higher pressure, where pipework turns a corner for instance, these bubbles collapse. Constant bombardment of a surface by collapsing bubbles causes cavitation erosion.

Corrosion prevention

Coatings

Coatings such as paint or powder help to prevent corrosion by forming a physical barrier over the surface of the material. This prevents any contact between the material and atmospheric oxygen, water or other chemicals that either cause or promote corrosion.

Paint and powder coating are available in a wide range of colours and enhance the aesthetics of a product as well as providing corrosion protection. However, these finishes are damaged easily, degrade over time and have to be renewed periodically to retain their effectiveness.

Galvanising

Galvanising not only forms a physical barrier to protect the surface of the material but also provides electrolytic (or galvanic) protection which continues to be effective even if the zinc coating is scratched or damaged. However, galvanised steel does not provide the improved aesthetics of a painted finish and so is restricted to industrial applications. It is also limited by the physical size of dipping tanks used to apply the zinc coating which cannot accommodate large components.

Electrolytic protection

When two dissimilar metals are in contact with one another in the presence of a liquid electrolyte such as water, an electrochemical cell is formed. At the anode of this cell, the material will lose mass as it releases positively charged ions into the electrolyte that then travel to the cathode. The electrons liberated when positive ions form move to the cathode through the metal. These combine with the electrolyte to form negatively charged ions, which in turn combine with the positive ions to form the by-product of corrosion.

When steel and zinc are in the presence of an electrolyte, it is the zinc that forms the anode and the steel (iron) that forms the cathode of an electrochemical cell. This means that the zinc becomes a **sacrificial anode** which corrodes in preference to the steel and by doing so protects the steel from corrosion.

This is the mechanism by which galvanising continues to protect steel from corrosion, even if the outer coating of zinc is damaged.

> **Key term**
>
> **Sacrificial anode:** sacrificial metal forming the anode in electrolytic protection; this corrodes in preference to the metal forming the cathode, which is therefore protected from corrosion.

6.6 Materials testing methods and interpretation of results

Visual inspection

Visual inspection of engineering components is limited to the observation of qualitative characteristics or features. It can spot the early signs and extent of corrosion, surface cracking or plastic deformation, which are all factors that should prompt further investigation.

Visual inspection is a basic form of non-destructive testing that can be used to monitor the condition of components in service.

Tensile testing

Tensile testing measures the response of a material to loading in tension.

Tensile testing machines use clamps to secure each end of a prepared test sample of known length and cross-sectional area. Hydraulic rams then apply tension to the sample by moving the clamps apart until the material fractures.

During the test, sensors measure the load on the material and any associated elongation of the sample.

Tensile testing is a destructive test carried out in the laboratory on expendable material samples. It cannot be used to carry out component tests in situ or to monitor the condition of components in service.

Load-extension graphs

Test data is used to plot a load-extension graph, as shown in Figure 6.8, which can be used to determine several important mechanical properties of the test material.

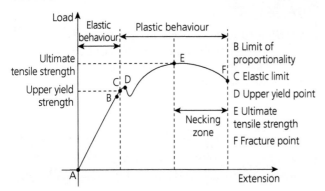

▲ Figure 6.8 Load-extension graph typical of mild steel

Hooke's law states that, in an elastic material, elongation is directly proportional to load. This can be seen by the straight-line part of the graph between A and B. Any loading in this region causes a proportional increase in elongation that is then recovered when the load is removed.

Point B is the limit of proportionality of the material, after which the proportionality between load and extension defined by Hooke's law no longer applies.

Point C is the elastic limit of the material, after which additional loading will cause permanent deformation of the material.

It is difficult to determine the exact moment when the elastic limit is reached and permanent plastic deformation of the material starts. So, engineers define this point as the load that causes a 0.05 per cent permanent extension in the sample and call it the upper yield strength, shown at point D (which you can see is close to the theoretical elastic limit at point C).

After the yield point, the extension caused by plastic deformation is non-linear. At this point, movement of large numbers of dislocations start to get in each other's way, causing pinning and subsequent work hardening of the material. This continues until the maximum load or ultimate tensile strength (UTS) is reached, as shown at point E.

After the elastic, plastic and work hardening phases, the material begins to fail as grains and grain boundaries start to cleave. Any small defect or stress raiser, such as a microcrack or inclusion, in the material becomes the weakest point in the structure and initiates localised straining called necking that will become the site of the final fracture (see Figure 6.9). Necking leads to a decrease in the cross-sectional area of the material supporting the load and a corresponding increase in localised stress. This increases the rate of localised deformation even further, causing rapid failure.

In the necking zone of the load-extension graph shown in Figure 6.8, there is an apparent reduction in load. However, as necking causes a local decease in cross-sectional area, the stress in that region actually continues to increase up until the material fractures at point F.

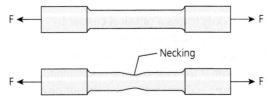

▲ Figure 6.9 Necking formed during the ductile failure of mild steel

Stress–strain characteristic graphs

The mechanical characteristics of different materials can be determined from their stress–strain graphs.

Figure 6.10 shows typical stress–strain curves for a range of different materials:
▶ Crystalline ceramics have extremely high ultimate tensile strength (UTS) and high stiffness (indicated by the steep gradient corresponding to a high value of Young's modulus, see next page) but are brittle, with little or no plastic deformation at fracture.

- Amorphous ceramics (or glasses) have lower UTS and lower stiffness (indicated by the shallower gradient) but are also brittle, showing little or no plastic deformation at fracture.
- High-carbon steels have high UTS, high stiffness and low ductility, showing a small amount of plastic deformation at fracture.
- Low-carbon steels have lower UTS, lower stiffness but much higher ductility, with large plastic deformation at fracture.
- Copper has even lower UTS, lower stiffness and even higher ductility, with very large plastic deformation at fracture.

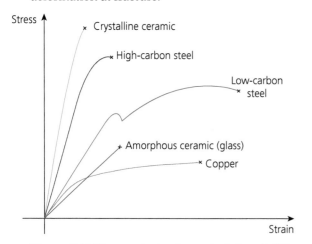

▲ Figure 6.10 Stress–strain characteristics of different materials

Stress calculations

Stress (σ) is the load force (F) per unit of cross-sectional area (A). This can be calculated using the formula:

$$\sigma = \frac{F}{A}$$

The units of stress are Nm^{-1} or Pa.

Strain calculations

Strain (ε) is the ratio of extension (Δl) and the original length of the material sample under test (l). This can be calculated using the formula:

$$\varepsilon = \frac{\Delta l}{l}$$

Strain is a ratio and therefore has no unit.

Calculating Young's modulus

Young's modulus or elastic modulus (E) is a proportionality factor, like the spring constant found in Hooke's law. It describes the stiffness of a material in its elastic region as the ratio of stress (σ) over strain (ε) and is calculated using the formula:

$$E = \frac{\sigma}{\varepsilon}$$

The units of Young's modulus are Nm^{-1} or Pa.

> **Improve your maths**
>
> 1. An M12 bolt has a cross-sectional area of $0.0113\,m^2$ and supports an axial load of $1200\,N$. Calculate the stress on the bolt.
> 2. A brittle test specimen undergoes a tensile test. Figure 6.11 shows the test results plotted as a stress–strain graph. From the graph, determine the ultimate tensile stress of the material.
>
>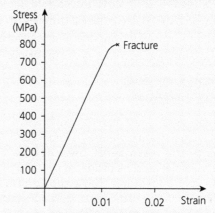
>
> ▲ Figure 6.11 Stress–strain graph of a brittle material that fractures with no plastic deformation

Toughness testing

Impact toughness is the amount of energy a material absorbs when it fractures and is measured using an impact test.

The Izod test is designed to measure the energy absorbed by a specially prepared material sample of known dimensions with a notch machined at its centre. The notch creates a weakness in the sample so the position of the fracture points can be predicted, allowing safe, repeatable testing.

The test sample is clamped in position so that it will be struck by a weighted pendulum at the bottom of its swing. The gravitational potential energy contained in the raised pendulum will have transformed completely into kinetic energy at the bottom of the swing when it impacts the test sample. After the impact, any remaining kinetic energy will be turned back into potential energy as the pendulum rises again on the other side. The difference between the potential energy in the pendulum at the start and end of the test is equal to the impact energy absorbed by the test sample during fracture.

The Izod test procedure is as follows:
▶ Raise the pendulum sufficiently to allow access to the test sample holder.
▶ Clamp the test sample into the holder.
▶ Raise the pendulum to the required test height.
▶ Release the pendulum so that it impacts and fractures the test sample.
▶ Measure the height the pendulum reaches as it continues to swing past the fractured sample.

Toughness testing is a destructive test carried out in the laboratory on expendable material samples. It cannot be used to carry out component tests in situ or to monitor the condition of components in service.

Health and safety

The apparatus used to test material toughness involves moving parts, a high-energy impact and the fracturing of the test material. These all pose significant risks that must be mitigated by using appropriate control measures. These should include training on the safe use of the apparatus, adequate guarding of the moving parts and the use of personal protective equipment (PPE).

Test yourself

What quantity is measured during an impact toughness test?

Hardness testing

The surface hardness of metal is commonly measured using a Rockwell test. This involves pressing an indenter into the surface of the material and measuring the size of the indentation it leaves when the load is removed.

There are three Rockwell test scales (HRA, HRB and HRC) meant for different metal types and anticipated hardness ranges. Each scale uses a different indenter shape and loading conditions.

For example, HRC is typically used on steel. This uses a diamond cone indenter, a preload of 10 kg and a test load of 100 kg.

The Rockwell HRC test procedure is as follows:
▶ Ensure the test surface is smooth, flat and free of scale or other contaminants.
▶ Place the test sample on the anvil of the testing machine.
▶ Bring the indenter into contact with the prepared surface.
▶ Apply the preload of 10 kg.
▶ Set the dial gauge on the tester to zero.
▶ Apply the test load of 100 kg.
▶ Release the test load after five seconds.
▶ With the preload still in place, read the Rockwell hardness from the tester dial gauge.

Hardness testing marks the surface of a component but is not a true destructive test. It can be carried out in the laboratory or in the field using portable equipment to measure the hardness of components in service.

▲ Figure 6.12 Rockwell hardness testing machine

Corrosion resistance testing

Corrosion resistance testing establishes the likelihood that a material will suffer problematic corrosion in a given working environment. It involves exposing material samples to certain conditions of temperature, humidity or other relevant factors in a controlled environment and monitoring them for signs of corrosion.

Corrosion can take a long time to develop in normal operating environments. To accelerate testing, especially where the material has a protective surface coating, a more aggressive test environment is required. The salt-spray test is widely used in these circumstances. It is inexpensive, repeatable and quick to produce results. The test is carried out in a sealed chamber that exposes the material to a fine spray of salt water in controlled conditions of temperature, spray pressure and salt concentration.

Corrosion resistance testing is a destructive test carried out in the laboratory on expendable component samples. It cannot truly reflect actual in-service conditions and is limited to providing a comparative indication of corrosion resistance between samples.

> ### Research
>
> Carry out your own simple corrosion resistance testing on a range of metal samples. Ensure that each sample is clearly identified, cleaned and degreased. Place the samples in a tray on a windowsill and spray them with a solution of 5 g of salt in 1 l of water. Respray them once a day.
>
> Take a photograph of the samples once a day so you can see any changes and monitor how corrosion progresses.

Wear resistance testing

The most common type of wear encountered in engineering is between two metal components in repeated sliding contact with each other.

The two most common types of wear resistance test are:
- ▶ Pin-on-disc – a stationary pin is pressed onto the surface of a rotating plate or disc. The wear between the two is measured as the movement caused by a reduction in the length of the pin and corresponding reduction in the thickness of the disc where the two have been in contact.
- ▶ Reciprocating – a reciprocating pin is pressed onto the surface of a stationary plate. The wear is measured in a similar way to pin-on-disc testing.

Wear resistance testing is a destructive test carried out in the laboratory on expendable material samples. It cannot be used to carry out component tests in situ or to monitor the condition of components in service.

Fatigue testing

The Wohler test is commonly used to determine the fatigue characteristics of a material sample. During the test, load reversal is achieved by rotating a test piece held in a chuck at one end, with a load applied at the other. With every 360° rotation of the chuck, one loading cycle is completed (see Figure 6.13).

The test continues until the specimen fails and the number of cycles before failure is used to measure fatigue resistance.

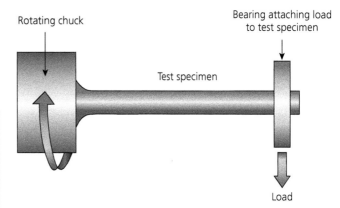

▲ Figure 6.13 Fatigue testing using the Wohler test method

Tests are repeated on several identical material samples with different loads. The results are plotted on a graph of stress versus number of cycles at failure, to provide a complete picture of the fatigue behaviour of the material.

S–N curves indicate the maximum number of loading cycles that can safely be made at any specified load.

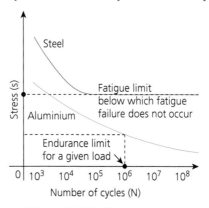

▲ Figure 6.14 S–N curves for typical steel and aluminium alloys

In steel and most ferrous metals, experiments have shown that there are load levels below which fatigue failure does not occur. This is known as the fatigue limit for that material and is shown on the S–N curve in Figure 6.14.

In aluminium and most non-ferrous metals, it has been shown that there is no fatigue limit and eventual fatigue failure can occur at any load. In this case, engineers can specify materials that must endure a minimum number of cycles at a given load. This is known as the endurance limit of the material at that load, shown in Figure 6.14.

Fatigue testing is a destructive test carried out in the laboratory on expendable material samples. It cannot be used to carry out component tests in situ or to monitor the condition of components in service.

Test yourself

What is meant by the endurance limit of a material at a given load?

Electrical conductivity testing

Electrical conductivity (σ) of solids can be calculated by measuring the resistance (R) of a material sample of known length (l) and cross-sectional area (A) using the formula:

$$\sigma = \frac{l}{RA}$$

Measuring electrical conductivity is a non-destructive test carried out in the laboratory.

Assessment practice

1. A metal component has a volume of $1.25 \times 10^{-4} \, m^3$ and a mass of 339 g. Calculate the density of the metal in kgm^{-3}.

2. Define what is meant by the term 'smart material'.

3. Describe the main difference between crystalline and amorphous solids in terms of their microstructure.

4. Outline two issues caused by the welding process that can affect the strength and ductility of the fused parent metal at a welded joint.

5. Heat treatment is commonly used to change the mechanical properties of metals. Describe the process of case hardening low-carbon steel.

6. Explain why galvanising is more effective than painting when used to protect steel components from corrosion.

7. Explain the term 'necking' when used to describe the behaviour of a ductile test sample undergoing a tensile test.

8. The strain measured in a steel bar is 0.002. Calculate the stress in the bar. (For steel, $E = 205 \, GNm^{-1}$.)

9. Outline how material hardness is measured using a Rockwell hardness (HRC) test.

10. A material test sample has a length of 0.10 m, a cross-sectional area of $7.9 \times 10^{-7} \, m^2$ and resistance of $2.21 \times 10^{-3} \, \Omega$. Calculate the electrical conductivity of the material.

Project practice

You work as a design engineer for a welding and fabrication company that is building steel ladder sections for use on offshore wind-turbine installations.

Write a short report to recommend measures to protect the ladders from corrosion and how the effectiveness of these measures can be tested.

You should include:
▶ an explanation of the different measures that can be used to prevent corrosion
▶ a fully justified recommendation of the most appropriate measures to take in this application
▶ a description of how the effectiveness of the recommended measures can be tested.

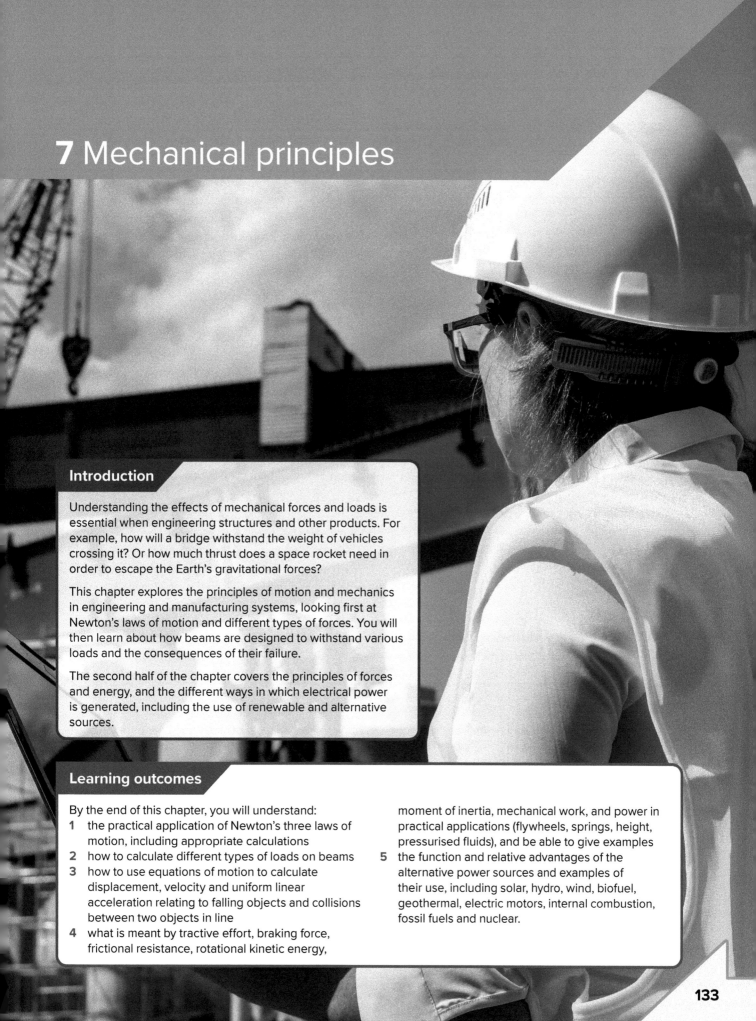

7 Mechanical principles

Introduction

Understanding the effects of mechanical forces and loads is essential when engineering structures and other products. For example, how will a bridge withstand the weight of vehicles crossing it? Or how much thrust does a space rocket need in order to escape the Earth's gravitational forces?

This chapter explores the principles of motion and mechanics in engineering and manufacturing systems, looking first at Newton's laws of motion and different types of forces. You will then learn about how beams are designed to withstand various loads and the consequences of their failure.

The second half of the chapter covers the principles of forces and energy, and the different ways in which electrical power is generated, including the use of renewable and alternative sources.

Learning outcomes

By the end of this chapter, you will understand:
1   the practical application of Newton's three laws of motion, including appropriate calculations
2   how to calculate different types of loads on beams
3   how to use equations of motion to calculate displacement, velocity and uniform linear acceleration relating to falling objects and collisions between two objects in line
4   what is meant by tractive effort, braking force, frictional resistance, rotational kinetic energy,

moment of inertia, mechanical work, and power in practical applications (flywheels, springs, height, pressurised fluids), and be able to give examples
5   the function and relative advantages of the alternative power sources and examples of their use, including solar, hydro, wind, biofuel, geothermal, electric motors, internal combustion, fossil fuels and nuclear.

7.1 Principles of motion and mechanics in engineering and manufacturing systems

Newton's three laws of motion

Newton's laws of motion are fundamental laws that describe the relationship between the motion of a body and the forces that are acting on it. They provide the basic principles upon which modern physics is built.

Newton's first law of motion

Newton's first law of motion, sometimes referred to as the law of inertia, states that a body will remain at rest, or will continue moving at a constant speed in a straight line, unless a force is acting upon it. For example, a ball sitting at rest on a table will not move until it is pushed.

It is because of this law that seatbelts and airbags are required safety equipment in motor vehicles. If a car comes to a sudden stop or crashes into an object, the passengers will continue to move at the speed that the car was previously moving because of inertia. The same would be the case for any unsecured load or cargo placed on top of the vehicle.

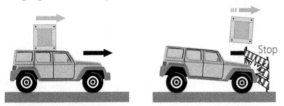

Stop

▲ Figure 7.1 Newton's first law of motion

Newton's second law of motion

Newton's second law of motion, sometimes referred to as the law of acceleration, states that the acceleration of an object depends on the mass of the object and the net force acting upon it. This is represented mathematically as:

$$Force\ (F) = mass\ (m) \times acceleration\ (a)$$

For example, to produce an acceleration of $9\,ms^{-2}$, an object of mass $5\,kg$ would need a force of $45\,N$ to act upon it ($F = 9 \times 5$).

> **Improve your maths**
>
> A rally car has a mass of 1320 kg. Calculate the acceleration of the car when a force of 4500 N is applied to it.

Newton's third law of motion

Newton's third law of motion, sometimes referred to as the law of opposing forces, states that for every action or force, there is an equal and opposite reaction.

An excellent example of this is gravity. When the Earth exerts a force of attraction on an object, then the object will also exert an equal force in the opposite direction, back on the Earth. This information is used when designing aircraft and space rockets, such as when calculating how much thrust is needed from the engines.

> **Key terms**
>
> **Newton's laws of motion:** fundamental laws that describe the relationship between the motion of a body and the forces that are acting on it.
>
> **Newton's first law of motion:** a body will remain at rest, or will continue moving at a constant speed in a straight line, unless a force is acting upon it.
>
> **Newton's second law of motion:** the acceleration of an object depends on the mass of the object and the net force acting upon it.
>
> **Newton's third law of motion:** for every action or force, there is an equal and opposite reaction.

▲ Figure 7.2 A Saturn V rocket launching

Another example is magnetic forces, such as when paperclips are attracted to a magnet. Maglev trains use the repelling forces of electromagnets to lift up the train and push it forward.

> **Test yourself**
>
> Describe two applications of Newton's third law of motion.

Types of forces

Concurrent, nonconcurrent and co-planar forces

With **concurrent forces**, the **lines of action** all meet at the same point (see Figure 7.3).

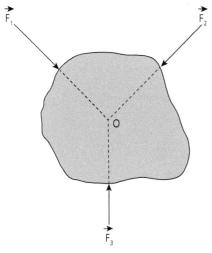

▲ Figure 7.3 Concurrent forces

By contrast, with **non-concurrent forces**, the lines of action do not meet at the same point. In this case, the sum of the forces has a non-zero resultant, meaning

an object will either speed up, slow down or change direction.

Co-planar forces are forces that are all acting in the same plane.

> **Key terms**
>
> *Concurrent forces:* forces where the lines of action all meet at the same point.
>
> *Lines of action:* geometric representations of how forces are applied.
>
> *Non-concurrent forces:* forces where the lines of action do not meet at the same point.
>
> *Co-planar forces:* forces that are all acting in the same plane.

> **Test yourself**
>
> Explain the difference between concurrent and non-concurrent forces.

Non-contact forces

Non-contact forces are forces acting between objects that do not physically touch each other. Examples include magnetic, electrostatic and gravitational forces.

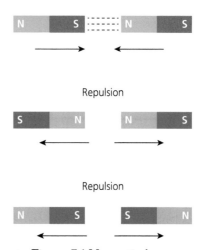

▲ Figure 7.4 Magnetic forces are non-contact forces

> **Key term**
>
> *Non-contact forces:* forces acting between objects that do not physically touch each other.

Simply supported beams

Beams are long structural elements that are designed to withstand various loads. A **simply supported beam** is one that rests on two supports and is free to move horizontally.

Simply supported beams are used in a wide range of structural engineering contexts, from small home extensions to huge bridges and skyscrapers.

> ### Key term
>
> **Simply supported beam:** beam that rests on two supports and is free to move horizontally.

Loading

Loads are the forces acting on or being exerted on a beam. They can either be vertical or horizontal, depending on the direction in which they are applied. For example, the weight of an object sitting on a bridge is a vertical load, whereas wind forces would act horizontally on the bridge.

Load distribution

When a load is spread out over a length, area or volume, it is said to be a distributed load.

- A **point load** is a force applied at a single point along a beam. It is calculated by working out the total load over a beam's length and then attributing it to its centre.
- A **uniformly distributed load** is applied evenly over the entire area or length of a beam. It is calculated by multiplying the load acting on the beam by the beam's length.

> ### Key terms
>
> **Point load:** force applied at a single point along a beam.
>
> **Uniformly distributed load:** load applied evenly over the entire area or length of a beam.

Some loads can be a combination of the two. However, if a point load is applied to a beam that is designed only to support uniformly distributed loads, there is a high risk of structural failure of the beam.

Beam failure can be catastrophic, for example in the case of a tall building or vehicle bridge, as it can lead to the collapse of the entire structure.

▲ Figure 7.5 A failed beam on a bridge, resulting in its collapse

> ### Industry tip
>
> Make sure you understand the different types of loads that beams need to be able to withstand and apply this knowledge when designing structures.

Reaction forces

Reaction forces act in opposition to action forces, in line with Newton's third law of motion. For example, friction is a reaction force to an object sliding against another object (see Figure 7.6).

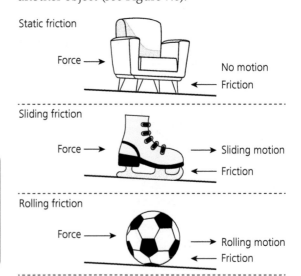

▲ Figure 7.6 Different types of friction

> ### Key term
>
> **Reaction forces:** forces that act in opposition to action forces, in line with Newton's third law of motion.

Loaded components, shear forces and bending moments

Structural components need to be able to resist the loads that are placed on them so they do not fail.

Shear forces act in a parallel direction to the surface of an object. They push the top part of the object in one direction and the bottom part in the opposite direction. This can result in deformation and cracks appearing in a structural component.

▲ Figure 7.7 The potential effects of shear forces on a component

A **bending moment** is a measure of the bending effect that occurs when a force is applied to a beam or other structural component. This is important to consider when designing structures, because too much bending can cause a beam or component to fail.

Key terms

Shear forces: forces that act in a parallel direction to the surface of an object.

Bending moment: measure of the bending effect that occurs when a force is applied to a beam or other structural component.

7.2 Principles of forces and energy

Conservation of momentum

Momentum cannot be created or destroyed. This is because the total momentum of two or more objects acting on each other remains constant unless an external force is applied. This law is based on Newton's third law of motion and can be represented mathematically as:

$$m_1u_1 + m_2u_2 = m_1v_1 + m_2v_2$$

where m_1 and m_2 are the masses of the objects, u_1 and u_2 are the initial velocities of the objects, and v_1 and v_2 are the final velocities of the objects.

Conservation of energy

As with momentum, it is also not possible for energy to be created or destroyed – it can only be changed from one form to another.

Unless energy is added from outside a system, the energy within the system remains the same, which can be represented mathematically as:

$$K_1 + U_1 = K_2 + U_2$$

where K_1 is the initial kinetic energy, U_1 is the initial potential energy, K_2 is the final kinetic energy and U_2 is the final potential energy.

D'Alembert's principle

D'Alembert's principle is another way of representing Newton's second law of motion.

Newton's second law states that *Force = mass × acceleration*, or $F = ma$. D'Alembert's principle, on the other hand, states that the force plus the negative of the mass multiplied by the acceleration is equal to zero: $F - ma = 0$.

Test yourself

Express each of the following principles mathematically:
▶ conservation of momentum
▶ conservation of energy
▶ D'Alembert's principle.

Potential and kinetic energy

Potential energy is the energy that is stored by an object due to its position, for example when a ball is held at height or when water is held behind a dam, ready for release.

When potential energy is released, it is changed into **kinetic energy**. Kinetic energy is therefore the energy that an object possesses because of its motion. An example would be a turning motor or a charged particle in an electric field.

Key terms

Potential energy: energy that is stored by an object due to its position.

Kinetic energy: energy that an object possesses because of its motion.

Gravitational force

Gravitational force is the force that attracts all objects with mass towards each other. As it is a relatively weak force, it only becomes noticeable when dealing with objects of particularly large mass, such as planetary bodies or stars.

> ### Key term
>
> **Gravitational force:** force that attracts all objects with mass towards each other.

For example, the Earth exerts a gravitational force that pulls objects towards its centre. It is this force that gives people and objects their weight. The larger the mass, the larger the gravitational force that is exerted, so people would weigh less on a planet with less mass than the Earth. At the surface of the Earth, the acceleration of gravity is approximately $9.81\,\text{ms}^{-2}$.

Gravity is an important force to consider when designing structures, tall buildings and air/spacecraft.

Frictional resistance

Friction is a force that acts in opposition to an object sliding along a surface, for example when pushing a book across a table.

> ### Key terms
>
> **Friction:** force that acts in opposition to an object sliding along a surface.
>
> **Tractive force:** force needed to overcome the resistance caused by friction.
>
> **Braking force:** force needed to slow down an object or bring it to a stop.

Friction often generates heat, which can cause the failure of components in engineering and manufacturing equipment. For example, machine gears meshing together in a gear train could overheat.

Friction is calculated using the following formula:

$$friction\ (F) = coefficient\ of\ friction\ (\mu) \times normal\ force\ (N)$$

The force needed to overcome the resistance caused by friction is called the **tractive force**.

Braking force is the force needed to slow down an object or bring it to a stop, for example when a driver operates a car's brake pedal.

Mechanical work and power

Mechanical work is the amount of energy transferred by a force, measured in joules (J). It is also described in its simplest form as the product of the strength of force applied to an object and the resulting distance travelled by it. It can therefore be calculated using the following formula:

$$Work\ (W) = force\ (F) \times distance\ (d)$$

Power is the rate at which energy is transferred or converted:

$$Power\ (P) = \frac{work\ done\ (E)}{time\ taken\ (t)}$$

> ### Key terms
>
> **Mechanical work:** amount of energy transferred by a force, measured in joules (J).
>
> **Power:** rate at which energy is transferred or converted.
>
> **Electrical power sources:** power sources that generate energy in the form of electricity.

Power applications

- ▶ Flywheels are used to smooth the transfer of power from a motor to a machine, resulting in continuous output. They consist of a wheel attached to a rotating shaft.
- ▶ Springs are used to store mechanical energy. The most common type is a coil spring.
- ▶ Fluid power is the use of pressurised fluids to control and transmit power throughout a system.

Types of power sources

Electrical power sources

Electrical power sources generate energy in the form of electricity.

Electrical power is an essential part of the modern world. Without it, there would be no electric lighting, heating, kitchen appliances, automated machinery or computers. Most electrical energy in the world is still generated using fossil fuels, although there is more and more development of **renewable power sources**.

▲ Figure 7.8 Solar panels and wind turbines

> ## Key terms
>
> **Renewable power sources:** natural power sources used to generate energy that are replenished at a higher rate than they are consumed, for example wind and solar.
>
> **Mechanical power sources:** power sources that generate energy in the form of electricity through mechanical motion, vibration or pressure.

Mechanical power sources

Mechanical power sources also generate energy in the form of electricity, through mechanical motion, vibration or pressure. For example:

▶ Piezoelectricity uses crystals to convert mechanical energy into electrical energy.
▶ Generators take kinetic energy in the form of movement and convert it into an electrical current.

> ## Health and safety
>
> When installing, repairing or maintaining wind turbines, a number of potential health and safety issues need to be considered, for example hazards associated with working at height.
>
> When working at height, workers must use the correct PPE, for example harnesses connected to a suitable anchor point and safety helmets.

▼ Table 7.1 Power sources

Power source	Function	Advantages
Solar	Solar panels create a current when sunlight strikes them, because of the ejection of electrons from the silicon material	Renewable and sustainable source of power Zero carbon emissions Clean source of power Produces a lot of electricity in sunny environments
Hydro	Water is slowly released from behind a dam to turn turbines. These then turn generators to produce electricity	Renewable and sustainable source of power Low carbon emissions Clean source of power Can produce large quantities of power
Wind	Turbines are turned by the power of the wind. These then turn generators to produce electricity	Renewable and sustainable source of power Low carbon emissions Clean source of power Produces a lot of electricity in windy environments, such as on top of hills
Biofuels	Fuels are made from biomass, which is produced using plant and animal waste. These can then be used to replace fossil fuels in power-generation applications	Renewable and sustainable source of power Reduced carbon emissions Reduces reliance on fossil fuels
Geothermal	Steam created from hot water below the surface of the Earth is used to turn turbines. These then turn generators to produce electricity	Renewable and sustainable source of power Clean and quiet Has relatively little impact on the environment compared to the use of fossil fuels

Power source	Function	Advantages
Electric motors/generators	Electric motors use the principle of electromagnetism to produce rotary movement when an electric current is applied. Electric generators work in reverse to motors in order to produce electricity from rotary movement.	Generators are a key component of most power-generation systems, e.g. fossil-fuel power stations, wind turbines and hydro systems Relatively cheap and widely understood technology
Internal combustion	Heat energy from burning fuel (such as petrol or diesel) is converted into movement. Internal combustion engines consist of fixed cylinders and moving pistons.	Relatively small and portable Relatively simple to maintain Relatively cheap and widely understood technology
Fossil fuels	Coal, oil and/or gas are burned to create steam that turns turbines. These then turn generators to produce electricity.	Still readily available Relatively cheap source of energy Can be used to generate large quantities of power
Nuclear	Nuclear fission creates steam which is used to turn turbines. These then turn generators to produce electricity.	Reduced carbon emissions Can be used to generate large quantities of power Reduces reliance on fossil fuels

Research

Investigate different methods of producing electrical power using renewable sources. Explain their relative advantages and disadvantages.

Case study

The UK is to build a new nuclear power plant in Sizewell, a small fishing village on the Suffolk coast, with construction expected to begin in 2024. The UK government claims that this is vital in helping the country to meet its net-zero carbon emission commitments in future decades. It also argues that it will reduce reliance on non-renewable fossil fuels.

However, many local people have voiced opposition about the proposal, saying the village is already the site of two other nuclear power plants and it should therefore be built elsewhere. They also argue that the government should be concentrating on developing renewable sources of power.

Questions

1 What are the advantages and disadvantages of building a new nuclear power plant at Sizewell? Do you think the project should go ahead? If so, why? If not, why not?
2 What are the advantages and disadvantages of nuclear power in general?

Assessment practice

1 State Newton's first law of motion.
2 Define the term 'co-planar forces'.
3 Give three examples of non-contact forces.
4 Describe what is meant by a simply supported beam.
5 Name two types of load that can be applied to a beam.
6 Explain the potential effects of shear forces on a beam.
7 Define D'Alembert's principle.
8 Explain the difference between potential and kinetic energy.
9 Describe two examples of mechanical power sources.
10 State two advantages of solar power.

Project practice

You are an engineer working for a structural engineering company. You have been asked to design a beam footbridge to allow people to cross a river. The river is 10 metres wide at the point where the bridge will be placed.

1 Produce a labelled sketch of your proposed design.
2 Explain how the design will withstand the loads and forces that it will be subjected to. Use calculations to support your response.

8 Electrical and electronic principles

Introduction

Electrical and electronic engineering is a rapidly growing sector of the engineering industry. It involves the application of knowledge in order to design and maintain systems, components and devices that make use of electricity and electron theory.

This chapter explores the underpinning laws and theories that allow engineers to design circuits and systems, and also examines the components and devices that make up DC circuit networks. You will also learn about the types of signals used in electrical and electronic systems and how they are processed.

Learning outcomes

By the end of this chapter, you will understand:
1. the physical principles underpinning electrical and electronic systems and devices
2. the basic properties and principles of magnetism and electromagnetism and their common applications
3. the relationship between flux density and field strength
4. the definitions of terms used in electric circuit theory and their applications
5. the use of Ohm's law and electric circuit theories to calculate values in circuits, such as voltage, current and resistance
6. how differential protection schemes work to protect transmission lines
7. how transformer protection schemes work for common faults
8. the characteristics of the different concepts related to signals
9. the characteristics of analogue and digital systems, including their waveforms and applications
10. the characteristics of DC circuit networks comprising resistors, capacitors and inductors in various arrangements, including time constants
11. the relationship between voltage, current and power in AC circuits and how to represent them in graphs and phasor diagrams
12. the properties and applications of semiconductor diodes and transistors
13. factors affecting the operation and applications of high-power electrical equipment and electronic devices.

8.1 Principles of electrical and electronic systems

Basic principles of electricity and electronics

Electron flow, charge and current

The first step in understanding how electrical and electronic circuits work is understanding electron theory.

All matter is composed of atoms, which contain particles known as **electrons**, protons and neutrons:
- electrons have a negative electrical charge
- protons have a positive electrical charge
- neutrons have no overall electrical charge.

The nucleus of an atom contains protons and neutrons, and this is surrounded by moving electrons. A body becomes either positively or negatively charged through the giving or gaining of electrons. Opposite charges attract each other, whereas like charges repel.

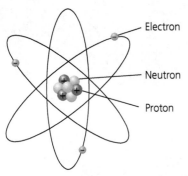

▲ Figure 8.1 Structure of an atom

Conventional current theory assumed that electrical **current** was created by the flow of protons in a circuit, from the positive to the negative terminal. However, we now know that electrical current occurs due to the flow of electrons from the negative to the positive terminal. Despite this, electrical circuit diagrams still use conventional current, as what is moving inside a wire makes no difference to electrical calculations.

Key terms

Electrons: particles of an atom which have a negative electrical charge.

Current: flow of electrons through a conductive material.

When electrical current flows through a material, it is said to be conducting electricity. Materials vary in their conductivity; for example, copper is a good conductor and is therefore used for wiring and circuit boards. Some materials do not allow current to flow through them and these are known as insulators. Good insulators include rubber, plastic and glass.

Current is measured in amperes, often shortened to amps and represented by the symbol A.

Test yourself

1 What are electrons?
2 What causes an electrical current?

Electrical energy, power and force

Electrical **energy** is caused by the movement of electrons from one atom to another, i.e. when current flows. It is described as the capacity for an electrical circuit to do work. Like other forms of energy, it is measured in joules (J).

Electrical energy is calculated using the following formula:

Energy (E) = Power (P) × Time (t)

Electrical **power** is the rate at which electrical energy is transferred. It is measured in watts (W), with one watt equivalent to one joule transferred per second.

Electrical power can be calculated using the following formulae:

Power (P) = Energy (E)/Time (t)

Power (P) = Current (I) × Voltage (V)

Power (P) = Current (I)^2 × Resistance (R)

Electrical **force** is the attractive or repulsive interaction between any two charged objects. As with all forces, it is governed by Newton's laws.

Key terms

Energy (electrical): capacity for an electrical circuit to do work.

Power (electrical): rate at which electrical energy is transferred.

Force (electrical): attractive or repulsive interaction between any two charged objects.

Test yourself

Explain the relationship between electrical energy and power.

Electrical networks and elements

When electrical or electronic components are connected together, they form a **network**. If the network is a closed loop, with a return path for the current, it is known as a circuit. Active networks contain an energy source, passive networks do not.

> ### Key term
>
> **Network (electrical):** arrangement of connected electrical or electronic components.

Networks with constant parameters (such as inductance and resistance) are referred to as linear. Networks with parameters whose values change with time are described as non-linear. Linear networks are much easier to analyse than non-linear networks and can be described using Ohm's law.

Electrical elements are conceptual abstractions that approximately correspond to actual electrical components in a network. They appear in schematics and circuit diagrams and are helpful for analysing and estimating how a circuit will work. Elements represent the ideal properties and behaviours of their physical counterparts; in reality, almost all physical components present a degree of uncertainty in terms of their values, tolerances and performance.

Capacitance

Capacitance is the ability of a circuit or component to store electrical charge. It is measured in farads (F).

Capacitors are components designed to store charge for various applications, such as timing, filtering and supplying electrical energy. Most capacitors have values measured in microfarads (μF), as they can only store small amounts of charge. However, supercapacitors can store much greater amounts and are sometimes used to replace batteries when powering circuits.

> ### Key term
>
> **Capacitance:** ability of a circuit or component to store electrical charge.

Magnetism and electromagnetism

Magnetism is a force that occurs between magnets or magnetic materials when they attract or repel each other. The magnetic forces are strongest at the north

and south poles of a magnet. Unlike poles attract each other, whereas like poles repel each other.

All magnets are surrounded by a **magnetic field** – an area where magnetic forces are observable. Magnetic fields can be represented using lines that indicate their strength and direction.

▶ The closer, or more dense, the lines appear, the stronger the magnetic field.
▶ The direction of the magnetic field is shown using arrows.

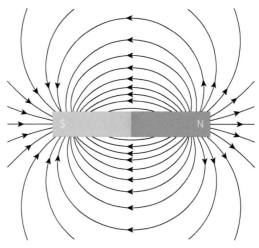

▲ Figure 8.2 Magnetic field of a bar magnet

The total magnetic field that passes through a given area is called the **magnetic flux**. The amount of magnetic flux that passes through a given area at right angles to the magnetic field is known as the **magnetic flux density**, measured in teslas (T).

> ### Key terms
>
> **Magnetic field:** area surrounding a magnet where magnetic forces are observable.
>
> **Magnetic flux:** total magnetic field that passes through a given area.
>
> **Magnetic flux density:** amount of magnetic flux that passes through a given area at right angles to the magnetic field.

When a magnet moves within a coil of wire, it produces a voltage. This is called electromagnetic induction. The direction of the voltage can be reversed by moving the magnet back out of the coil, or by moving the other pole of the magnet into the coil. This is useful for a variety of engineering applications, such as maglev trains and electromagnetic door locks. It is also the principle that allows transformers and generators to work.

Inductance is the property of a component that opposes, through its changing magnetic field and the subsequent creation of an electromotive force (emf), a change in the current flowing through it.

Measurements of electricity in systems

Electrical parameters, such as voltage (volts), current (amps), resistance (ohms) and power (watts), can be measured within circuits. This helps engineers to understand how different circuits work and to test that they are functioning as expected.

The most common instrument used to take electrical measurements is a **multimeter**. It has probes that are placed at the appropriate test points and a digital screen to read the numerical results.

▲ Figure 8.3 A digital multimeter

Multimeters are popular because they can be used to measure a number of different parameters, thereby reducing the need to carry separate meters.

Current is measured in series, whereas voltage/potential difference is measured in parallel.

For both accuracy and safety, it is important that there is no power supplied to a circuit when measuring its resistance. This is because a multimeter cannot actually measure resistance directly, so it applies a small current, measures the resulting voltage and uses this to calculate the resistance. If there is already a voltage across the test point, this not only renders the measurement meaningless but also can cause damage to the circuit or the meter itself.

The **SI** units of measurement for the main parameters in electrical and electronic circuits are summarised in Table 8.1 on the next page.

▼ Table 8.1 SI units of measurement for the main electrical parameters

Parameter	Unit of measurement	Unit abbreviation
Voltage	volt	V
Current	ampere (amp)	A
Resistance	ohm	Ω
Power	watt	W
Capacitance	farad	F
Inductance	henry	H
Energy	joule	J
Time	seconds	s
Frequency	hertz	Hz
Magnetic flux	weber	Wb
Magnetic flux density	tesla	T

Electric circuit theories

Voltage, current, resistance and power

Voltage is the difference in electrical potential between two points in a circuit. It is measured in volts (V). It is sometimes described as the 'push' that results in a current flowing around a circuit. The bigger the push, the larger the current.

Current is the flow of electrons through a circuit. It is measured in amps (A). There are two types of current, alternating (AC) and direct (DC):

▶ AC changes direction periodically and is used for mains electricity applications and in high-voltage power distribution.
▶ DC only flows in a single direction and is typically used in low-voltage electronic circuit boards and programmable microcontroller-based applications.

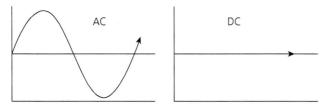

▲ Figure 8.4 AC and DC transmission systems

Resistance is a measure of the opposition to the flow of electrical current and is measured in ohms (Ω).

Power is the rate at which electrical energy is transferred by a circuit, measured in watts (W).

Voltage, current, resistance and power are all related to each other. Ohm's law describes the relationship between voltage, current and resistance as follows:

Voltage (V) = Current (I) × Resistance (R)

The triangle shown in Figure 8.5 can be used to help calculate values. By covering up the required parameter, it leaves the formula for calculating it.

▲ Figure 8.5 Ohm's law triangle

For example, if the current flowing through a 1.5 kΩ resistor is 20 mA, then the voltage across it would be calculated as follows:

$$V = I \times R$$

$$V = 0.020 \times 1500$$

$$V = 30 \text{ V}$$

Ohm's law can be used to calculate values in series and parallel circuit arrangements, and also within circuits that are a combination of both. There is more information on these types of circuits later in this chapter.

Watt's law describes the relationship between power, current and voltage as follows:

$$P = I \times V$$

Again, a triangle can be used to help calculate the required values, and this is shown in Figure 8.6.

▲ Figure 8.6 Watt's law triangle

Potential difference and dividers

Voltage is also known as potential difference. A **potential divider** is a circuit arrangement that divides the initial supply voltage, resulting in a smaller output voltage. It does this by using two resistors connected in series, R_1 and R_2, and taking advantage of the way the voltage drops across them.

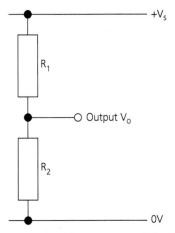

▲ Figure 8.7 A potential divider circuit

Potential dividers are often used in sensing circuits. For example, R_1 or R_2 could be replaced with a light-dependent resistor to form a light/dark-sensing circuit. Or a thermistor could be used to produce a temperature-sensing circuit.

Kirchhoff's current and voltage laws

It is not always possible to model the behaviour of more complex circuits and networks using Ohm's law. This is where Kirchhoff's current and voltage laws come in:

▶ Kirchhoff's current law is concerned with the conservation of charge within a circuit. It states that the total current or charge entering a node is equal to the charge leaving the node. This is because the charge has nowhere to go, as none of it is lost within the node.
▶ Kirchhoff's voltage law is concerned with the conservation of energy within a circuit. It states that in any closed-loop network, the total voltage around the loop is equal to the sum of all the voltage drops within the loop. This means that the sum of all the voltages within the loop must be equal to zero.

Phasor diagrams

In AC circuits, phase differences can exist between different sine waves with the same frequency. **Phasor diagrams** are used to show the phase relationships between two or more of these alternating quantities, in terms of both their magnitude and direction. For example, the sine wave in Figure 8.8 can be represented in a phasor diagram, as shown.

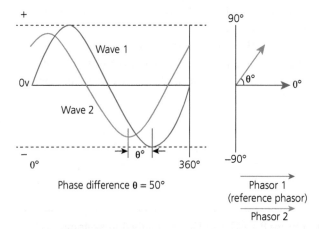

▲ Figure 8.8 Waveform diagram and phasor diagram

Circuit protection systems

Circuit protection systems are designed to safeguard electrical and electronic circuits from damage. They usually provide a weak link in a circuit, which is broken in the event of a fault, or a diversion to the flow of current. This then prevents further damage or safety issues.

Lightning arrestors are used to protect power transmission lines, systems and components from the effects of lightning strikes. They do this via high-voltage and ground terminals, which divert the current from the strike to earth. Differential protection systems are also used to protect transmission lines. They compare the current entering the line with the current leaving it and trip the power if the difference is too great.

Too much current flowing through an electrical or electronic system, known as **overcurrent**, can result in damage to components, fires and even explosions. Two of the main protection systems that guard against this are fuses and circuit breakers.

> ### Key term
>
> **Overcurrent:** when the safe load current for a circuit is exceeded.

Fuses have an internal metal strip of wire that melts when too much current flows through it. This breaks the circuit, preventing any further flow. Once the fuse has blown, it must be replaced before the repaired circuit can be used again. Fuses are, however, cheap and easy to source.

▲ Figure 8.9 A fuse

Circuit breakers use contacts that open when too much current is detected, to interrupt the flow of current.

The advantage of these over fuses is that they can be reset and used again without replacement, but they do cost more to purchase and install. Special types of circuit breaker are used to protect transformers from overcurrent damage.

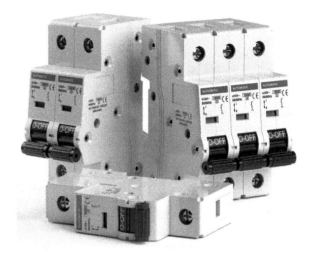

▲ Figure 8.10 Examples of circuit breakers

Distance protection systems measure the impedance (opposition) to the AC current between a relay and a fault location, and compare this with a set value. If the measured value is less than this set value, the relay actuates an electrically operated switch and isolates the fault.

Residual current devices

Residual current devices (RCDs) are often installed in consumer units (fuse boxes) and are designed to protect people from coming into contact with electricity. If there is an imbalance between the neutral and live wires in a circuit, the RCD trips and the power supply is cut off.

> ### Research
>
> Research and analyse the function of different circuit protection systems. For each, give examples of why, where and how it is used.

DC circuit networks

Types of DC circuit networks

DC circuit networks are arrangements of components that use low-voltage and direct-current power supplies, such as batteries. The components are generally arranged in series, in parallel or in a combination of both.

▶ In a **series circuit**, the components are arranged in a single line. There is only one path for the current to flow through.

▶ In a **parallel circuit**, the components are arranged in loops or branches. There are several paths for the current to flow through.

> ### Key terms
>
> **Series circuit:** circuit where the components are arranged in a single line.
>
> **Parallel circuit:** circuit where the components are arranged in loops or branches.

Resistors in series and parallel

Resistors are components that reduce the flow of current. The higher the value of the resistor in ohms, the less current can pass through it.

The total resistance of resistors connected in series is calculated using the following formula:

$$R_{tot} = R_1 + R_2 + \dots$$

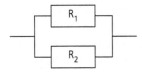

▲ **Figure 8.11 Resistors in series**

The total resistance of resistors connected in parallel is calculated using the following formula:

$$(1/R_{tot}) = (1/R_1) + (1/R_2) + \dots$$

▲ **Figure 8.12 Resistors in parallel**

Some circuits combine both series and parallel arrangements. For example, the total resistance of the circuit in Figure 8.13 would be calculated as follows.

First, calculate the resistance of the series section of the circuit:

$$R_s = 100 + 330 = 430 \ \Omega$$

Next, calculate the resistance of the parallel section of the circuit:

$$(1/R_p) = (1/1200) + (1/5600)$$

$$R_p = 988.2 \ \Omega$$

Finally, add the two values together:

$$R_{tot} = 430 + 988.2 = 1418.2 \ \Omega$$

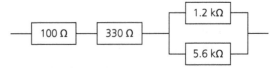

▲ **Figure 8.13 Resistors in series and parallel**

Capacitors in series and parallel

The total capacitance of capacitors connected in series is calculated using the following formula:

$$(1/C_{tot}) = (1/C_1) + (1/C_2) + \dots$$

And in parallel:

$$C_{tot} = C_1 + C_2 + \dots$$

Inductors in series and parallel

The total inductance of inductors arranged in series or parallel can be calculated in a similar manner to resistance.

In series:

$$L_{tot} = L_1 + L_2 + \dots$$

In parallel:

$$(1/L_{tot}) = (1/L_1) + (1/L_2) + \dots$$

> ### Improve your maths
>
> 1. Two 47 µF capacitors are connected in parallel. What is the total capacitance of this arrangement?
> 2. A circuit consists of two resistors of 680 Ω and 1 kΩ connected in parallel. Calculate the total resistance of the circuit.

Semiconductors

Semiconductors are materials that conduct current better than insulators but not as well as conductors. Examples include germanium, silicon and gallium arsenide. They are fundamental in the construction and function of electronic components such as diodes, transistors and integrated circuits.

> ### Key term
>
> **Semiconductors:** materials that conduct current better than insulators but not as well as conductors.

Semiconductors are either intrinsic or extrinsic. Intrinsic semiconductors are chemically pure, whereas extrinsic semiconductors have impurities added to them during a process called doping. The amount and type of impurities added affect the materials' electrical properties, and can be altered to control the flow of electrical current.

There are two main types of extrinsic semiconductor, N-type and P-type, which are differentiated according to the direction in which current flows through them. In N-type semiconductors, electrons are the majority charge carriers. In P-type semiconductors, electrons are the minority carriers.

P- and N-type semiconductors can be put together to form a P–N junction. The most common application of this in electronic circuits is a diode. A standard diode allows current to flow from the positive end (anode) to the negative end (cathode) when a positive voltage is applied to the P-type side of the junction. Current cannot flow in the opposite direction. This is called forward bias. In reverse bias mode, a diode is connected the opposite way around, with a positive voltage applied to the N-type side of the junction. In this configuration, no current can flow until the electric field intensity is so high that the diode breaks down.

> **Test yourself**
>
> 1 Explain the difference between N- and P-type semiconductors.
> 2 Explain how reverse bias operation of a diode differs from forward bias.

Hierarchical circuit design

In hierarchical circuit design, parts and components are divided into different blocks and sub-blocks, grouped according to their function. This makes it easier to keep track of the different inputs and outputs for each part of the circuit. The approach is therefore well suited to large and complex circuit designs. New blocks can be created for new components, whereas old blocks can be reused as necessary, saving time and increasing efficiency.

Signals

Types of signal

There are two main types of signal used in electrical and electronic systems:

▶ **Analogue signals** are continuous and are usually represented using sinusoidal (sine) waves.

▶ **Digital signals** are discrete and are usually represented using square waves.

> **Key terms**
>
> *Analogue signals:* continuous signals, usually represented as sine waves.
>
> *Digital signals:* discrete signals, usually represented as square waves.

Figure 8.14 shows an analogue signal (top) and a digital signal (bottom).

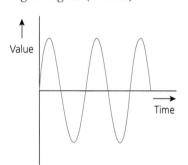

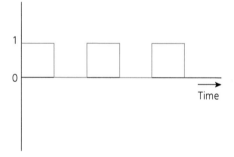

▲ Figure 8.14 (top): analogue (sinusoidal) and (bottom): digital (square) signal waveforms

> **Test yourself**
>
> Explain the differences between analogue and digital signals.

Waveforms

Although sinusoidal and square are the most commonly used signal waveform shapes, signals can also be represented as rectangular, triangular or sawtooth waves (see Figures 8.15 and 8.16 on the next page).

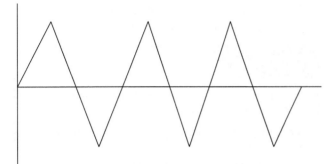

▲ Figure 8.15 Triangular wave

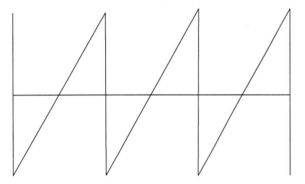

▲ Figure 8.16 Sawtooth wave

Signal processing and conditioning

Signal processing is the name given to a broad range of techniques used to synthesise, modify and analyse different types of electrical and electronic signals. Applications of signal processing include:

▶ filtering noise from an audio signal
▶ interpretation of image and video signals
▶ digital signal processing (DSP).

Signal conditioning is the preparation and manipulation of signals so that they meet the requirements of the next stage of processing. A common example of this in electrical and electronic systems is analogue to digital signal conversion.

Test yourself

Sketch examples of sinusoidal, square, triangular and sawtooth waveform shapes.

Research

Investigate the operation of a noise filter for an audio system. Find out its purpose, how it works, and its benefits and limitations. If a physical device is available, investigate the different input and output connections and how these link with the rest of the audio system. Draw a systems diagram to help show this.

Fan in and fan out

The terms fan in and fan out are used to describe the function of logic gates with regard to their inputs and outputs.

Logic gates are digital devices that produce output signals that are either high (1) or low (0), depending on the input signals received. Fan in refers to the number of inputs that a logic gate is capable of handling safely. Fan out is the number of logic gate inputs that are driven by the output of another logic gate.

Assessment practice

1. Explain the difference between an electron and a proton.
2. Explain the purpose of electrical elements in schematics and circuit diagrams.
3. State what is meant by magnetic flux.
4. Give two examples of the application of electromagnetic induction.
5. Two 330 Ω resistors are arranged in series. The arrangement has a current of 15 mA flowing through it. Calculate the voltage across the arrangement.
6. Explain how a potential divider works.
7. State two circuit protection systems that safeguard against overcurrent.
8. Describe how a diode operates and behaves in reverse bias mode.
9. Describe the purpose of signal conditioning.
10. Explain the difference between fan in and fan out.

Project practice

You work for an electrical installation and maintenance company. You have been asked to complete the handover of a recently installed circuit-breaker system for an outdoor light at a domestic property.

Produce a set of notes for the handover discussion with the client that includes:
▶ an explanation of the function of the system
▶ the key electrical parameters/signals and how they are measured/calculated
▶ information about how to use the system safely
▶ any ongoing maintenance requirements.

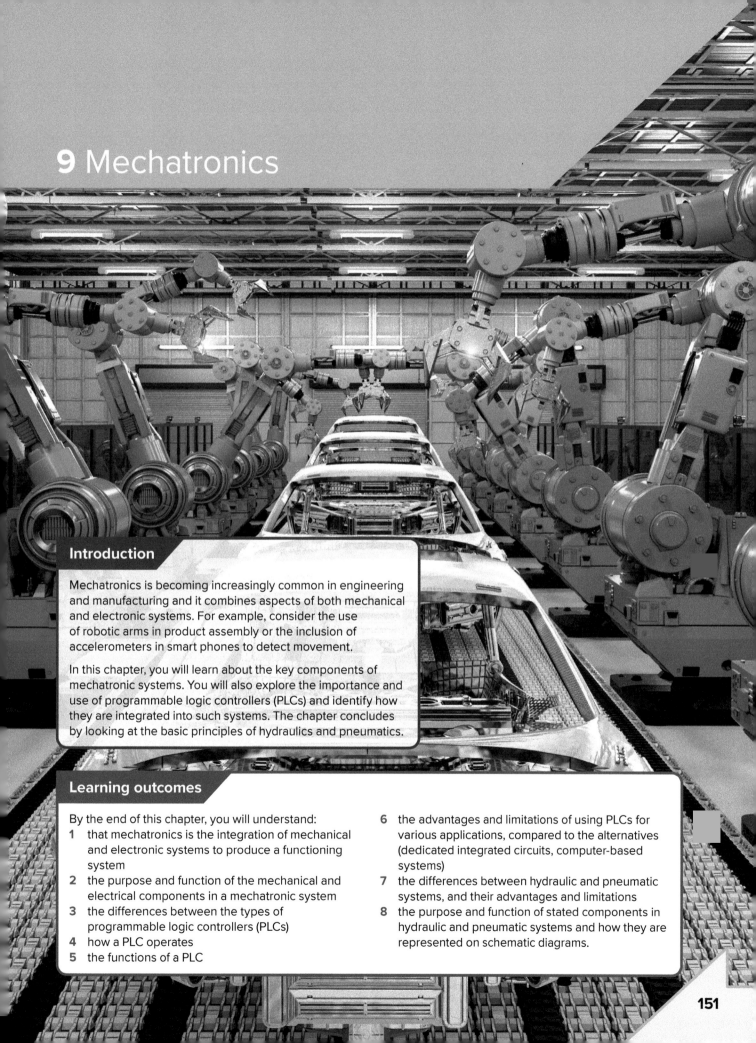

9 Mechatronics

Introduction

Mechatronics is becoming increasingly common in engineering and manufacturing and it combines aspects of both mechanical and electronic systems. For example, consider the use of robotic arms in product assembly or the inclusion of accelerometers in smart phones to detect movement.

In this chapter, you will learn about the key components of mechatronic systems. You will also explore the importance and use of programmable logic controllers (PLCs) and identify how they are integrated into such systems. The chapter concludes by looking at the basic principles of hydraulics and pneumatics.

Learning outcomes

By the end of this chapter, you will understand:
1 that mechatronics is the integration of mechanical and electronic systems to produce a functioning system
2 the purpose and function of the mechanical and electrical components in a mechatronic system
3 the differences between the types of programmable logic controllers (PLCs)
4 how a PLC operates
5 the functions of a PLC

6 the advantages and limitations of using PLCs for various applications, compared to the alternatives (dedicated integrated circuits, computer-based systems)
7 the differences between hydraulic and pneumatic systems, and their advantages and limitations
8 the purpose and function of stated components in hydraulic and pneumatic systems and how they are represented on schematic diagrams.

9.1 The key components of a mechatronic system

Mechatronics is the integration of mechanical and electronic systems and sub-systems in order to create fully functional systems.

> **Key term**
>
> **Mechatronics:** the integration of mechanical and electronic systems and sub-systems in order to create fully functional systems.

Mechanical components

Gears

Gears are toothed wheels that are linked together to transmit drive. Two or more gears linked together are referred to as a gear train. For a simple two-gear train, there is a driver gear and a driven gear.

> **Key term**
>
> **Gears:** toothed wheels that are linked together to transmit drive.

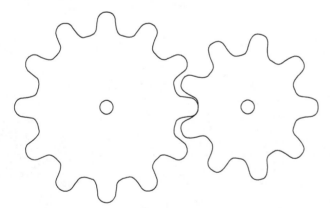

▲ Figure 9.1 A simple gear train

A small gear driving a larger gear results in an increase in torque (rotational force) but a decrease in speed. A large gear driving a smaller gear achieves the opposite – a decrease in torque and an increase in speed. This can be represented mathematically by calculating the gear ratio:

Gear ratio = number of teeth on driven gear/number of teeth on driver gear

For example, if the driver gear has 20 teeth and the driven gear has 10 teeth, then:

Gear ratio = 10/20 = 1:2

Compound gear trains have multiple stages. The gear ratio for a compound gear train is the product of each of these stages.

> **Improve your maths**
>
> In a two-gear train, the driven gear has 18 teeth and the driver gear has 6 teeth. Calculate the gear ratio of this gear train.

> **Research**
>
> Research different examples of compound gear trains and explain how they work.

Cams

A **cam mechanism** is used to turn rotary motion into linear or reciprocating motion. The cam is rotated, either manually or using a motor. This causes the follower to move up and down. A guide is used to ensure the movement is in a straight line.

> **Key term**
>
> **Cam mechanism:** mechanism that uses a cam and a follower to turn rotary motion into linear or reciprocating motion.

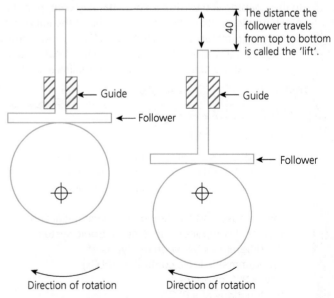

▲ Figure 9.2 A cam and follower mechanism

Eccentric cams are circular but have an off-centre rotating shaft, which produces a symmetrical rise and fall. Different shapes can be used when asymmetrical movement is required, such as pear-shaped cams.

Linkages

Linkages are systems made up of levers or rods connected by pivots. The pivots can be either fixed or moveable, depending on whether they allow or restrict movement. They are used to change the size of a force and/or the direction or type of motion. The exact function is dependent on the type of linkage:

▶ **reverse-motion linkage** – the output direction of motion is the reverse of the input direction of motion

▶ **parallel-motion linkage** – two or more parts of the linkage move in the same direction but parallel to each other

▶ **bell-crank linkage** – the direction of motion is converted between horizontal and vertical; the output force can be increased or decreased compared to the input force, depending on the positioning of a fixed pivot.

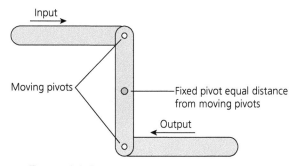

▲ Figure 9.3 A reverse-motion linkage

Key terms

Linkages: systems made up of levers or rods connected by pivots.

Reverse-motion linkage: type of linkage where the output direction of motion is the reverse of the input direction of motion.

Parallel-motion linkage: type of linkage where two or more parts of the linkage move in the same direction but parallel to each other.

Bell-crank linkage: type of linkage where the direction of motion is converted between horizontal and vertical; the output force can be increased or decreased compared to the input force, depending on the positioning of a fixed pivot.

Test yourself

Describe the function of a parallel-motion linkage.

Levers

A **lever** is a rigid rod that turns on a pivot or fulcrum. Levers can be used to exert a large force using a smaller effort force.

▶ With first-class levers (also known as first-order levers), the fulcrum is in the middle, between the load and the effort. Examples include seesaws and scissors.

▶ With second-class levers (also known as second-order levers), the load is between the fulcrum and the effort. Examples include wheelbarrows and car doors.

▶ With third-class levers (also known as third-order levers), the effort is between the fulcrum and the load. Examples include fishing rods and brooms.

Key term

Lever: rigid rod that turns on a pivot or fulcrum.

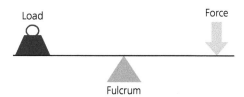

▲ Figure 9.4a A first-class, or first-order, lever

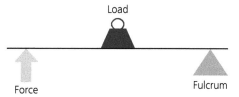

▲ Figure 9.4b A second-class, or second-order, lever

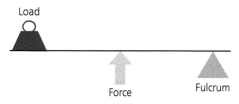

▲ Figure 9.4c A third-class, or third-order, lever

Test yourself

Explain what is meant by a second-class lever and give an example.

Pulleys

A **pulley** comprises wheels, an axle and a rope. While pulleys can be used to transmit drive, their main function is to reduce the effort force needed when lifting loads. For example, a crane uses a pulley to lift heavy materials on a construction site.

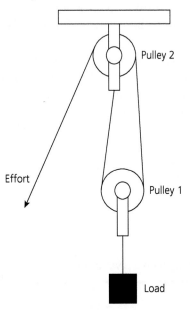

▲ Figure 9.5 A two-wheel pulley system

Case study

The Burj Khalifa in Dubai is the tallest building in the world, reaching over 800 metres in height.

An incredible 330,000 m³ of concrete was used in its construction. However, none of this would have been possible without the use of mechanical systems to lift the materials into place as the structure grew in size.

In addition, the building makes use of the world's third fastest lift system to get people from floor to floor. This requires the use of a complex modern pulley system powered by electric motors.

Questions

1 Give three examples of mechanical components and systems used in the civil engineering industry.
2 What other applications are pulley systems used for?

Industry tip

Mechatronics engineers need a good understanding of how both mechanical and electronic components work. They also need to be able to think creatively and have good problem-solving skills.

Electrical and electronic components

Transducers

Transducer is an umbrella term used to describe both electrical/electronic sensors (input devices) and actuators (output devices). These devices are used to convert one form of energy, or signal, into another.

Sensors

Sensors convert signals from the physical environment, such as light, sound or temperature level, into an electrical or electronic signal, such as a voltage, current or resistance. For example, light-dependent resistors have a resistance that decreases as the light level increases, and vice versa. Thermistors work in a similar way but respond instead to temperature levels.

▲ Figure 9.6 A light-dependent resistor

Key terms

Transducer: umbrella term used to describe both electrical/electronic sensors (input devices) and actuators (output devices) that convert one form of energy, or signal, into another.

Sensors: electronic components that detect changes in the physical environment around them.

Test yourself

Name a sensor that can detect changes in temperature level and describe how it functions.

Actuators

Actuators convert electrical, electronic or mechanical signals into physical movement.

Two types of actuator widely used in mechatronic systems are motors and solenoids. Motors produce rotary motion, whereas solenoids produce a pushing or pulling motion. Both of these actuators work using the principle of electromagnetism.

> ### Key term
>
> **Actuators:** devices that convert electrical, electronic or mechanical signals into physical movement.

▲ Figure 9.7 A small DC motor

Microprocessors and microcontrollers

Microprocessors and **microcontrollers** are examples of programmable components. They replace discrete components and hardware with programming, thus reducing the size of circuits and offering greater flexibility of design.

Microprocessors house a powerful central processing unit (CPU) on an integrated circuit (IC), with all other devices connected externally. They are mainly used in computing applications, such as computer-controlled robot systems.

Microcontrollers are basically small computers, contained in microchips, that include processing, memory and other functions such as analogue to digital conversion. They are often used to replace traditional analogue and digital hardware ICs in engineering applications such as timing, counting, responding to sensor inputs and controlling actuators.

> ### Key terms
>
> **Microprocessors:** powerful central processing units mainly used in computing applications.
>
> **Microcontrollers:** small computers, contained in microchips, that respond to input devices and control output devices.
>
> **Motor:** drive device that turns electrical energy into rotary movement.

▲ Figure 9.8 A microcontroller circuit board with ports for connecting different input and output devices

Drive devices

Drive devices turn an electrical or electronic signal into a mechanical signal in the form of movement.

▼ Table 9.1 Purpose and function of drive devices

Drive device	Purpose	Function
Standard **motor**	To create rotary motion when current flows through it, for example turning wheels on electric vehicles	Velocity/speed is controlled by varying the supply voltage signal Direction is controlled by reversing the direction of the supply current
Servo motor	To create rotary motion through a precise angle or velocity, for example when moving the cutting tool in a CNC machine	A series of electronic pulses controls the movement of the motor. These pulses can be provided by a microcontroller working alongside a suitable motor driver circuit
Stepper motor	To create rotary motion through a series of precise steps, for example when moving digital cameras into specific positions	Each full 360° rotation is divided into a sequence of equal steps A suitable process device (such as a microcontroller) and motor driver circuit are required

9.2 The operation, function and applications of programmable logic controllers (PLCs) in mechatronic systems

PLCs and their types

Programmable logic controllers (PLCs) are programmable devices used in manufacturing and industrial control systems. They control different production processes by responding to sensors and activating actuators. Their exact function depends on their programming, making them extremely flexible. They are also very robust and reliable, so are well suited to sometimes harsh manufacturing environments.

Key term

Programmable logic controllers (PLCs): programmable devices used in manufacturing and industrial control systems.

PLCs are process devices and are therefore shown as such in systems block diagrams. They generally consist of a power supply, a CPU, a programming device and ports for connecting input and output devices. Inputs and outputs can be hardwired or connected using wireless technology.

▲ Figure 9.9 An industrial PLC

Unitary PLCs contain all the different parts and components within a single housing. Modular PLCs have different parts, or modules, that are connected together to form a customisable device. This allows greater flexibility of design and functionality, but increases cost

and complexity. Some PLCs have characteristics of both unitary and modular types, and therefore offer some of the benefits and limitations of each.

PLC operation

PLCs must be programmed using an appropriate language or system, including, for example:
- ladder logic
- sequential function charts
- function block diagrams
- structured text
- instruction lists.

Ladder logic is most commonly used, which presents instructions as a graphical diagram based on relay logic hardware. This makes it highly visual and easy to follow, which is especially useful when fault finding.

Signal conditioning

It is sometimes necessary to modify a signal from a sensor or other device so that it meets the requirements for processing by a PLC. This is called **signal conditioning**. Typical examples include:
- analogue to digital conversion – changing a continuous analogue signal into a discrete digital signal
- amplification – increasing the size, or gain, of a signal
- attenuation – reducing the strength of a signal
- filtering – removing unwanted features from a signal, such as certain frequencies.

Key term

Signal conditioning: modifying the signal from a sensor or other device so that it meets the requirements for processing by a PLC.

PLC function

Motor drivers and interface devices

A **motor driver IC** is used when a programmable device cannot provide the required output current to drive a motor on its own. A simple option is a Darlington driver IC, such as the ULN2803, but this can only provide current in one direction. If a motor needs to be able to reverse as well as rotate forwards, then a driver capable of providing bidirectional current is required. One such option is the L293D IC. This is a 16-pin IC that can provide up to 600 mA each for up to two motors.

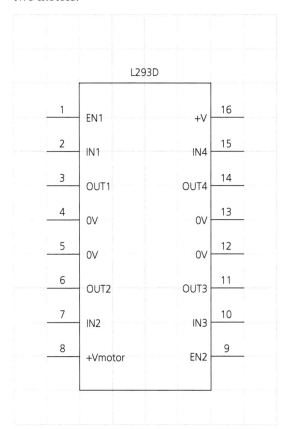

▲ Figure 9.10 L293D pinout diagram

Interface devices ensure accurate communication can take place between a PLC and the input and output devices, such as sensors and actuators. These are usually provided as modules for PLCs that can be connected to the main controller base.

> **Key terms**
>
> **Motor driver IC:** integrated circuit used to ensure that there is enough current for motors to operate correctly.
>
> **Interface devices:** modules that ensure accurate communication takes place between PLCs and input/output devices.

Applications of PLCs

PLCs are used in a wide range of industrial, engineering and manufacturing applications. One of the most common is robotics. In manufacturing, robots can be used to assemble products more quickly and efficiently than humans, and also to perform operations that would be hazardous to humans. PLCs are also used in control systems for conveyor belts and automated packaging systems.

Animatronics is the use of mechanical and programmable electronic components to produce lifelike robots, typically for films or other entertainment. Animatronic arms and hands can also be used to perform operations that previously would have had to be completed manually.

Crucial to the application of PLCs is **supervisory control and data acquisition (SCADA) software**, which is used to supervise, monitor and control industrial engineering processes. In addition to the PLCs connected to input and output devices, SCADA systems consist of supervisory computers, remote terminal units, communications devices and a human–machine interface (HMI).

> **Key term**
>
> **Supervisory control and data acquisition (SCADA) software:** software used to supervise, monitor and control industrial engineering processes.

> **Research**
>
> Research the use of PLCs at your placement.

9.3 The basic principles of hydraulics and pneumatics

Principles of fluid power systems

Fluid power systems can be either hydraulic or pneumatic:
- **Hydraulic systems** use a liquid, such as oil or water, as the power transmission medium.
- **Pneumatic systems** use compressed air to transmit power.

> **Key terms**
>
> **Hydraulic systems:** systems that use a liquid as the power transmission medium.
>
> **Pneumatic systems:** systems that use compressed air to transmit power.

Hydraulic vs pneumatic systems

Due to the risk of leakage of the power transmission medium, hydraulic systems are not generally used where there is the potential for contamination, for example in food processing. However, they are used where greater power is required, such as in heavy-lifting applications.

Pneumatic systems tend to be used where greater speed of operation is required, such as when operating CNC machine tools.

Components of hydraulic and pneumatic systems

Table 9.2 outlines the purpose and function of some of the main components used in fluid power systems.

As with electrical and electronic circuits, fluid power systems can be represented using schematics.

▼ Table 9.2 Components of fluid power systems

Component	Type of fluid power system typically used in	Purpose and function
Valve	Hydraulic and pneumatic	Controls the direction of the fluid and hence the movement of other components, such as cylinders and actuators
		Arranged in different layouts to create AND and OR systems
Pump	Hydraulic	Creates the flow of fluid by overcoming the pressure induced by the resistive load
Actuator	Hydraulic and pneumatic	Turns the hydraulic or pneumatic energy back into mechanical energy
		Acts as an output device to the system
Cylinder	Hydraulic and pneumatic	An example of an actuator within a fluid power system
		Forces a piston to move in a certain direction
		Creates movement by a rod moving back and forth in a barrel
Compressor	Pneumatic	Converts electrical or mechanical energy into pressurised air

Test yourself

Describe one example of an actuator within a fluid power system.

Assessment practice

1   Describe, using an example, a first-class (or first-order) lever.
2   Explain the purpose of a pulley mechanism.
3   Name a sensor that can detect changes in light level.
4   Explain, using examples, the purpose of an actuator in a mechatronic system.
5   Explain what is meant by a microcontroller.
6   Describe the architecture of a PLC.
7   Describe two examples of signal conditioning.
8   Give an example of a motor driver IC capable of providing bidirectional current.
9   Explain when pneumatic systems are typically used instead of hydraulic systems.
10 Describe how a pneumatic cylinder functions.

Project practice

You work as a mechatronic engineer for a manufacturing company. The company manufactures different workshop tools on a production line, using conveyor belts to move workpieces from station to station. The production line is currently monitored and controlled using a non-programmable, hardware-based system. The company wishes to move to a programmable system for greater flexibility of operation.

Produce a report detailing how this change will be implemented. Your report should include:
▶   an explanation of the different programmable system options available and their relative advantages and disadvantages
▶   the selection of a programmable system to use and justification as to why it has been chosen
▶   a description of how the programmable system will be implemented and how it will function.

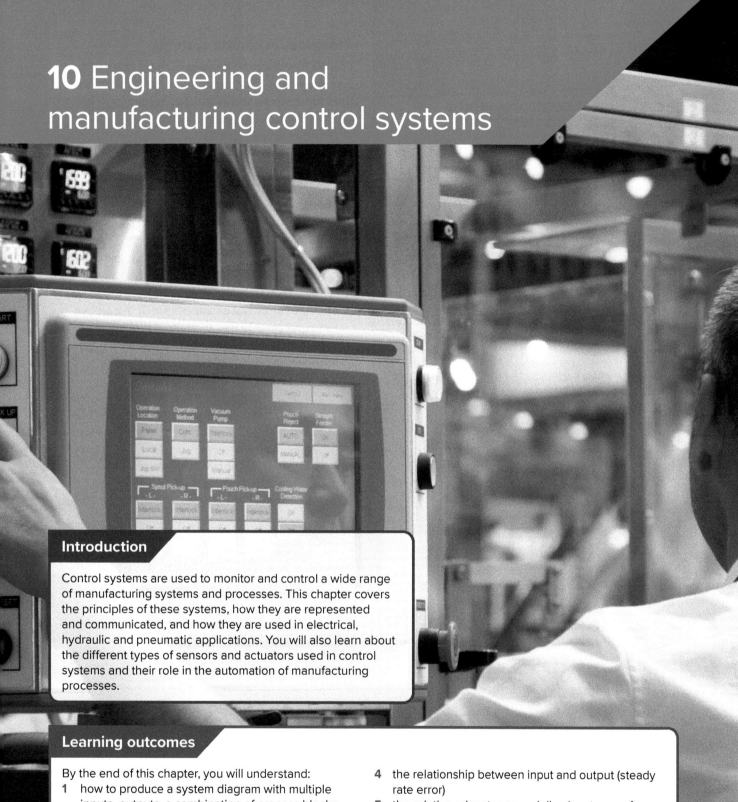

10 Engineering and manufacturing control systems

Introduction

Control systems are used to monitor and control a wide range of manufacturing systems and processes. This chapter covers the principles of these systems, how they are represented and communicated, and how they are used in electrical, hydraulic and pneumatic applications. You will also learn about the different types of sensors and actuators used in control systems and their role in the automation of manufacturing processes.

Learning outcomes

By the end of this chapter, you will understand:
1 how to produce a system diagram with multiple inputs, outputs, a combination of process blocks and feedback, and explain its operation
2 applications of open- and closed-loop control systems (under- or over-damped, and time dependency)
3 the advantages and disadvantages of open- and closed-loop control systems

4 the relationship between input and output (steady rate error)
5 the relative advantages and disadvantages of analogue and digital signals in control systems
6 applications of control systems in industry, including effective and efficient networked communication and data transmission
7 the purpose and function of the different types of sensors and actuators
8 applications and uses of sensors and actuators.

10.1 Principles and applications of control system theory

Principles of control systems

Control systems are made up of:
▶ **input blocks**
▶ **process blocks**
▶ **output blocks**.

> **Key terms**
>
> **Input blocks:** system blocks that take signals from the real-world environment and change them into signals that process blocks can understand.
>
> **Process blocks:** system blocks that respond to signals they receive from input blocks and alter them in some way, before sending them to the output blocks.
>
> **Output blocks:** system blocks that take signals from process blocks and turn them back into real-world environmental signals.

Inputs in control systems

Input blocks take signals from the real-world environment and change them into signals that process blocks can understand.

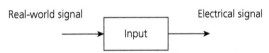

Real-world signal   Electrical signal

Input

▲ **Figure 10.1 An input block in an electrical control system**

Input devices within control systems tend to be electrical sensors that convert real-world signals into a voltage or current. They are used to monitor different environmental parameters, for example:
▶ Temperature sensors monitor temperature levels.
▶ Light sensors monitor light levels.
▶ Flow sensors measure the flow rate of liquids or gases.
▶ Pressure sensors measure pressure levels.

Switches are also examples of input devices.

> **Test yourself**
>
> Describe the function of input blocks in control systems.

Processes in control systems

Process blocks can be thought of as the brain of a control system. They respond to signals they receive from input blocks and alter them in some way, before sending them to the output blocks.
▶ Timers take an input signal and keep it high or low for a set time period. This can often be adjusted using different component values, for example by changing resistor or capacitor values, or by using variable resistors and capacitors.
▶ Comparators compare two different signals, often a measured signal and a reference signal, and indicate which is the highest. An example of this is temperature control, where the desired temperature is the reference signal and this is compared to the actual measured temperature. If the desired temperature is higher than the actual temperature, a heater can be turned on.
▶ Pulse units produce a continuous sequence of digital pulses. As with timers, the duration that the signals are high and/or low can be changed by altering component values or by using variable resistors. An application of this is in electronic circuit test equipment, where a series of pulses can be used to act as a stimulus for taking measurements.
▶ Counters add up the number of digital signals or pulses received. This figure can then be outputted to a display. An example of this is counting the number of manufactured products that pass by on a conveyor belt.
▶ Latches take an input signal and keep it high or low until it is reset. An example of this is a security alarm system which sounds when triggered but which can be reset using a keypad or key switch.

> **Test yourself**
>
> Explain how different types of process blocks work in control systems.

Logic gates

Logic gates are process devices that receive digital input signals (1s and 0s) and produce digital outputs depending on those inputs. How they respond to the signals received depends on the type of logic gate. Three of the most common are AND, OR and NOT gates.

> **Key term**
>
> **Logic gates:** process devices that receive digital input signals (1s and 0s) and produce digital outputs depending on those inputs.

NOT gates are the simplest type of logic gate, as they have just one input and one output. They act as inverters, that is, the output signal is always the inverse, or opposite, of the input signal. The way in which a NOT gate works is shown using its truth table (see Table 10.1). This gives the output signal states for any given combination of inputs.

▼ Table 10.1 Truth table for a NOT gate

Input	Output
0	1
1	0

AND gates and OR gates are more complex, as they respond to two digital input signals.

An AND gate only produces a high (1) output signal when both of the input signals are high. Otherwise, the output is low (0). The truth table for an AND gate is shown in Table 10.2.

▼ Table 10.2 Truth table for an AND gate

Input A	Input B	Output
0	0	0
0	1	0
1	0	0
1	1	1

An OR gate produces a high output signal when either or both of the input signals are high. Otherwise, the output is low. The truth table for an OR gate is shown in Table 10.3.

▼ Table 10.3 Truth table for an OR gate

Input A	Input B	Output
0	0	0
0	1	1
1	0	1
1	1	1

Test yourself

Draw the truth tables for AND, OR and NOT logic gates.

Outputs in control systems

Output blocks take signals from process blocks and turn them back into real-world environmental signals. For example:

▶ Actuators, such as those found in motors and solenoids, turn a mechanical, pneumatic, hydraulic or electrical signal into movement.
▶ Buzzers turn an electrical signal into sound.
▶ Light outputs, such as light-emitting diodes, turn an electrical or electronic signal into visible light.

▲ Figure 10.2 An output block in an electrical control system

Test yourself

Describe the function of output blocks in control systems.

Signals used in control systems

The two main types of signal used in control systems are analogue and digital.

As we learned in Chapter 8, analogue signals are continuous and usually represented using sinusoidal (sine) waves. Examples include temperature, sound and light levels. Analogue signals carry more information than digital signals and as such are more accurate. However, they are more susceptible to the effects of noise and interference.

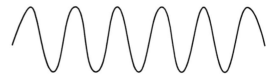

▲ Figure 10.3 An analogue signal

Digital signals are discrete and can only be high (1) or low (0). They are usually represented using square waves. Digital signals carry less information than analogue signals, so they take up less bandwidth and are easier to decode. However, this also makes them less detailed.

▲ Figure 10.4 A digital signal

Pulse width and amplitude modulation

Signals often need to be converted into a format that is suitable for transmission. This is called modulation. There are different methods of achieving this, depending on the signal characteristic that is modified or changed, for example:

▶ **pulse width modulation (PWM)** – the width of the pulsed carrier signal is varied according to the amplitude of the message signal

▶ **pulse amplitude modulation (PAM)** – the amplitude of the pulsed carrier signal is varied according to the amplitude of the message signal (this occurs with AM radio waves).

> ### Key terms
>
> **Pulse width modulation (PWM):** the width of the pulsed carrier signal is varied according to the amplitude of the message signal.
>
> **Pulse amplitude modulation (PAM):** the amplitude of the pulsed carrier signal is varied according to the amplitude of the message signal.

Open- and closed-loop control systems

There are two main types of control system:

▶ An **open-loop system** does not have a feedback loop.

▶ A **closed-loop system** has one or more feedback loops.

A feedback loop is created when the output signal from a system is also an input signal to the same system.

> ### Key terms
>
> **Open-loop system:** system without a feedback loop.
>
> **Closed-loop system:** system with one or more feedback loops.

▲ Figure 10.5 An open-loop system does not have a feedback loop

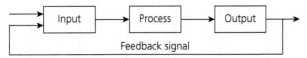

▲ Figure 10.6 A closed-loop system has at least one feedback loop

An example of a closed-loop system is a heating control system. A heater is turned on when a sensor detects that the temperature in the room is too cold. This output signal is then fed back, so that the system knows when it is warm enough for the heater to be turned off again. Robotics systems are often closed-loop systems, as they rely on sensor information being fed back about the movement of the robot.

Closed-loop systems are less likely to be affected by noise and are therefore more robust and accurate than open-loop systems. However, they are more complicated to produce and more costly to maintain.

Under-damped and over-damped systems

Systems that are under-damped return to their equilibrium point with only a few oscillations. Over-damped systems return very slowly to equilibrium, without any oscillations.

> ### Test yourself
>
> Explain the difference between open- and closed-loop control systems.

Transfer functions

Control systems, their constituent blocks and their input/output signals can be represented algebraically. This helps with system modelling, analysis and fault finding, as it shows the relationships between the different parts that make up a system.

The contents of each individual system block can be represented mathematically using a **transfer function**. For example, G(s) in Figure 10.7 can be expressed as:

$$G(s) = Y(s)/X(s)$$

or:

$$Y(s) = G(s)X(s)$$

where X(s) is the input signal and Y(s) is the output signal.

▲ Figure 10.7 Transfer function

> ### Key term
>
> **Transfer function:** algebraic representation of the contents of a system block.

Summing points

The **summing point** (represented as a circle with a cross inside) is the part of a system that produces the algebraic sum of the reference signal and the feedback signal. This results in the error signal which, for the example in Figure 10.8, can be written as:

$$E(s) = R(s)-B(s)$$

where E(s) is the error signal, R(s) is the reference signal and B(s) is the feedback signal (C(s) is the output signal).

> ### Key term
>
> **Summing point:** the part of a system that produces the algebraic sum of the reference signal and the feedback signal.

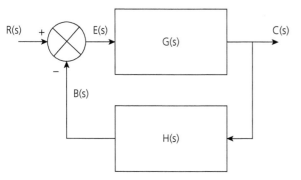

▲ Figure 10.8 Summing point

Steady state error is the difference between the input and output of a system, as time goes to infinity. It is usually better for a system to have a low steady state error.

> ### Improve your maths
>
> Draw a block diagram of a control system with feedback and a summing point.

Representing control systems

Block diagrams are the main method of representing control systems, and you have seen lots of examples of these in this chapter. They provide a top-down overview of a system in terms of its inputs, processes and outputs. Blocks are used to represent the groups of components that make up each sub-system, and arrows are used to represent the signals that flow between them.

When individual components need to be shown, **schematic diagrams** or **wiring diagrams** can be used.

Schematic diagrams (also known as schematics) show components as standard symbols, joined together by straight lines that represent connecting wires. They usually follow the same layout as a block diagram, with input components to the left, process components in the middle, and output components towards the right of the diagram.

Wiring diagrams show components as pictorial representations joined together by wires. The layout indicates how the different parts will be physically arranged when installed or placed in situ.

> ### Key terms
>
> **Block diagrams:** diagrams that use blocks and arrows to provide a top-down overview of a system in terms of its inputs, processes and outputs.
>
> **Schematic diagrams:** diagrams that show a system in terms of its individual components, represented by standard symbols.
>
> **Wiring diagrams:** diagrams that show a system in terms of its physical wiring layout, with components shown as pictorial representations.

> ### Industry tip
>
> Make sure you can locate and interpret any diagrams (block, schematic or wiring) for systems you will be working on in the workplace.

Applications of control systems

Electrical control systems

Electrical control systems make use of electrical power supplies, components and signals.

> ### Key term
>
> **Electrical control systems:** control systems that make use of electrical power supplies, components and signals.

Typical examples of electrical control systems in engineering and manufacturing include:
▶ temperature-, light- and climate-control systems, when making products that are temperature sensitive or require a specific set of environmental conditions to be maintained

▶ liquid-flow and pressure-control systems, such as in hydroelectric power generation and heat-pump systems

▶ systems that make use of infrared-sensor technologies, such as production-line quality control and/or safety warning systems

▶ data-transmission systems that provide effective and efficient networked communication.

Health and safety

Always follow the relevant workplace safety procedures and legislation when working with electrical systems. This should include safe isolation procedures when taking measurements of electrical parameters.

Fluid power control systems

As we learned in Chapter 9, **fluid power control systems** use a fluid as the power transmission medium. In hydraulic systems, this is a liquid such as oil or water, whereas in pneumatic systems it is compressed air.

Hydraulic control systems are commonly used where a large amount of power is needed, for example to control the landing gear and brakes in aircraft, and in construction industry applications, such as heavy-lifting systems, excavators and drills.

Pneumatic systems are extremely accurate and precise. Examples include air-braking systems, pressure-control systems and CNC machine movement, speed and positioning controls.

Key term

Fluid power control systems: control systems that use a fluid, such as liquid or a compressed gas, as the power transmission medium.

▲ Figure 10.9 A pneumatic control system

Research

Research examples of electrical, hydraulic and pneumatic control systems used in different engineering and manufacturing contexts. What is their purpose? How do they work? What are their benefits and limitations? Present your responses to the class.

10.2 How sensors and actuators are used in automation control systems

Sensors

As we learned in Chapter 9, sensors are electronic components that detect changes in the physical environment around them, for example light and temperature levels. They then change this information into a signal that can be processed by a control device, such as a PLC or microcontroller. Sensors are therefore examples of input devices in systems.

Sensors can be analogue or digital, depending on the type of signal they measure and/or produce:

▶ Analogue sensors measure and/or produce continuous output signals. In the case of electrical sensors, this is usually a voltage, resistance or current that is either proportional or inversely proportional to the quantity being measured. For example, an NTC (negative temperature coefficient) thermistor has a varying resistance that decreases as temperature level increases, and vice-versa.

▶ Digital sensors only produce output signals that are discrete, that is, they can only be high (1) or low (0). A tilt switch is an example of a simple digital sensor.

Sensors can be active or passive:

▶ Active sensors work by sending a signal into the environment and measuring the responses they get back. An example is a proximity sensor that uses infrared signals to detect how close an object is to it. Another example is a laser sensor, such as those used in measuring applications.

▶ Passive sensors monitor changes in the environment around them without otherwise interfering with it.

Actuators

Actuators are examples of output devices in control systems. They produce physical movement in response to a signal received. One of the most commonly used actuators is a motor, which produces rotary movement.

As with sensors, actuators can be analogue or digital, depending on the type of signal that they output and/or respond to. They can also be active or passive:

▶ Active actuators require their own power supply and/or are capable of introducing new energy into a system.

▶ Passive actuators neither need their own power supply nor introduce any new energy into a system.

Uses of sensors and actuators in automation

Automated systems perform their jobs without the need for human operators. They are widely used when mass-producing items on production lines.

Sensors and actuators are integral to the function of automated systems. Sensors detect changes in the conditions around them and actuators then implement the required changes. For example, in an automated lifting system:

▶ a sensor would detect when an object is placed on a platform

▶ a motor actuator would power a gear or pulley system to physically lift the object

▶ another sensor would then detect when the object has been lifted to the required height and the motor would stop moving.

On a production line, infrared sensors could be used to detect the volume of manufactured products moving along a conveyor belt to a packaging station. The motor powering the belt would then stop when the required number of products had passed through.

▲ Figure 10.10 A laser sensor used for taking measurements on a conveyor belt

Case study

Airbus is a European company that designs and manufactures commercial jet aircraft, including the second-best-selling airliner in the world, the Airbus A320. The fuselage for this plane is currently made in Hamburg, Germany, using a recently introduced automated assembly line. The assembly line makes use of a number of robotic control systems that minimise the need for human input. Airbus believes that this is a demonstration of the future of aircraft manufacture.

Questions

1 Discuss the advantages and disadvantages of this approach to manufacturing aircraft.

2 As technology continues to improve, how do you think this will impact the design and manufacture of aircraft in the future?

Measurement applications

One stage of the manufacturing process that has benefitted immensely from automated systems is quality control. This is because highly accurate measurements can be taken using automated sensors, thus reducing the need for manual tools and equipment.

Examples of measurements that can be automated through the use of sensors include:

▶ electrical and electronic parameters, such as voltage, current and resistance

▶ mechanical outputs, such as force, torque and velocity

▶ chemical and biological measurements

▶ thermal output signals

▶ types and levels of radiation

▶ optical and acoustic signal waveforms and characteristics.

Automating measurement systems can result in improved accuracy, efficiency and quality, and a reduction in human error, labour costs and waste.

Research

Research ways in which sensors and actuators can be used to automate the control of different manufacturing processes. What sensors could be used? What are their benefits and limitations?

Assessment practice

1 Draw a labelled example of an input block in a control system.
2 Describe the function of a latch in a control system.
3 Describe how a NOT gate functions.
4 Explain why control systems are represented algebraically.
5 Explain the purpose of a wiring diagram.
6 Explain the main difference between electrical, hydraulic and pneumatic control systems.

7 Explain what is meant by an active sensor.
8 Give an example of an actuator used in control systems.
9 Explain what is meant by an analogue actuator.
10 Explain why companies are increasingly automating measurement systems in manufacturing environments.

Project practice

You work for a manufacturing company that makes parts for vacuum cleaners. Measurement of the physical dimensions of manufactured parts is currently completed using human workers and manual equipment, but the company is keen to use automation.

Produce a plan for how the company could use sensors and actuators to automate this process.

Your plan should include:
▶ how the existing manual operations will be automated
▶ the sensors and actuators to be used, how they function and their benefits/limitations
▶ the potential impact(s) on the company's workforce.

11 Quality management

Introduction

This chapter looks at the role of quality management in monitoring the operations needed to maintain high standards in the field of engineering. Quality management involves the creation and implementation of:
▶ an organisation's quality policy
▶ quality planning and assurance
▶ quality control
▶ quality improvement.

We will consider how quality standards and engineering bodies offer guidance to engineers, before comparing quality assurance and quality control, and investigating the methods available to improve quality within an organisation. To conclude, we will identify the types and applications of standard operating procedures and their purposes.

Learning outcomes

By the end of this chapter, you will understand:
1 the function, purpose and value of standards (safety, quality, compliance) and how to access this information
2 the roles and responsibilities of the engineering bodies
3 the main principles, purposes and outcomes of quality assurance, quality control, inspection and testing; the difference between quality control and quality assurance
4 the main requirements of quality standards
5 the reasons for document management and version control
6 the advantages and disadvantages of 100 per cent sampling compared to statistical process control (SPC)

7 the use of Six Sigma for high-volume manufacturing
8 the main principles, purposes, advantages and disadvantages of different approaches to quality improvement
9 the typical format and content of Standard Operating Procedures (SOPs)
10 how SOPs are used in different applications
11 the reasons for using SOPs (consistency, conformance to standards)
12 how SOPs are produced, implemented and evaluated.

11.1 Quality standards, assurance, control and improvement

Quality standards

Quality standards provide guidance on how to meet **legislation**. They:

▶ give recommendations on how products should be designed and manufactured, and set minimum requirements

▶ are written and agreed by experts with specialist engineering knowledge

▶ are updated on a regular basis to ensure **compliance** with the latest legislation and to take advantage of technological developments

▶ provide guidance on design and manufacturing processes to ensure high levels of quality and safety.

> **Key terms**
>
> **Legislation:** a law or set of laws passed by Parliament.
>
> **Compliance:** being in accordance with commands, rules or requests.

British standards

The British Standards Institution (BSI) is the UK's national standards body. It develops standards for use in the UK and also works with other international standards organisations.

The standards are **technical specifications** that can be used as guidance within engineering, particularly in the manufacturing of products.

Manufacturers need to pay close attention to the standards that are applicable to what they produce. For example, a company that manufactures bicycles needs to consider safety requirements by referencing *BS EN ISO 4210 – Cycles. Safety requirements for bicycles*.

The BSI Kitemark is a quality trademark that shows that a product or service meets the appropriate national standards. Products that comply with the BSI Kitemark scheme can display the trademark on them.

> **Key term**
>
> **Technical specifications:** documents that define the requirements for projects, products or systems.

European conformity (CE)

The letters CE appear on many products to signify that they conform with the health, safety and environmental protection standards for products sold within the European Economic Area (EEA). CE is the abbreviation for *conformité européenne* – European **conformity**.

Conformity is assessed by the European Union (EU) and the CE marking is recognised worldwide. It does not mean that a product was made in the EEA, but indicates that it satisfies the legal requirements to be sold there.

> **Key term**
>
> **Conformity:** compliance with standards, rules or laws.

> **Test yourself**
>
> Who assesses the conformity signified by the CE mark?

International Organization for Standardization (ISO)

The International Organization for Standardization (ISO) develops standards to ensure the quality, safety and efficiency of products and services in the international marketplace. The standards cover a huge range of areas, from product manufacture and technology, to agriculture, food safety and healthcare.

ISO certification is recognised worldwide and confirms that an organisation is meeting the requirements for quality **assurance**.

> **Key term**
>
> **Assurance:** certainty about something.

Roles and responsibilities of engineering bodies

Engineering bodies oversee the registration of graduate and professional engineers and govern the conduct and practice of registered engineers to ensure public safety.

There are numerous engineering bodies in the UK.

Engineering Council

The Engineering Council is a UK regulatory body that establishes and upholds internationally recognised standards of professional engineering **competence**. It holds a register of engineers and technicians who have been assessed against these standards, so that employers can be confident in the knowledge and experience of those who have been awarded one of the Engineering Council's professional titles.

> **Key term**
>
> **Competence:** the ability to do something effectively, with sufficient knowledge, judgement and skill.

Institution of Engineering and Technology (IET)

The Institution of Engineering and Technology (IET) provides specialist advice to the UK government and other organisations that require engineering expertise. It issues reports on engineering-related policy issues and explains the impacts of engineering and technology on society.

Institution of Mechanical Engineers (IMechE)

The Institution of Mechanical Engineers (IMechE) supports and oversees mechanical engineers across the world, equipping them with mechanical engineering knowledge and skills. To ensure that technical workforces are trained to the highest standards, it develops and maintains strategic partnerships with leading multinational organisations.

Society of Operations Engineers (SOE)

The Society of Operations Engineers (SOE) is the regulatory body for maintenance engineers and technicians who inspect, maintain and manage equipment and machinery to ensure safety in the workplace and for the public. It works to improve maintenance and inspection processes and to ensure sustainable engineering practices.

Chartered Institution of Building Services Engineers (CIBSE)

The Chartered Institution of Building Services Engineers (CIBSE) helps building services engineers to achieve professional registration and membership by providing courses at all levels of study. Buildings are responsible for almost 50 per cent of harmful carbon emissions, so CIBSE works with engineers to help design and construct more energy-efficient buildings.

Institution of Agricultural Engineers (IAgrE)

The Institution of Agricultural Engineers (IAgrE) is a professional membership organisation with interests in agriculture, forestry, environment, horticulture, amenity and all other land-based industry. It brings together academics, engineers and industry to promote the application and improvement of technology in this sector.

Institute of the Motor Industry (IMI)

The Institute of the Motor Industry (IMI) develops skills benchmarks and qualifications for engineers and technicians working in the automotive industry. This includes training in new technologies, such as electric vehicles (EVs) and hydrogen fuel cell electric vehicles (FCEVs).

The Welding Institute (TWI)

The Welding Institute (TWI) ensures engineering organisations understand the importance of welding and joining, and that they value the registered professionals in those trades. It is committed to professional development and the safety of those employed in welding and joining.

Quality assurance and quality control

Culture of quality

A culture of quality can be defined as an environment in which everyone is responsible for, and contributes to, quality. An organisation should constantly discuss its quality at all levels and not just simply follow its quality guidelines. Quality can only be ensured if everyone is committed to it, and it cannot be guaranteed by a single tool, test or person.

▲ Figure 11.1 A culture of quality is an environment in which everyone is responsible for, and contributes to, quality

The difference between quality assurance and quality control

Quality assurance (QA) is a management **methodology** used by organisations to prevent defects in manufactured products. It ensures the achievement of both the organisation's quality standards and the customer's requirements through the use of process planning and documentation such as quality plans and inspection and test plans.

ISO 9000:2015 defines QA as 'part of quality management focused on providing confidence that quality requirements will be fulfilled'.

> ### Key term
>
> **Methodology:** a system of methods or rules used for a particular activity.

Quality control (QC) is a corrective tool used to ensure that manufactured products meet accepted standards. It implements the inspection and testing requirements of quality assurance in order to detect and rectify defects.

ISO 9000:2015 defines QC as 'part of quality management focused on fulfilling quality requirements'.

▲ Figure 11.2 Quality assurance and quality control

▼ Table 11.1 Quality assurance (QA) and quality control (QC) comparison

Quality assurance (QA)	Quality control (QC)
A proactive management tool	A reactive corrective tool
Focuses on the process	Focuses on the product
Prevents defects	Detects defects
Responsibility of everyone in an organisation	Responsibility of the testing department
Carried out during the manufacturing process	Carried out once the product has been manufactured

Quality assurance approaches

Total quality management (TQM)

Total quality management (TQM) is a continual process that aims to:

- detect and reduce errors in manufacturing
- improve supply-chain management
- enhance customer experience
- ensure employees are adequately trained.

TQM can be divided into four basic categories, as shown in Figure 11.3.

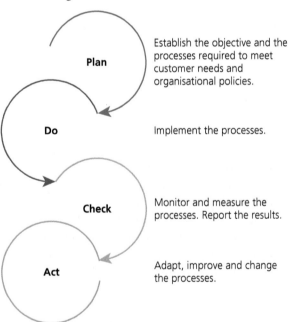

Plan	Establish the objective and the processes required to meet customer needs and organisational policies.
Do	Implement the processes.
Check	Monitor and measure the processes. Report the results.
Act	Adapt, improve and change the processes.

▲ Figure 11.3 Four categories of TQM

Right first time

This is the concept of making sure that all procedures are performed correctly the first time and every time, in order to improve production efficiency and decrease waste. To be successful, right first time needs to follow the problem-solving **DMAIC** method shown in Figure 11.4, which originates from **Six Sigma**:

▶ **D**efine the problem: identify the problem that needs to be fixed. For example, why are products being rejected for one particular machining process?

▶ **M**easure the problem: collect data in order to make an informed decision about what is causing the problem. For example, all the testing data would be collected for the machining process.

▶ **A**nalyse the root cause of the problem: examine and interpret the collected data in order to identify the underlying reason for the problem, for example worn-out tools being used for the machining process.

▶ **I**mprove the process: fix the root cause of the problem, for example by fitting new cutting tools to the machining process.

▶ **C**ontrol the process: put control procedures in place to ensure the machining process is maintained and reviewed on a regular basis.

> **Key terms**
>
> **DMAIC:** a data-driven, problem-solving method used to help identify and fix problems within a process and improve future output.
>
> **Six Sigma:** a set of quality management techniques that aim to improve processes within an organisation and in turn reduce the amount of manufacturing errors and product defects.

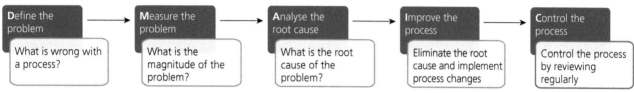

▲ Figure 11.4 Right first time DMAIC method

ISO 9001

ISO 9001 is an international standard that specifies criteria for a quality management system within an organisation. Conformity with the standard will:

▶ improve organisational productivity and efficiency, thereby reducing expenditure

▶ increase customer satisfaction, thereby encouraging repeat business and increasing turnover

▶ improve an organisation's reputation, thereby attracting new customers.

> **Test yourself**
>
> Which international standard sets out the requirements for a quality management system?

> **Improve your English**
>
> Write a short report on the benefits of ISO 9001 and why it would help an organisation.

Inspection and testing

Inspection and testing are key requirements of a quality assurance system, ensuring both confidence in the manufacturing process and customer satisfaction.

Inspection is used to measure conformity to a standard, such as the quality of raw materials and components purchased by an organisation before they enter the production line.

Testing is used during quality control checks to ensure that the manufactured products meet required **tolerances**. Any product that fails to do so will be rejected, and corrective measures will be implemented to ensure that future products conform with the specified standards.

> **Key term**
>
> **Tolerances:** allowable amounts of variation for a specified quantity, especially in the dimensions of a product or part.

Traceability

Traceability means being able to identify and track all the processes within an organisation, from the purchase of raw materials through to the manufacture of the product and, finally, to disposal of the product.

To manage this process successfully:

- the source of (and other information about) each component should be easily and fully identifiable, for example by using batch or lot numbers
- a transaction record should be attached to each component or material as it is used, which will include a date/time stamp, a batch number and any other relevant product information.

If a problem occurs with the quality of a product which could affect public safety, traceability allows the manufacturer to respond quickly and effectively, for example by recalling faulty products.

Document management and version control

A document management system is an essential part of quality assurance, making sure that documents are current, accessible and secure.

Documents may need to be modified when an organisation develops new or updated processes. Effective version control can help to keep track of amendments, record key decisions and prevent the use of out-of-date documents.

Industry tip

Always know where to find policy documents and, when requested, implement any quality changes and monitor the results. Accept changes to working practices and always listen to your co-workers.

Quality control approaches

Purposes of inspection

- To differentiate good material batches from bad material batches: before a production run, batches of materials are inspected to ensure they conform to customer requirements. If the material is of good quality, it is accepted; if it is of bad quality, it is rejected.
- To establish if a process is changing over time: a manufacturing process may change over time as issues are identified and dealt with. For example, if an adhesive is failing to join two materials together, the process may be changed to use permanent fixings like rivets or bolts.

- To establish if a process is nearing **specification limits**: machinery tools wear out during the manufacturing process, which will affect the dimensions of the products being made and eventually cause specification limits to be exceeded.
- To assess the quality of the manufactured product: this involves testing the product to ensure that all required dimensions and finishes are correct.
- To assess the accuracy of inspectors: inspectors periodically check each other's testing data to ensure their testing is accurate and that they are all operating in a standardised way.
- To verify the precision of measuring instruments: all measuring instruments are tested regularly using a **measurement gauge**, to ensure they are taking accurate readings.
- To measure the overall effectiveness of each process: inspectors review each manufacturing process through discussions with technicians and engineers, and verify that the methods used are still producing high-quality products.

Key terms

Specification limits: boundaries that define where a product works and does not work; they assess how capable a process is of meeting customer requirements.

Measurement gauge: an instrument used to check that measuring equipment is still recording precise measurements.

▲ Figure 11.5 Measuring with a micrometer

▲ Figure 11.6 Inspection is important for verifying the precision of measuring instruments

Stages of inspection

1 Raw materials and components are inspected on receipt for suitability, before they are put into stock or used on the production line.
2 Samples are taken and inspected at critical points in the production process. If errors are found, the stage of production can be identified and corrective action can be taken.
3 Finished products are inspected before being released. If products are of a lower quality than required, they may be rejected or sold at reduced prices.

Types of inspection

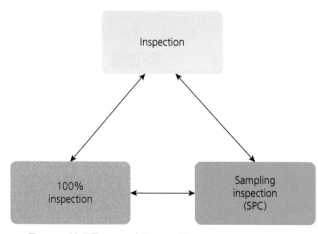

▲ Figure 11.7 Types of inspection

100 per cent sampling/testing

100 per cent **sampling/testing** is a non-destructive inspection method where every component or product is inspected separately.

Due to the amount of detailed inspection required, this is an expensive and time-consuming technique, so it is only used where either an exceptionally high standard of quality is required or there is only a small quantity of parts available to the manufacturer. For example, it may be used when inspecting aircraft, nuclear equipment and medical apparatus.

The risks of 100 per cent sampling/testing include errors arising due to fatigue, negligence and the difficulty of supervising such a specialised task.

> **Key terms**
>
> **Sampling:** the process of selecting batches of products for testing.
>
> **Testing:** the process of assessing product sample batches to check they are within customer specifications for the final product; this determines whether corrective actions are needed in the manufacturing process.

▲ Figure 11.8 100 per cent sampling/testing is required when inspecting aircraft components

Process capability

Process capability measures how the manufacturing process produces products to meet customer requirements. It defines specification limits by using a lower specification limit (LSL) and an upper specification limit (USL). Any products that do not fall between the specification limits would be rejected.

For example, in Figure 11.9 on the next page, a customer has specified the manufacturing limits for a product. The height measurement has a nominal value of 4 mm, an LSL value of 3.98 mm and a USL value of 4.02 mm. After a sample of 100 products was tested and the data recorded in a **histogram**, it was found that 96 per cent of the products fell within the specification limits, so the process was deemed capable.

Test yourself

Name one method of inspection.

Key term

Histogram: a graphical chart that displays data using bars of different heights; it groups the data into ranges and the height of each bar shows the amount of data in each range.

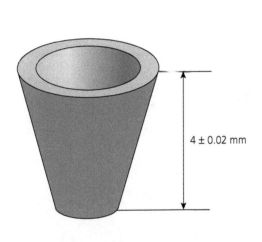

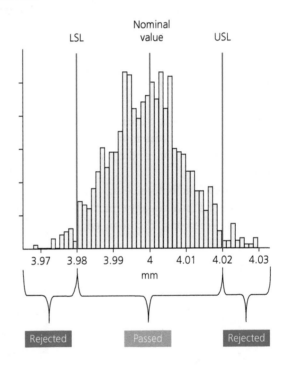

▲ Figure 11.9 Process capability

Statistical process control (SPC)

Statistical process control (SPC) uses statistical methods to measure, monitor and control a process. It ensures the efficiency of manufacturing processes, resulting in less waste, and also allows for the early detection of problems, avoiding the need to make changes after products have been manufactured.

One typical SPC method is to use a control chart. Figure 11.10 shows how product sample dimensions vary over time. The centre line is the average output.

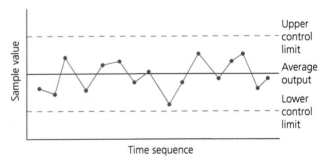

▲ Figure 11.10 SPC control chart

100 per cent sampling is used for high-value and safety-critical products. SPC is typically used for products manufactured at high speed.

▼ Table 11.2 Comparison of 100 per cent sampling and SPC

	100 per cent sampling	SPC
Cost	Expensive to carry out	Cheaper to carry out
Time	More inspection time required	Can be automated at high speed
Fatigue	Operators may suffer from fatigue	Operator fatigue mainly eliminated
Staff	More staff required	Fewer staff required as can be automated
Damage	Damaged products due to increased handling	Fewer damaged products due to minimal handling
Destructive testing	Components/products cannot be destroyed	Components/products can be destroyed
Product lot	Lot may be saved as each piece is inspected	Whole lot may be rejected if a sample is found to be defective

Improve your maths

If a 20 per cent inspection sample is required from a batch of 10,000 components, how many components are needed for testing?

Six Sigma for high-volume manufacturing

Six Sigma is a method used to reduce the occurrence of defects in the manufacturing process. It aims to:

▶ reduce the manufacturing time, by making processes more efficient

▶ reduce wastage, by producing products with fewer defects

▶ avoid variations in the manufacturing process, ensuring that products are identical and meet the customer's requirements.

This will all contribute to making a business more successful.

Six Sigma uses the DMAIC methodology, which you will recognise from right first time.

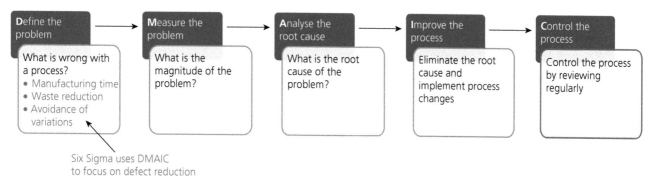

Define the problem
What is wrong with a process?
• Manufacturing time
• Waste reduction
• Avoidance of variations

Measure the problem
What is the magnitude of the problem?

Analyse the root cause
What is the root cause of the problem?

Improve the process
Eliminate the root cause and implement process changes

Control the process
Control the process by reviewing regularly

Six Sigma uses DMAIC to focus on defect reduction in the manufacturing process

▲ Figure 11.11 DMAIC methodology used in Six Sigma

Test yourself

Which quality control method suits high volume production?

Quality improvement

Quality improvement is a systematic approach that uses specific methods and tools to achieve measurable improvements in quality. The model for quality improvement is the PDSA (plan, do, study, act) cycle, shown in Figure 11.12.

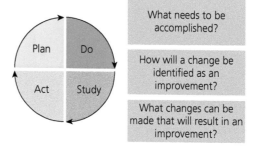

Plan | Do | Act | Study

What needs to be accomplished?

How will a change be identified as an improvement?

What changes can be made that will result in an improvement?

▲ Figure 11.12 PDSA model for quality improvement

Failure mode and effects analysis (FMEA)

Failure mode and effects analysis (FMEA) is a method of identifying and eliminating potential problems or failures in a manufacturing process during the design stage of a project. It can support continuous improvement, as it allows organisations to document actions taken to mitigate risks of failure.

Failure	Potential
Mode	Types, ways, possibilities
Effects	Negative effect on process
Analysis	Assess risk and reduce it

▲ Figure 11.13 FMEA

▼ Table 11.3 Advantages and disadvantages of FMEA

Advantages	It can be used for both existing and new products
	It enables actions to be prioritised
	It is a good communication tool
	It clarifies the magnitude of risks and actions
Disadvantages	Evaluation is often subjective
	Evaluation can vary between teams
	It is time consuming
	Solutions can be so large and complex that they become impossible to manage

Quality circles

Quality circles (also known as quality control circles) are small groups of employees from all areas of an organisation who meet to discuss and solve manufacturing problems. They aim to improve processes by analysing problems and proposing remedies. The groups are run by a manager or supervisor, but all participants are able to offer opinions based on their expertise.

▼ Table 11.4 Advantages and disadvantages of quality circles

Advantages	They result in increased productivity and product quality
	They boost employee morale and improve teamwork skills
Disadvantages	They require time to implement, as well as training for participants
	They often lack management support
	They are often not empowered to make decisions

▲ Figure 11.14 Quality circle

Pareto analysis

Pareto analysis is a tool that helps to identify different ways that a manufacturer could solve a problem, before selecting the most effective solution.

The 80/20 rule (also known as the Pareto principle) states that, for many incidents, roughly 80 per cent of consequences come from 20 per cent of the underlying causes.

For example, 80 per cent of car accidents are caused by 20 per cent of motorists, or 20 per cent of tools in the toolbox are used for 80 per cent of tasks.

Pareto analysis chart

A Pareto analysis chart shows the frequency of any defects and the cumulative percentage of them happening. An 80 per cent line is drawn to work out the 20 per cent defects that need attention.

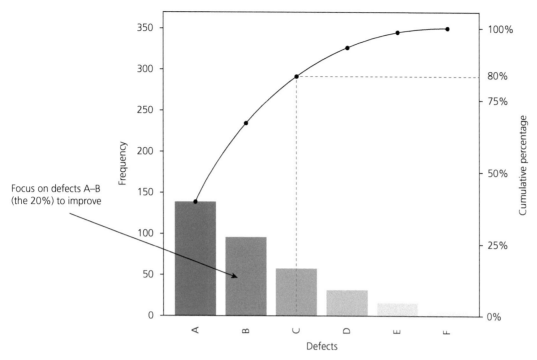

▲ Figure 11.15 Pareto analysis chart

▼ Table 11.5 Advantages and disadvantages of Pareto analysis charts

Advantages	They clearly present significant and insignificant variables
	They highlight which problem to prioritise
	Hierarchical presentation of data is useful when working with limited resources
Disadvantages	More than one chart may be required for data collection
	They only use qualitative data, which cannot be used in statistical calculations
	They are unable to specify the cause of the problem being analysed

Cause and effect diagrams

A cause and effect diagram (also known as an Ishikawa or fishbone diagram) shows the possible causes of a problem in graphic form. It divides the causes into separate branches (or fishbones) which can then be analysed when developing a solution.

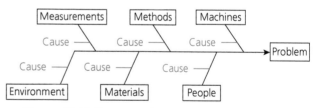

▲ Figure 11.16 Cause and effect diagram

▼ Table 11.6 Advantages and disadvantages of cause and effect diagrams

Advantages	They provide a graphical structure for analysing the possible causes of a problem
	They are easy to draw and understand
	They enable the development of other improvement strategies
	Decision makers can exchange ideas in an organised way in order to solve a problem
Disadvantages	The graphical structure is often too simplistic for real-world problems
	Some people find the format difficult to read
	Unverified data may lead to unsuitable improvement strategies being carried out
	They must be used in conjunction with other decision-making tools, such as FMEA, to ascertain the root causes of a problem

11.2 Types and applications of standard operating procedures (SOPs) and their purposes

Standard operating procedures (SOPs)

Standard operating procedures (SOPs) are instructions drawn up by an organisation to help employees carry out required processes to required standards. They are used for manufacturing, quality and maintenance.

SOPs are made up of step-by-step routines and checklists that are clear and straightforward to follow. The aim is to reduce the likelihood of mistakes being made.

SOPs ensure:
▶ consistency in the way tasks are performed, to keep an organisation running effectively
▶ conformity to standards and industry regulations.

> **Key term**
>
> **Standard operating procedures (SOPs):** step-by-step instructions for routine operations. These are put together by an organisation and all workers must follow them.

▲ Figure 11.17 SOPs are made up of step-by-step routines and checklists

Types and applications

▶ Manufacturing SOPs are provided for workers performing both manual and automated manufacturing tasks. They can include practices that are essential for health and safety.

▶ Quality assurance and quality control SOPs are used to ensure the results of quantitative testing are accurate. The use of step-by-step guides also ensures the data collected is consistent and precise.

▶ Maintenance SOPs are documented instructions to help workers perform routine maintenance tasks. They ensure machinery and equipment are kept running efficiently.

Typical formats and content

SOPs typically come in three different formats: step-by-step, hierarchical and flowchart.

Step-by-step SOPs

Step-by-step SOPs break down processes into numbered lists of detailed steps. The steps should be clear enough that someone can follow them without supervision. For example, a step-by-step SOP for a milling machine might include:

1 Locate the start and emergency stop buttons.
2 Keep the work table clear of materials and tools.
3 Keep all cutting tools sharpened and in good condition.
4 Tighten the collet chuck using the correct spanner.
5 Ensure the work-holding device is secured to the table.

Hierarchical SOPs

Hierarchical SOPs are used for more complex procedures with a greater number of steps. These may cover:
▶ policy – why the process is required
▶ procedures – the steps and the people who will perform them
▶ guidelines – advice on how to meet the requirements of the product
▶ documentation – records required to ensure the process is correctly documented.

▲ Figure 11.18 Hierarchical SOP

Flowchart SOPs

Flowchart SOPs represent a process or workflow from start to finish in diagrammatic form. They provide a visual step-by-step approach to completing a process.

Test yourself

What is a hierarchical SOP used for?

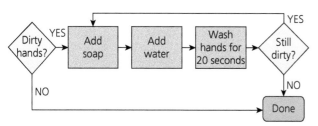

▲ Figure 11.19 Flowchart SOP

Production of SOPs

Figure 11.20 outlines how SOPs are produced.

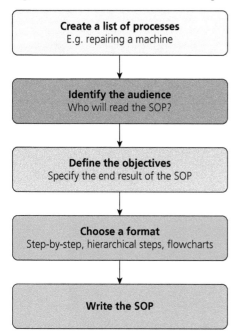

▲ Figure 11.20 SOP production

Implementation of SOPs

When writing a SOP, it is useful to create a rough draft first, so that it can be edited.

A SOP for a maintenance task might use the following format:
▶ title and document number
▶ equipment and materials required
▶ potential hazards
▶ skills required

▶ employees required
▶ time required to complete the task
▶ how often the task should be completed
▶ the steps to follow
▶ what feedback will be required.

Once the document has been completed and reviewed, some space should be left for approval signatures and dates.

Evaluation of SOPs

Once completed, SOPs need to be evaluated. They should be reviewed and approved by employees from all stages of the production process.

SOPs should be tested, and any feedback should be used to review the SOPs and amend them if necessary.

▲ Figure 11.21 SOPs should be reviewed by employees from all stages of the production process

Purposes

SOPs aim to achieve:
▶ standardisation of activity, by allowing processes to be completed in a prescribed way and eliminating the need for any guesswork
▶ customer satisfaction, by ensuring products are produced to the same specification in the same way
▶ safety and training, by ensuring employees know the correct methods and procedures to reduce accidents
▶ quality output, by guaranteeing the work is always done in a predefined and optimised way
▶ reduction of miscommunication, by ensuring employees do not need to ask around or look through documentation to get answers.

Case study

A producer of industrial injection-moulding equipment wanted to improve product quality. Management decided to implement a quality policy that focused on customer satisfaction.

The quality policy plan involved making improvements to the production line by using statistical methods to better measure, monitor and control the manufacturing processes. To improve the quality culture within the organisation, each employee played a particular role to support the initiatives.

The quality initiatives included:
- improving measurement precision when testing
- reducing production cycle times for each manufactured component
- lowering wastage costs, particularly for scrap and rework
- making inspections faster by using sampling methods.

To achieve these objectives, the organisation needed to monitor quality data continually and use data analysis to make any required process adjustments. Management therefore introduced statistical process control (SPC) into the organisation's culture, with every employee using data to improve the manufacturing processes.

The organisation was able to measure significant quality improvements. For example, instead of machinists making their own personal adjustments to the machines, the SPC data allowed machine adjustments to be limited, thereby saving time. The organisation lowered wastage costs and this decreased production cycle times by an average of 20 per cent per component, generating annual savings of approximately 3034 production hours.

The organisation reported 51 per cent annual cost savings from lower wastage and up to an 11 per cent reduction in overall scrap and rework costs. More importantly, this affected customer satisfaction, as the quality improvements reduced warranty claims by 51 per cent.

Questions
1 What benefits did the company see from using SPC data instead of machinists making their own adjustments?
2 How does a culture of quality improve an organisation's attitude to product improvement?

Assessment practice

1 Which engineering body oversees welding?
2 Describe the difference between quality assurance and quality control.
3 What are the four stages of total quality management (TQM)?
4 What is ISO 9001?
5 Describe what is meant by traceability.
6 Identify three purposes of document version control.
7 What does the acronym FMEA stand for?
8 What are two advantages of quality circles?
9 What is the Pareto principle?
10 Explain how a step-by-step SOP is prepared.

Project practice

On receipt of several new orders, an engineering business wants to review how to improve the quality of its products. The new products will be manufactured in high volume and sold in Europe.

You are part of a new team that has been tasked with investigating and reporting on quality assurance and quality control requirements.
1 In a group, discuss the key differences between quality assurance and quality control.

2 Produce a report by:
  - researching which standards are relevant for exporting goods to Europe
  - researching the available approaches and methods to ensure quality improvement
  - considering 100 per cent sampling compared to statistical process control (SPC)
  - making a recommendation to the management team about your findings.

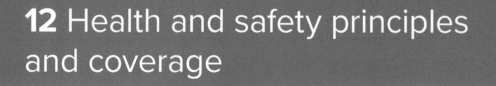

12 Health and safety principles and coverage

Introduction

This chapter focuses on the importance of health and safety in the engineering sector and how it might affect your activities in the workplace. It identifies key legislation and its application in order to manage risks and protect workers and businesses. We will look at the responsibilities of employees and employers, and explore how risk assessment is critical for preventing accidents and injuries.

The chapter closes by examining specific health and safety considerations for a range of engineering contexts and identifying how an organisation must respond to environmental legislation.

Learning outcomes

By the end of this chapter, you will understand:
1 the main requirements of the current key health and safety legislation, how to access it and how it affects your own activities in the workplace
2 that the legislation should be satisfied by your company's safe systems of work and other procedures, and you therefore do not need to know every detail of the law
3 the purpose of health and safety legislation within the engineering industry
4 how health and safety legislation affects the frequency of accidents and related incidents
5 the importance of mental health and wellbeing in the workplace
6 the persons responsible for ensuring compliance
7 the implications of non-compliance
8 the health and safety responsibilities of employees
9 the health and safety responsibilities of employers
10 the differences between local, national and global requirements
11 the hazards associated with engineering and manufacturing contexts
12 common industrial injuries that can occur when the appropriate precautions are not taken
13 methods of identifying hazards
14 how to evaluate risks
15 the hierarchy of control for control measures
16 the types of control measures typically used in engineering
17 the different health and safety considerations appropriate to a range of engineering contexts
18 the main requirements of the current key environmental legislation, how to access it and how it affects your own activities in the workplace
19 that the legislation should be satisfied by your company's environmental policies and other procedures, and you therefore do not need to know every detail of the law
20 the purpose of environmental legislation within the engineering industry
21 the aims, benefits and consequences of ISO 14001
22 methods of waste disposal and their implications.

12.1 The main requirements of key health and safety legislation applicable to engineering activities

Health and safety legislation exists to safeguard the health, safety and welfare of employers, employees and anyone else who may be exposed to **hazards** due to work activities. It ensures that the **risk** of harm is eliminated or minimised by placing obligations on various **duty holders**.

According to the Health and Safety Executive (HSE), in the UK, on average, over 200 people are killed each year in workplace **accidents**, and over one million are injured. In 2020/21, when millions of workers were furloughed or working from home during the Covid-19 pandemic, the figures reported by HSE were lower than average:

▶ 142 workers were killed at work
▶ 441,000 workers sustained a non-fatal injury at work
▶ 51,211 non-fatal injuries to employees were reported under the Reporting of Injuries, Diseases and Dangerous Occurrences Regulations (RIDDOR).

▲ Figure 12.1 On average, over one million people are injured in workplace accidents in the UK each year

Key terms

Hazards: anything with the potential to cause harm.

Risk: the likelihood that a hazard will cause harm.

Duty holders: people with a legal responsibility.

Accidents: unplanned, undesired occurrences which may result in harm.

The UK engineering industry is subject to a number of laws which cover many different areas, including health, safety and welfare. Legislation comprises Acts of Parliament and regulations (statutory legislation) which are mandatory and must be complied with in order to avoid prosecution.

However, there are also many guidance documents which are not legally enforceable but offer advice on best practice and how to comply with the law. These are usually created by authorities or government agencies to achieve a particular outcome.

Legislation

Companies must follow health and safety legislation to protect their employees. Changes to legislation are published via government agencies, such as www.legislation.gov.uk, and regulatory bodies, such as HSE (www.hse.gov.uk). Companies can subscribe to an **RSS feed** to receive notifications about new legislation as it is published on these websites.

Key term

RSS feed (Really Simple Syndication feed): a web-based application that notifies users of any updates on a selected website.

According to the HSE, a company's safe system of work must ensure that:

▶ all risks that affect employees and other people that visit the company are assessed
▶ all preventative and protective measures are planned, controlled, monitored and reviewed on a regular basis
▶ all employees are consulted about the risks within their work areas and about the preventative and protective measures in use
▶ a written health and safety policy is provided if five or more people are employed by the company.

Health and Safety at Work etc. Act 1974

Under the Health and Safety at Work etc. Act (HASAWA) 1974, everyone is responsible for health and safety at work, including employers, employees, self-employed people and those in control of work premises. Specific responsibilities are discussed later in this chapter.

The main objectives of HASAWA are to:

▶ secure the health, safety and welfare of people at work

▶ protect people other than those at work (for example visitors or the general public) from risks to health or safety caused by work activities

▶ control the possession and use of explosive, highly flammable or otherwise dangerous substances.

▲ Figure 12.2 HASAWA aims to control the use of highly flammable substances, such as those used in welding

Regulations

Legislation made under HASAWA is divided into a number of different regulations, which aim to cover every aspect of the workplace. The main regulations that control health, safety and welfare in the engineering sector include:

▶ Management of Health and Safety at Work Regulations (MHSAWR) 1999

▶ Provision and Use of Work Equipment Regulations (PUWER) 1998

▶ Personal Protective Equipment (PPE) Regulations 1992

▶ Control of Noise at Work Regulations 2005

▶ Manual Handling Operations Regulations (MHOR) 1992 (amended 2002)

▶ Lifting Operations and Lifting Equipment Regulations (LOLER) 1998

▶ Work at Height Regulations (WAHR) 2005 (amended 2007)

▶ Electricity at Work Regulations 1989

▶ Control of Electromagnetic Fields at Work (CEMFAW) Regulations 2016

▶ Reporting of Injuries, Diseases and Dangerous Occurrences Regulations (RIDDOR) 2013

▶ Control of Substances Hazardous to Health (COSHH) Regulations 2002 (amended 2004).

Management of Health and Safety at Work Regulations (MHSWR) 1999

The Management of Health and Safety at Work Regulations 1999 apply to every work activity and outline the duties of employers and employees for maintaining workplace health and safety.

Under this legislation, the main requirement for employers of five or more employees is to undertake a **risk assessment** and record significant findings. They must also:

▶ implement any health and safety measures required by the risk assessment

▶ provide appropriate health surveillance for employees

▶ appoint competent people to support them in implementing health and safety measures

▶ establish emergency procedures

▶ ensure employees are provided with adequate health and safety information and training

▶ cooperate with other employers sharing the same workplace and coordinate approaches for managing health and safety.

Employees are required to:

▶ use machinery, equipment, dangerous substances, transport equipment, means of production and safety devices in accordance with training and instruction

▶ report work situations which pose a serious and imminent danger to health and safety

▶ report shortcomings in their employer's protection arrangements for health and safety.

> **Key term**
>
> **Risk assessment:** systematic process of identifying and evaluating hazards in order to determine whether sufficient control measures are in place to prevent harm.

Provision and Use of Work Equipment Regulations (PUWER) 1998

These regulations place duties on people and companies who own, operate or have control over work equipment.

According to the HSE, PUWER requires that equipment provided for use at work is:

▶ suitable for the intended use

▶ safe for use, maintained in a safe condition and inspected to ensure it is correctly installed and does not subsequently deteriorate

- used only by people who have received adequate information, instruction and training
- accompanied by suitable health and safety measures, such as protective devices and controls. These will normally include guarding, emergency stop devices, adequate means of isolation from sources of energy, clearly visible markings and warning devices
- used in accordance with specific requirements, for mobile work equipment and power presses.

Personal Protective Equipment at Work Regulations 1992

These regulations require that **personal protective equipment (PPE)** is provided by the employer when health and safety risks cannot be controlled by other means. They also state that PPE must be:
- fit for purpose
- correctly stored
- properly maintained
- used properly by employees.

▼ Table 12.1 Examples of personal protective equipment (PPE)

Workplace hazards	PPE required
Chemicals	Gloves, overalls, respirator, safety glasses, safety boots
Welding	Welding mask and filtered lens, fire resistant gloves, leather apron, overalls, safety boots
Equipment with moving parts	Safety glasses and hearing protection where needed

Key term

Personal protective equipment (PPE): clothing or equipment designed to protect a user from workplace hazards.

Control of Noise at Work Regulations 2005

Exposure to excessive noise in the workplace can lead to temporary or permanent hearing damage, including tinnitus (ringing in the ears) and hearing loss. The Control of Noise at Work Regulations 2005 require employers to carry out a risk assessment if noise levels reach 80 decibels, in order to eliminate or control exposure.

Control measures may include:
- provision of PPE (for example earplugs or ear defenders)
- provision of noise-control equipment
- the creation of hearing protection zones if noise levels exceed 85 decibels (daily or weekly average exposure).

Employers should provide hearing checks, information and training for employees at risk.

▲ Figure 12.3 Employees should be provided with appropriate PPE if there is a risk of hearing damage

Manual Handling Operations Regulations 1992 (amended 2002)

The Manual Handling Operations Regulations (MHOR) 1992 (as amended) define manual handling as 'any transporting or supporting of a load ... by hand or bodily force'. This covers activities such as lifting, lowering, pushing, pulling, moving or carrying a load.

Poor manual handling can cause injuries such as joint and musculoskeletal problems. Typical injuries include:
- cuts and bruises
- broken bones
- muscular sprains and strains.

More serious accidents may occur when loads are not correctly handled, including crush injuries, slips, trips, falls and even fatalities.

Where possible, employers should avoid the need for hazardous manual handling tasks. If they cannot be avoided, employers must conduct a risk assessment that takes into consideration the task to be carried out, the capability of the individual performing the task, the load being handled and the environment in which the handling takes place. It is then important to implement appropriate control measures, such as planning tasks in advance and using correct posture.

Employees are responsible for making use of any system of work provided by their employer, for example mechanical aids.

Lifting Operations and Lifting Equipment Regulations (LOLER) 1998

Under this legislation, employers are responsible for controlling risks associated with **lifting operations** and lifting equipment in the workplace. Lifting equipment includes overhead cranes and their supporting runways, motor vehicle lifts, vehicle tail lifts and cranes fitted to vehicles, goods and passenger lifts, and telehandlers and forklifts.

> **Key term**
>
> **Lifting operation:** an operation concerned with the lifting or lowering of a load.

▶ Every lifting operation should be properly planned by a competent person, appropriately supervised and carried out in a safe manner.
▶ Employers must ensure that lifting equipment is of adequate strength and stability for each load, and that the load itself and anything attached to it for lifting purposes is of adequate strength.
▶ Equipment for lifting people must protect them from being crushed, trapped or struck, or from falling.
▶ All lifting equipment must be installed in such a way as to eliminate or reduce the risk of the equipment or load striking a person or of a load drifting, falling or being released unintentionally.
▶ Machinery and accessories for lifting loads must be clearly marked to indicate their safe working loads.

Lifting equipment can also be classified as work equipment, so PUWER will apply, requiring:
▶ inspection, maintenance and record keeping
▶ statutory periodic 'thorough examination'.

Work at Height Regulations (WAHR) 2005 (amended 2007)

The HSE defines working at height as working in any place where, if there were no precautions in place, a person could fall a distance liable to cause personal injury.

If possible, work at height should be avoided. If it cannot be avoided, an employer should:
▶ carry out a risk assessment and identify control measures to minimise the risk of injury
▶ plan the task carefully

▶ choose appropriate access equipment
▶ select competent people to carry out the task
▶ implement control measures to reduce the risk of falling, for example guard rails, toe boards and barriers
▶ ensure the use of appropriate PPE, such as harnesses
▶ inspect equipment regularly.

▲ Figure 12.4 If work at height cannot be avoided, it should be risk assessed to minimise the risk of injury

Electricity at Work Regulations 1989

All uses of electricity within the workplace are regulated under the Electricity at Work Regulations 1989. The regulations are designed around the prevention of danger and, in order to comply with the legislation, employers must ensure that:
▶ only trained and competent employees work on or with electrical equipment
▶ all electrical systems are designed to prevent danger
▶ electrical systems are maintained and fixed installations are inspected every 5 years
▶ any electrical work is carried out safely.

▲ Figure 12.5 Employers and employees must follow strict guidelines when working on or with electrical equipment

Control of Electromagnetic Fields at Work (CEMFAW) Regulations 2016

These regulations require employers to make a suitable and sufficient assessment of the levels of electromagnetic fields (EMFs) to which their employees may be exposed, and to ensure that exposure remains below exposure limit values (ELVs). If a risk assessment identifies that employees are exposed to EMFs in excess of ELVs, employers must implement control measures to eliminate or minimise risks to health.

Some engineering testing equipment uses EMFs:
▶ X-rays (for example x-ray machines are used to test the internal parts of aircraft engines)
▶ gamma rays (for example gamma rays are used to find weak spots in welded structures)
▶ some higher-energy ultraviolet (UV) rays (for example UV rays are used to carry out detailed surface inspections of materials).

Reporting of Injuries, Diseases and Dangerous Occurrences Regulations (RIDDOR) 2013

Under RIDDOR regulations, employers, self-employed individuals and the 'responsible person' in charge of work premises are required to report and keep records of workplace injuries, diseases and 'dangerous occurrences', including:
▶ deaths or serious injuries resulting from work-related accidents (reportable injuries)
▶ diagnosed diseases associated with the occupation, such as Carpal Tunnel Syndrome, occupational asthma, tendonitis and cancer
▶ 'dangerous occurrences' (incidents that could cause death or serious injury).

Control of Substances Hazardous to Health (COSHH) Regulations 2002 (amended 2004)

COSHH regulations are designed to protect workers from illness while working with certain hazardous substances and materials. Employers or employees who fail to comply with COSHH regulations risk an unlimited fine.

A hazardous substance is any substance that poses a direct or indirect risk of harm to humans. Within engineering and manufacturing, hazardous substances may be found in fumes, gases, dusts and products containing chemicals.

COSHH requires employers to:
▶ carry out risk assessments for work involving hazardous substances

▶ prevent or control exposure to hazardous substances by implementing and maintaining control measures
▶ monitor exposure to hazardous substances
▶ provide health surveillance for employees who may be exposed to hazardous substances
▶ provide information, instruction and training to employees who may be exposed to hazardous substances
▶ make arrangements for responding to accidents, incidents or emergencies involving hazardous substances.

Employees must safeguard the health and safety of themselves and others by following any procedures or control measures put in place by their employer when working with hazardous substances, including wearing the correct PPE and managing contaminated PPE appropriately. All accidents, spillages and breakages must be reported immediately. It is important for employees to keep up to date with any training provided.

▲ Figure 12.6 The COSHH Regulations protect employees from ill health when working with hazardous substances

Test yourself

1 Which Act of Parliament makes health and safety everyone's responsibility?
2 Give three examples of regulations that control health, safety and welfare in the engineering sector.

Health and safety culture

The overall standard of health and safety in a workplace is defined by both the safety management system and the health and safety culture – that is, the shared values, beliefs, expectations and attitudes about how to behave safely. The ACSNI Human Factors Study Group found that this culture is crucial in order to implement, maintain and improve an organisation's health and safety management system.

> Organisations with a positive safety culture are characterised by communications founded on mutual trust, shared perceptions of the importance of safety and by confidence in the efficacy of preventive measures.
>
> Source: ACSNI Human Factors Study Group (HSC, 1993)

The Health and Safety at Work Act 1974 states that health and safety training must be provided by the employer to ensure that employees carry out their roles safely and effectively.

12.2 The importance of health and safety practices within the workplace

The impact of legislation on accidents and related incidents

Following health and safety legislation will lead to:
- a better company reputation
- increased levels of productivity
- less staff turnover
- lower employee absence
- fewer accidents (and in turn a reduced risk of legal claims)
- reduced insurance premiums.

The most recent data from the European Statistical Office (Eurostat) shows that in 2018 (before the UK left the EU), the UK had one of the lowest rates of fatal workplace injury compared with other large EU economies, with a standardised incidence rate of 0.61 per 100,000 employees. This demonstrates the effectiveness of the UK's health and safety legislation.

RIDDOR reports allow the HSE to record statistics of the kind of accidents, fatal and non-fatal, that commonly occur in the workplace.

▼ Table 12.2 Non-fatal injuries by accident type, 2020/21

Kind of accident	Percentage of all accidents reported by employers (%)
Slips, trips or falls on same level	33
Injured while handling, lifting or carrying	18
Struck by a moving object	10
Acts of violence	8
Fall from a height	8

Source: Kind of accident statistics in Great Britain (HSE, 2021)

This legislation allows the HSE to assess the frequency of accidents and related injuries within organisations, and to review whether current regulations are still effective.

Standardised incidence rate per 100,000 employees

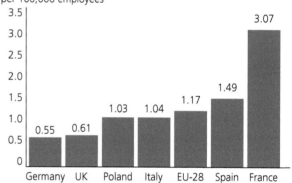

▲ Figure 12.7 Fatal workplace injuries in large EU economies (Eurostat 2018)

The importance of mental health and wellbeing in the workplace

Mental health is about emotional, psychological and social wellbeing. It affects how we think, feel and behave.

The most common mental health problems are anxiety and depression. These can be caused by:
- traumatic events, such as serious injury or the threat of serious injury or death
- challenging life events, such as bereavement or divorce
- work-related stress.

▼ Table 12.3 Causes of stress in the workplace

Area	How employees might feel
Demands	Employees are not able to cope with the demands of their jobs
Control	Employees are unable to control the way they do their work
Support	Employees don't receive enough information and support
Relationships	Employees are having trouble with relationships at work or are being bullied
Role	Employees don't fully understand their roles and responsibilities
Change	Employees are not engaged when a business is undergoing change

Source: www.hse.gov.uk/stress/causes.htm

According to HSE statistics, in 2020/21 822,000 workers suffered from work-related stress, depression or anxiety. Employers have a duty of care to manage and prevent stress by improving conditions at work. Under equality legislation, they are also required to make reasonable adjustments to manage an employee's mental health, for example by allowing them to take more rest breaks.

Ensuring compliance with health and safety legislation

HASAWA relates to all people at work. Duty holders include employers, employees, the self-employed, suppliers of goods and substances for use at work, persons in control of work premises, persons who manage and maintain work premises, and persons in general. The legislation is specifically aimed at people and their activities at work, rather than premises or processes. It includes provisions for both the protection of people at work and members of the general public who may be at risk as a consequence of workplace activities.

The two main branches of law in the UK are criminal law and civil law, and these both apply to health and safety:

▶ Criminal law is concerned with maintaining law and order and protecting society. The HSE or a local authority may take action against an employer who breaches health and safety legislation.
▶ Civil law is concerned with upholding the rights of individuals who have suffered harm. For example, someone injured in a workplace accident may make a claim for compensation against an employer.

Enforcing authorities

The Health and Safety Executive (HSE) and local authorities (LAs) are responsible for enforcing health and safety legislation in Great Britain. Local authorities have responsibility for retail, hotel and catering sites, wholesale distribution and warehousing, offices and consumer/leisure sites.

Health and Safety Executive (HSE)

HSE inspectors have powers to enforce health and safety legislation, for example:

▶ carrying out workplace inspections
▶ issuing informal cautions and formal legal notices to duty holders
▶ initiating prosecutions.

Inspection visits can be unannounced, as long as they take place at a reasonable time. During a visit, the inspector may:

▶ examine the workplace
▶ speak to employees
▶ offer guidance and advice on work or processes
▶ check equipment and machinery
▶ take samples, photographs and measurements
▶ make copies of documentation pertinent to their investigation
▶ seize evidence, equipment or materials.

▲ Figure 12.8 An HSE inspector may check equipment and machinery

Implications of non-compliance

When health and safety risks are not managed and legislation is breached, the HSE can take enforcement action against individuals or companies.

If a minor problem is identified during an inspection, the inspector may offer informal verbal or written advice and allow the duty holder a chance to resolve the issue before formal action is taken.

Where the issue is more serious and HASAWA has been breached, an inspector can:

▶ issue an improvement notice, which outlines what action must be taken in order to comply with the law, why and in what timeframe (usually a minimum of 21 days, to allow the duty holder enough time to appeal to an industrial tribunal)

▶ issue a prohibition notice to stop (either immediately or after a specified period of time) an activity that presents a risk of serious personal injury, which will not be lifted until corrective action has been taken

▶ initiate legal proceedings in the criminal court.

Prosecution may be considered necessary in cases where a duty holder fails to comply with a legal notice. The penalties are unlimited fines, imprisonment for up to two years, or both.

In the case of a death resulting from a work activity, a manslaughter investigation will be considered and investigated by the police, with assistance from the HSE.

▲ Figure 12.9 Prosecution may be considered necessary in cases where a duty holder fails to comply with a legal notice

Test yourself

What powers does the HSE have to enforce HASAWA?

Research

Visit the HSE website at **www.hse.gov.uk/simple-health-safety/law/index.htm**.

Find, read and check that you understand the information provided by the HSE for guidance on health and safety: criminal and civil law in the workplace.

What would happen if a company did not comply with the law?

Is it a criminal offence for an employer not to have employers' liability insurance?

Case study

Poor lighting in a loft space resulted in a gas engineer falling through the ceiling. The employer was found guilty of not providing adequate equipment for the employee to work safely, which led to him sustaining several serious soft-tissue injuries.

The claim was settled for £10,000, of which the employer only paid £250 due to having insurance.

Questions

1 Why was the employer found guilty of not protecting its employee?
2 Which regulations had the employer breached?
3 Investigate other examples of times when engineering companies have been fined for not following health and safety legislation. How does this affect a company's reputation?

Improve your maths

An engineering company has been prosecuted under HASAWA for not providing adequate hearing protection for its workers. It has been fined 20 per cent of its turnover of £250,000 and ordered to pay £10,366 in costs.

How much has the company been fined in total?

12.3 Responsibilities for health and safety

Employee responsibilities

Under HASAWA, while at work, employees are required to:

▶ take reasonable care of the health and safety of themselves and others who may be affected by their acts or omissions at work

► cooperate with their employer regarding health and safety policy
► not to intentionally or recklessly interfere with, or misuse, anything provided in the interests of health, safety or welfare.

▲ Figure 12.10 Employees are required to take reasonable care of the health and safety of themselves and others affected by work activities

Employer responsibilities

Under HASAWA, every employer must ensure, so far as is reasonably practicable, the health, safety and welfare at work of all their employees.

They must also:
► provide and maintain equipment and work systems that are safe and without risks to health
► make arrangements for the safe use, handling, storage and transport of substances
► provide information, instruction, training and supervision to ensure the health and safety of employees
► maintain a safe and healthy workplace with safe entrances and exits
► provide a safe working environment and adequate welfare facilities
► prepare a written health and safety policy if the company has more than five employees.

In addition, MHSWR 1999 tells us that all employers must:
► carry out risk assessments of all work activities
► identify and implement adequate control measures
► inform all employees of risk assessments and associated control measures
► review risk assessments at regular intervals
► make a record of risk assessments if there are five or more employees
► ensure the safety of visitors, contractors and members of the public.

Test yourself

What are the key responsibilities of employers and employees with regards to workplace health and safety?

Local, national and global requirements

Health and safety responsibilities vary locally and also at national and global levels.

Local

In the UK, health and safety legislation is enforced locally by either the HSE or local authorities, depending on the work activity:
► The HSE tends to be responsible for the manufacturing, agricultural, construction, medical and educational sectors, for example factories, farms, building sites, hospitals, nursing homes and schools.
► Local authorities tend to be responsible for the retail, wholesale distribution, hospitality, catering, beauty and leisure sectors, for example shops, offices, warehouses, hotels, pubs, restaurants, hotels, hairdressers, nail salons, health clubs, theatres and cinemas.

National

The HSE is the UK's national regulator for workplace health and safety, and provides advice, information and guidance to help manage risks at work. It engages with various stakeholders in order to develop strategies to improve health and safety awareness and to encourage changes in behaviour.

Global

According to the International Labor Organization, worldwide there are around 340 million workplace accidents and 160 million cases of occupational illness annually, corresponding to over 6000 deaths every day. Approaches to health and safety differ across the world.

Workplace health and safety is often not a priority in developing countries and often there is no enforceable legislation. Without legal protections, occupational injuries and diseases are commonplace, with those who live in poverty enduring unsafe working conditions. By contrast, in Europe, the US and Canada, health and safety tends to be strongly implemented.

▲ Figure 12.11 Workplace health and safety is often not a priority in developing countries

12.4 Risk assessment

Hazards associated with engineering and manufacturing

The main hazards associated with engineering and manufacturing relate to:
▶ equipment and tools: improper manual handling, unguarded machinery, excessive noise and vibration from tools and equipment
▶ stored energy: risk of electrical shock from non-isolated equipment
▶ electricity: defective or poorly maintained electrical equipment, improper use of electrical equipment
▶ harmful substances, including gases: exposure to substances such as metalworking fluids, degreasing solvents and dust or fumes from welding, brazing, soldering, coating and painting
▶ environments: poor housekeeping leading to slips, trips and falls; dangerous working environments (for example working at height or in confined spaces).

▲ Figure 12.12 Contact with electricity is a main cause of injury in the engineering industry

Improve your English

Your company is auditing its health and safety provision. Write a short report on the most common industrial injuries that could occur in your workshop.

A risk assessment is essential in order to identify which hazards exist in the workplace and to put in place control measures to eliminate or reduce the risk of harm.

As an engineer, you must assess the risk to yourself and others when work is being carried out. This is at the heart of working safely. After a risk assessment has been completed, it will, as a minimum, result in the use of personal protective equipment (PPE) and the presence of a first-aid kit. However, the more risk that is identified in the risk assessment, the more mitigation processes that need to be put in place to protect the operative and anyone else that may be affected by the work being completed.

Stages of risk assessment

The HSE recommends a five-step risk assessment process:
1 Identify hazards.
2 Decide who might be harmed and how.
   Who will have access to the workplace, equipment and processes and what hazards will they be exposed to?
3 Evaluate the risks and decide on control measures.
4 Record any findings (when there are five or more employees).
   Use a risk assessment document that contains information on hazards, risks and how risks are controlled.
5 Review the risk assessment and update if necessary. This is to ensure that the risk assessment is still effective and not out of date. For example, changes have been made to staff or equipment used.

Hazard identification (HAZID)

HAZID is the first stage in risk assessment. It is used to identify the activities that have the potential to cause harm.

The most common methods used to spot hazards in the workplace are:
▶ inspections
▶ talking to people involved in activities
▶ analysing accident records
▶ looking at manufacturer's instructions for machines
▶ reviewing Material Safety Data Sheets (MSDS).

Hazard and operability study (HAZOP)

HAZOP is an approach to managing risk which looks at specific processes or operations, such as a machine or automated processes, to identify potential hazards and operational issues, including those which could affect the quality of the final product.

▲ Figure 12.13 HAZOP process diagram

For each hazard identified, consideration should be given to injuries that might be sustained by employees, contractors, visitors or the general public. Some employees may need special consideration, such as young people, migrant workers, pregnant workers and disabled workers.

Evaluation of risks

Once hazards have been identified, it is important to consider the level of risk each one presents.

A common method for assessing the level of risk is to attribute scores to the likelihood of harm being caused and the severity of that harm. These scores are multiplied together to produce an overall risk rating, which can then be read on a risk matrix.

For example, likelihood may be scored 1–5, where:
▶ 1 represents highly unlikely
▶ 2 represents unlikely
▶ 3 represents possible
▶ 4 represents likely
▶ 5 represents highly likely.

Consequences may also be scored 1–5, where:
▶ 1 represents insignificant (for example an injury where no first aid is required)
▶ 2 represents minor (for example an injury where minor first aid is required)
▶ 3 represents moderate (for example a serious injury that requires medical assistance and must be reported)
▶ 4 represents major (for example a serious injury requiring urgent medical assistance or hospitalisation)
▶ 5 represents catastrophic (usually a fatality).

These figures can then be used in the risk matrix in Figure 12.14 to determine the level of risk as low (green), medium–low (yellow), medium–high (orange) or high (red).

The Management of Health and Safety at Work Regulations 1999 legally require employers to make decisions about all hazards faced in the workplace, whether or not they are significant.

Consequences/impact	Catastrophic	5	5	10	15	20	25
	Major	4	4	8	12	16	20
	Moderate	3	3	6	9	12	15
	Minor	2	2	4	6	8	10
	Insignificant	1	1	2	3	4	5
			1	2	3	4	5
			Rare	Unlikely	Possible	Likely	Almost certain
			Likelihood/probability				

▲ Figure 12.14 A risk matrix

Hierarchy of control

A risk assessment should identify control measures that either eliminate a hazard or reduce the risk of it causing harm. Reference should be made to the hierarchy of control, which organises control measures according to how effective they are at reducing risk.

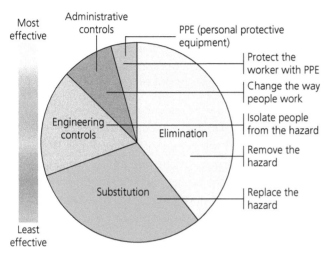

▲ Figure 12.15 Hierarchy of control

▲ Figure 12.16 PPE should only be used as a last resort, when hazards cannot be controlled by other means

Typical control measures in engineering

Typical control measures used in engineering include:
► machine guarding
► machine isolation
► PPE:
  – eye protection
  – safety shoes
  – ear protection
  – gauntlets
  – helmets.

Test yourself

Explain the importance of risk assessment in engineering workplaces.

Research

Visit **www.hse.gov.uk/simple-health-safety/risk**.

Read and check you understand the information provided by the HSE on managing risks and risk assessment at work.

What considerations should be made for vulnerable workers?

Recording findings

If a business employs five or more people, it must record the findings of its risk assessments. The HSE recommends that records show:
► who is at risk from the hazard and in what way are they at risk
► how the business is currently controlling the risk
► additional action needed in order to control the risk
► who is responsible for performing this additional action
► when the additional action should be completed.

▼ Table 12.4 Excerpt from a risk assessment

Hazard	Who might be harmed and how	Existing control measures	Risk rating		
			Likelihood	Consequence	Rating
Work piece/waste/chuck key ejected from machine	Fitter and others Cuts to hands, bruises, eye damage	Work is adequately secured (small work is mechanically secured) Machine vice and hand vice are kept on machine Safe working area is demarcated Adequate guarding Eye protection is worn	2	3	6

Reviewing and revising the risk assessment

It is important to review a risk assessment to verify that the controls in place are still working. Workers may have identified new issues since the risk assessment was carried out, and accidents or incidents may have occurred which mean further controls need to be implemented. Changes to staff, processes or the equipment used may all affect the level of risk. The risk assessment record should be updated accordingly following a review.

12.5 Health and safety considerations in specific engineering contexts

Engineering has many processes and operations that require consideration to ensure that health and safety requirements are met.

Safe systems of work

A safe system of work (SSoW) is a set of procedures designed to eliminate or minimise the risks involved in specific operations. It is essential that everyone on site adheres to these systems, so that they are all working to the same high safety standards at all times.

An SSoW may be necessary if it is not possible to completely eliminate the hazards associated with a particular activity, if there have been safety issues relating to an activity, or if staff have raised safety concerns. Changes to the law regarding the performance of an activity may also bring about the need for an SSoW.

The following information should be recorded in an SSoW:
▶ a detailed description of the steps involved in the activity
▶ a list of the potential hazards at each stage of the activity
▶ an explanation of the controls in place at each stage of the activity
▶ the name of the person responsible for the creation and implementation of the SSoW
▶ a date for review of the SSoW.

All staff involved in the performance of the activity should be trained to follow the SSoW. As with a risk assessment, the SSoW should be reviewed to make sure the procedures in place are still appropriate and that they are being followed correctly.

Permit-to-work

According to the HSE:

> Where proposed work is identified as having a high risk, strict controls are required. The work must be carried out against previously agreed safety procedures, a 'permit-to-work' system.
>
> The permit-to-work is a documented procedure that authorises certain people to carry out specific work within a specified time frame. It sets out the precautions required to complete the work safely, based on a risk assessment. It describes what work will be done and how it will be done; the latter can be detailed in a method statement.
>
> The permit-to-work requires declarations from the people authorising the work and carrying out the work.

Source: www.hse.gov.uk/coshh/basics/permits.htm

Examples of when a permit-to-work is required include:
▶ when completing 'hot works', for example welding/plasma cutting
▶ when working in confined spaces, for example in a tank
▶ when lone working, for example in a sewer/tank
▶ when maintaining machinery and equipment, for example maintenance of a disc cutter.

Test yourself

Explain, using examples, when a permit-to-work would be required in the engineering workplace.

Research

Visit **www.hse.gov.uk/coshh/casestudies/welder.htm**.

Read and check you understand why the welding company was not complying with regulations.

What should have been included in the company's risk assessments?

Oxygen use in the workplace

Oxygen is a very reactive gas that poses fire and explosion hazards in the workplace. When pure oxygen is at high pressure, for example from a cylinder, it can react destructively with materials such as oil and grease. Fires also burn hotter and faster in the presence of oxygen.

Some materials, such as oily rags, may catch fire spontaneously due to oxidation.

According to the HSE, the main causes of fires and explosions when using oxygen in the workplace are:
▶ oxygen enrichment (oxygen in the air increasing) from leaking equipment
▶ use of materials not compatible with oxygen
▶ use of oxygen in equipment not designed for oxygen service
▶ incorrect or careless operation of oxygen equipment.

Test yourself

Describe the risks of using oxygen in the workplace.

Fire and explosion hazards

In engineering, the following potential causes of fire and explosion need to be considered:
▶ incorrect equipment or material specification
▶ defective material or equipment
▶ pressure outside design limits
▶ temperature outside design limits
▶ vibration
▶ corrosion
▶ erosion
▶ human error.

It is also important to consider the storage and use of flammable or explosive substances, for example:
▶ grease, oil and gas
▶ process additives (for example methanol)
▶ fuels (for example diesel and aviation fuel) and lubricants
▶ bottled gas
▶ explosives and detonators
▶ chemicals
▶ ordinary combustibles.

Asphyxiation hazards

Asphyxiation is when the body has an insufficient supply of oxygen. It can lead to unconsciousness, brain damage and death.

Some gases used in engineering can cause asphyxiation:
▶ Simple asphyxiants, such as hydrogen, methane and nitrogen, are not poisonous but are often odourless and pose a hazard to health because they replace the oxygen in the air.
▶ Chemical asphyxiants, such as carbon monoxide and hydrogen sulphide, are toxic gases that can be fatal if breathed in. They are often colourless and sometimes odourless.

It is essential to have suitable ventilation systems in areas where there is a risk of asphyxiating gases accumulating. For example, in MIG (metal inert gas) welding, there must be a clean air supply to the work area to replace extracted air.

Heat

Some engineering processes involve heat, so it is important to consider the effects of heat on the body.

Heat stress is when the body is unable to regulate its internal temperature. It may be caused by increased air temperature, increased work rate, increased humidity and excess work clothing.

The HSE gives examples of workplaces where people might suffer from heat stress due to a hot working environment:
▶ glass and rubber manufacturing plants
▶ mines
▶ compressed air tunnels
▶ conventional and nuclear power stations
▶ foundries and smelting operations
▶ brick-firing and ceramics plants
▶ boiler rooms.

Moving parts

Before using machinery with moving parts, it is important to consider what the hazards may be and how they can be controlled. The HSE gives examples of how moving machinery parts can cause injuries:
▶ People can be struck and injured by moving parts of machinery or ejected material.

- Parts of the body can be drawn in or trapped between rollers, belts and pulley drives.
- Sharp edges can cause cuts and severing injuries.
- Sharp-pointed parts can cause stabbing or puncture injuries to the skin.
- Rough surface parts can cause friction or abrasion injuries.
- People can be crushed, both between parts moving together and towards a fixed part of the machine, wall or other object.

Injuries can also occur due to machinery being used by inexperienced or untrained operators, or due to machines being badly maintained and developing faults.

▲ Figure 12.17 Moving pulley drives can cause injury unless suitable control measures are in place

Fire safety

Fire safety is a major health and safety consideration in all engineering organisations. It is essential that fire risk assessments are carried out to make sure that any potential fire hazards are identified and controlled.

There are many possible sources of ignition of fire, including open flames, hot surfaces, friction, sparks and static electricity. The most common sources of fuel for fire are wood, paper, powder, foam, waste, chemicals or oil-based products. In addition to tanks of compressed oxygen, the air we breathe can also contribute to the ignition or fuelling of fires.

The fire triangle is a simple model for understanding the necessary ingredients for most fires.
- A fire naturally occurs when the elements are present and combined in the right mixture.
- A fire can be prevented or extinguished by removing any one of the elements in the fire triangle.

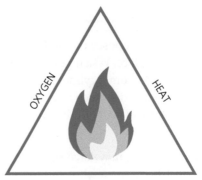

▲ Figure 12.18 The fire triangle

Some common causes of fire in engineering workplaces are:
- overloaded plugs, loose wires and faulty connections. Regular inspections should be carried out to avoid the risk of electrical fires
- dust and grease due to a build-up of clutter. Dirt and dust on machinery can also cause it to overheat. Regular cleaning and maintenance should be carried out
- combustible materials, such as paper, cardboard and wood, being incorrectly stored. These should be stored either in locked containers or off site, and they should be regularly disposed of.

Guarding

Guarding protects employees from the hazards presented by the operation of a machine where those hazards cannot be eliminated. It also minimises the chances of an accident happening due to electrical or mechanical failure, poor design or human negligence and malice.

The main types of machine guarding are:
- fixed: the guard is a part of the machine which is permanently installed and serves as a barrier. It does not require moving parts to function
- interlocking: the guard stops the machine from moving if it is opened or removed, causing the tripping mechanism and power to shut off
- adjustable: the guard can be adjusted to allow products of different sizes to be manufactured
- self-adjustable: the guard protects the operator by placing a barrier between the danger area and the operator. When the stock is moved into the danger area, the guard is pushed away, leaving only a small opening to allow the stock to pass through.

Manual handling

Most people working in the engineering industry will be involved in manual handling activities, where they are required to lift, lower, push, pull, carry or move a load. They must always receive training to ensure correct posture and handling techniques, thereby reducing their risk of injury.

Lock out, tag out (LOTO)

Lock out, tag out (LOTO) is a safety procedure that ensures the isolation of hazardous energy sources before maintenance work is carried out. These energy sources may be:

▶ electrical
▶ pneumatic
▶ mechanical
▶ hydraulic
▶ chemical
▶ radiation
▶ thermal energy.

Once the energy source is turned off, a safety lock is used to prevent it being turned on again (the energy source is 'locked out'). A tag is then attached to the device to warn others not to turn it on (the device is 'tagged out').

▲ Figure 12.19 Lock out, tag out is used to make sure hazardous energy sources are isolated before maintenance work begins

Maintenance

Under PUWER, employers must ensure that work equipment is 'maintained in an efficient state, in efficient order and in good repair' to ensure it does not deteriorate and put people at risk of harm. Safety-critical parts of machinery may also need more attention than other parts. Regular preventative maintenance tasks, such as inspecting, cleaning, lubricating and making minor adjustments, can ensure that machinery remains in a safe condition and avoids the need for costly production shutdowns.

Other health and safety considerations specific to engineering

▼ Table 12.5 Other health and safety considerations specific to engineering

Contexts	Employers should
Chemicals	Provide material safety data sheets (MSDSs) Train workers in safety procedures Provide PPE (gloves, overalls, respirator, safety glasses and boots)
Confined spaces	Provide adequate ventilation Provide PPE (welding mask or helmet with filtered lens, fire-resistant gloves, leather apron, overalls, boots) Establish safe working procedures Provide appropriate training and supervision
Electrical testing	Ensure equipment is regularly inspected, tested and tagged Use lock out, tag out (LOTO) Label faulty equipment, e.g. DANGER – DO NOT USE
High-voltage electrical (generation, distribution, isolation and storage)	Maintain recommended safety distance from exposed high voltage equipment Apply earthing devices only when danger from electrocution has been removed Establish safe working procedures Provide appropriate training and supervision

12.6 Principles and practices relating to environmental legislation and considerations

Environmental legislation

Environmental legislation seeks to protect the land, air, water and soil from the harmful effects of pollution. Employers are legally responsible for compliance.

In England, the main body responsible for developing environmental policy and legislation is the Department for Environment, Food and Rural Affairs (DEFRA).

Environmental Protection Act 1990

Waste created on site usually needs to be stored for an amount of time before it is collected, and there is a legal obligation to ensure this is done safely:

▶ Waste should be sorted into recyclable and non-recyclable materials, either on site or at a sorting facility.
▶ Waste must be collected by a registered waste carrier or taken directly to a permitted waste disposal facility.
▶ Waste transfer notes must be produced and retained for at least two years.

▲ Figure 12.20 Environmental legislation seeks to protect the land, air, water and soil from the harmful effects of pollution

Pollution Prevention and Control Act 1999

The Pollution Prevention and Control Act 1999 requires industrial activities with high pollution potential to have an environmental permit and to meet environmental conditions.

It covers such areas as:
▶ waste
▶ water quality
▶ air quality
▶ environmental noise.

Clean Air Act 1993

The Clean Air Act 1993 regulates smoke emissions, the height of chimneys and the content and composition of motor fuels.

Industrial premises are prohibited from emitting dark smoke from chimneys or flues. This also applies to the burning of materials on site.

▲ Figure 12.21 White smoke is permitted from chimneys as it is mostly harmless water droplets

Radioactive Substances Act 1993

The Radioactive Substances Act 1993 is concerned with the control of radioactive material and the disposal of radioactive waste.

Radioactive materials must be:
▶ accounted for so that the location is always known
▶ kept in suitable containers to protect against accidental damage or loss and to protect employees and the public
▶ inspected, and leak tests must be carried out every two years.

Controlled Waste (England and Wales) Regulations 2012

Household, industrial and commercial waste are classed as controlled waste and are subject to the Environmental Protection Act 1990. They are regulated due to their toxicity, hazardous nature, or ability to cause harm to human health or the environment. Sewage, sewage sludge and septic tank sludge are not controlled wastes.

Household and commercial waste includes litter and refuse while industrial waste may include building site and demolition waste. The waste must be collected or taken to local authority landfill or recycling plants.

Hazardous Waste (England and Wales) Regulations 2005

Waste is considered hazardous under environmental legislation if it contains substances or has properties that might make it harmful to human health or the environment. Examples of hazardous waste include asbestos, chemicals (such as brake fluid or printer-toner), batteries, fluorescent tubes and refrigerators.

The regulations ensure that any hazardous waste is only produced, moved, received and disposed of at registered sites. A registration process for producers of hazardous waste was introduced to control, track and record the movement of hazardous waste.

Dangerous Substances and Explosive Atmospheres Regulations (DSEAR) 2002

The Dangerous Substances and Explosive Atmospheres Regulations (DSEAR) 2002 set the minimum requirements to protect from the risks of fire, explosion and other similar events caused by dangerous substances.

According to the HSE:

> Dangerous substances are any substances used or present at work that could, if not properly controlled, cause harm to people as a result of a fire or explosion or corrosion of metal. They can be found in nearly all workplaces and include such things as solvents, paints, varnishes, flammable gases, such as liquid petroleum gas (LPG), dusts from machining and sanding operations, dusts from foodstuffs, pressurised gases and substances corrosive to metal.

Source: www.hse.gov.uk/fireandexplosion/dsear.htm

Employers are required to identify dangerous substances that are used or present in the workplace and put in place measures to control or remove the risks associated with them.

Employees must be properly trained in working with the dangerous substances and there must be plans in place in the event of accidents and emergencies relating to the substances.

Environmental considerations

ISO 14001

There has been an increase in environmental management in recent years. This is partly due to growing awareness of our impact on the environment but in business it may be primarily driven by cost savings and market opportunities.

ISO, the International Organization for Standardization, has published international standards to help businesses manage their environmental responsibilities.

> ISO 14001 is an internationally agreed standard that sets out the requirements for an environmental management system. It helps organisations improve their environmental performance through more efficient use of resources and reduction of waste, gaining a competitive advantage and the trust of stakeholders.

Source: International Organization for Standardization, ISO 14001

▲ Figure 12.22 Solar panels can help to improve an organisation's environmental performance

Waste disposal

▼ Table 12.6 Benefits and limitations of different methods of waste disposal

Method	Benefits	Limitations
Landfill (disposing of waste material by burying it)	Cheaper than recycling processes	Toxic chemicals can be produced Heavy metal ions can leach into the ground Non-biodegradable plastics may not be broken down Occupies a large area
Reuse (using all or parts of a product to make something else)	Cuts down single-use items, reduces waste	Reused products are often of lesser quality
Recycling (reprocessing a material or product to make something else)	Less waste goes to landfill Recycling materials can require less energy/carbon than producing new materials	Expensive Some wastes cannot be recycled Separation of useful material from waste is difficult
Controlled waste disposal (litter and refuse from household, commercial and industrial waste)	Cuts down on toxic waste going into landfill	Expensive Disposal of waste must take place at a licensed waste disposal facility

Industry tip

Health and safety documentation must be reviewed regularly and kept up to date. It is often referred to as controlled documentation. Always check with your supervisor that you have access to the latest versions of documents such as risk assessments.

Test yourself

Explain the benefits of different methods of waste disposal in the engineering industry.

Research

Visit **www.hse.gov.uk/fireandexplosion/dsear.htm**.

Find, read and check that you understand the information provided by the HSE for guidance on the DSEAR 2002 regulations.

What must an employer do when using dangerous substances?

Assessment practice

1. What are the main objectives of the Health and Safety at Work etc. Act (HASAWA) 1974?
2. Describe what is meant by a health and safety culture in the workplace.
3. What powers do HSE inspectors have to enforce health and safety legislation?
4. What are the three main responsibilities of employees under HASAWA?
5. Suggest three typical injuries in the engineering industry.
6. What are the five stages of risk assessment?
7. Describe what is meant by a safe system of work.
8. Why must oxygen use in the workplace be carefully controlled?
9. Describe how a lock out, tag out system works.
10. Suggest two benefits and two limitations of recycling waste.

Project practice

You have been asked to produce a safe system of work (SSoW) for one process in the manufacture of metal street signs. The process includes the drilling of six holes (8 mm Ø) into sheet steel (400 × 500 mm) on a pillar drill machine.

Your SSoW **must** include the following:
▶ a step-by-step description of the task (SOP)
▶ work holding devices, guarding and general machine safety requirements
▶ identification of hazards and control measures
▶ any PPE requirements.

13 Business, commercial and financial awareness

Introduction

Almost all engineering activity takes place in an industrial context. Engineering companies in the industrial sector work within a wide range of disciplines, including automotive, aeronautical, electrical, electronic and biomedical. Whatever the discipline, companies all follow similar business and financial practices.

To work effectively, a company needs managers and staff who understand how the world of business works. This chapter looks at the basics of the financial side of any business, with an emphasis on engineering contexts.

Learning outcomes

By the end of this chapter, you will understand:
1   the goals of commercial operations and how these are addressed
2   how organisations address the needs of different customers and markets
3   how organisations evaluate activities in terms of quality, cost and time
4   the role of research and development and innovation to address changing customer needs
5   how business practices influence the operation of engineering organisations
6   the legislation affecting tendering and contracts
7   the meaning of some financial concepts and their implications for the operation of a business.

13.1 Principles of commercial operations and markets

Some engineering companies are well known, such as Airbus, Aston Martin and the BT Group. However, there are tens of thousands of engineering companies in the UK, ranging in size from one or two people working in a small workshop to thousands of **employees** working in different locations around the world. The only qualification for being an engineering company is using, providing or producing engineering tools and products. Even Amazon can be regarded as an engineering company as it has a large engineering department and relies heavily on state-of-the-art engineering equipment.

Society tends to measure the value of everything in terms of money. This allows the movement of value in an agreed form from person to person, company to company, customer to customer and **employer** to employee. It is therefore important to understand the financial principles that underpin the operation of companies. While this knowledge may not be needed by an engineering apprentice in the early stages of their career, it becomes more important as their career progresses.

▲ Figure 13.1 It is important to understand the financial principles that underpin the operation of companies

Even if a company is set up for charitable or philanthropic purposes, its value is still measured in financial terms. Being an employer comes with certain responsibilities, and all businesses need to understand and abide by national and international legislation in order to be successful.

Key terms

Employees: people who work for a company in return for wages or a salary.

Employer: someone who sources the pay of an employee and controls the work they do.

Commercial priorities

To ensure success, a company also needs a clear understanding of its business purpose, that is, what it will make or provide. An important part of this is carrying out **market research** to gather information about customers' wants and needs. This may involve sending emails to customers, posting questionnaires, adding surveys to a website or carrying out focus groups. It is then critical to review the information and act on it.

If a product is popular with the market, it can be sold and generate **income**. As long as the **costs** of making a product or service are kept to a reasonable level, a company will make a **profit**. Some legal requirements for a company, for example the type of accounts that must be submitted, are based on **turnover**, which is the total amount of money received by a company from the sale of goods or services over a certain length of time, without taking costs into consideration.

A new business or **start-up** might seek funding from other organisations to help it purchase equipment, employ staff or market itself to customers. If the start-up is successful, it will grow and make a profit. It may need to take on more staff, increase its range of products or services, or even move to larger premises.

Key terms

Market research: collecting information about customers' wants and needs to help a business make decisions about design, production and marketing.

Income: money that comes into a company.

Costs: money required to make, store, sell or market a product.

Profit: what is left when costs are subtracted from income.

Turnover: total amount of money received by a company from the sale of goods or services over a certain period of time.

Start-up: a company that has been set up recently, often by someone with a new idea or product.

What are costs?

Costs are everything that a company has to spend money on. In business terms, they can be split into direct and indirect costs.

Direct costs include:

- raw materials used to make a product
- equipment or tools needed to make a product
- energy required to make a product
- costs related to selling a product
- wages of the personnel who make a product
- rent, if the factory is making only this product.

Indirect costs include:

- wages of personnel not directly involved with making the product, for example human resources, catering, cleaning, reception and office staff
- rent for a building for the whole company, if the company makes several products
- insurance, IT and IT support
- maintenance.

When budgeting, it can be easy to identify direct costs, but indirect costs must also be considered carefully and not overlooked.

▲ Figure 13.2 Direct costs include the wages of the personnel who make a product

What can a company do with the profit?

A company's profit could go directly to its owner. However, to ensure continued success, some or most of the profit should be used for the following:

- dividends for **shareholders**
- **bonuses** for key staff
- investment as **capital**
- investment in new tools and equipment
- innovation through a research and development budget.

Key terms

Shareholders: individuals or entities that have invested money in a company in exchange for shares in the company's stock.

Bonuses: extra money over and above the salary or wages of an employee.

Capital: money or valuable items that a business can use to generate income.

What is a stakeholder?

A stakeholder is someone who has an interest in a company and is impacted by its successes and failures.

When a company is new, it may only have a few stakeholders – the people who set up the business plus any suppliers or customers. As a company grows, it will take on more personnel and support staff (such as cleaners, caterers and receptionists) and gain more customers and suppliers. The number of stakeholders grows with the company.

A large company will have numerous stakeholders. Often, it will add several departments to the company structure and divide the business into smaller, more manageable sections.

Everyone with a connection to a company is considered a stakeholder. Some might be regarded as more important, such as the employees or the bank controlling the money, but in any business decision, all stakeholders should be taken into account, including the local county council and the people who live next door to the factory.

What is a customer?

A customer is anyone who buys (exchanges money for) a product or service. Engineering companies may be customers of other engineering companies who supply parts.

Market research should always identify the needs and wants of existing customers, as well as looking for new markets. Sometimes customers may not realise they need a company's new products, and the company will have to use skilled marketing to influence the existing client base.

▲ Figure 13.3 A customer is anyone who exchanges money for a product or service

Where are customers based?

In the past, customers for a new company would be found in the area local to the company. Since the introduction of the internet, customers can be found anywhere, nationally and internationally. This is, of course, good for the business, but customers from other cultures might have differing requirements, and this should be part of the market research that leads to product development.

▲ Figure 13.4 The internet has provided a platform for international trade

---

Case study

▲ Figure 13.5 Production of gears

Abdul, a machinist working at a machine tool manufacturer, found a gear hobbing machine for sale from a local company that was closing down. He bought the machine using a bank loan and set it up in a small industrial unit near his house.

He retired from his job at the machine tool company and then marketed his services back to the company, with the offer of making the gearbox gears needed for the lathes at a reduced cost but higher quality. His costs were lower because his indirect costs were

lower, but he had time to take more care when making the product.

Abdul's old company took him up on the offer and ordered gears from him. The larger company saved money because it could reduce the number of machines and staff working in its workshop. Abdul employed some of the staff from the old company and grew his business; soon he was able to start making gears for other machine tool manufacturers across the country.

In the end, Abdul was able to buy his old employer, and with the increased capacity he could make gears for companies across the world.

Questions

1 Choose an engineering company, maybe one you already have knowledge of, and investigate the way it started. See how it grew from small beginnings. Is the original owner or their family still involved in the company?

2 Abdul started his company with a bank loan. Research other ways he could have financed his start-up and think about how you could start a company of your own.

What is efficiency?

A company can measure its success in several ways, and one of the most important is efficiency.

Efficient companies make sure they are getting the best value for money from all aspects of their business. Here are some examples of how different departments contribute to efficiency.

An efficient sales department

A sales department would be wasting its time contacting people who would never be interested in a product. Instead, it should target potential customers who have:

- already registered an interest in a product by contacting the company directly
- previously bought a product
- been identified by market research.

These potential customers are known as leads. To be efficient at bringing in new business, a sales team should make sure it follows these leads.

An efficient production department

A production department must make sure it uses a company's resources in the best way. For example:

- A large piece of equipment that is expensive to purchase, run and maintain should be used as much as possible.
- Materials should be bought in the correct quantities and at a low price.

▲ Figure 13.6 Why would a curtain-sided truck help efficiency?

How can a transport department ensure it is being efficient when transporting finished products to a warehouse away from the manufacturing site?

Consider how different types of loads are transported, for example perishable items, electronic equipment or building materials. Does the quantity or weight of the goods require a variety of vehicles?

Efficiency and the avoidance of waste (the eight wastes)

Companies should strive to avoid wasting valuable resources and, in turn, money. The acronym TIMWOODS can be used to help identify and eliminate different types of waste found in processes:

- **T**ransportation: if goods or materials travel further than necessary, it wastes both fuel and time, and it causes wear and tear to vehicles.
- **I**nventory: keeping too much of anything in stock wastes space and possibly money.
- **M**otion: unnecessary movements of people or products result in wasted time or effort. The shorter the distance, the better and more efficient it is.
- **W**aiting: waiting for the next process to be carried out wastes time. Employees are paid even if they are not actually doing anything purposeful, so wasting time is wasting money.
- **O**ver-production: making too much of something suggests that production has not been planned, and there will not be a ready market for all the products. They will occupy space in the stores and may never be sold.
- **O**ver-processing: this is when a manufacturing process uses more materials, steps or activities than are actually required by the customer.
- **D**efects: mistakes cost time and money. An efficient company will spend time ensuring processes are as error-free as possible before production commences.
- **S**kills: this is about a waste of human knowledge and expertise. Employees should be fully utilised according to their skills, talents and abilities in order to avoid waste.

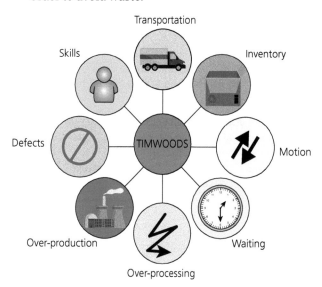

▲ Figure 13.7 TIMWOODS

What does 'value added' mean?

'Value added' refers to the increase in value of a product or service following development and production processes.

It could be argued that all engineering is the process of taking something simple and making it into something more complicated and more valuable.

A company such as Amazon does not really add value to its products, and the items are sold in the same state as they were purchased by Amazon in the first place. However, Amazon does add value by having efficient website and shipping processes that add convenience for the customer.

Car manufacturers take sheet steel and engineer car body panels. These panels are worth more than the raw material they are made of.

▲ Figure 13.8 A car manufacturer adds value to sheet steel by turning it into car body panels

Research

Look up the recipe for a cake. Add up the cost of the ingredients (don't forget to take the amount into account, for example two eggs are one third of the price of half a dozen eggs).

Compare your total cost with the price of an equivalent cake sold in a supermarket, after being made by a food manufacturing facility. How much value has been added by the manufacturer of the supermarket cake?

The role of research and development (R&D)

A company's research and development (R&D) department focuses on the future direction of products. It looks at the current range of products and suggests both changes to existing designs and new products for development.

There can be many reasons why a company might ask for a product to be redesigned:

- Certain materials or components are no longer available.
- A competitor has made a better, more popular product.
- New materials or technologies have become available.
- Customers no longer want or need the product.
- New regulations have come into force.
- Materials or components have increased in price.

A company may have to work fast to adapt its offering in circumstances such as these – circumstances over which it may have no control.

A failure to look to the future will result in obsolete products that no one will want to buy, and a company in this situation will go out of business.

Competition

When a company starts up, it may have a unique offering in terms of the goods it sells or the services it offers. Products might be covered by **copyright** or **patent**, but if they become popular, other companies are likely to find a way into the market and become competition.

Key terms

Copyright: a legal right to be identified as the creator of a literary or creative work and to control its reproduction.

Patent: a legal document that grants an owner exclusive rights over an invention.

The competition might even make improvements to a design, and a company will risk losing customers unless its R&D department keeps up and comes up with its own new ideas.

The Competition and Markets Authority is a non-ministerial government department responsible for the protection of copyrights, patents and trademarks.

Research

Visit the website for the Competition and Markets Authority and research how it controls patents. Then find the website for the Intellectual Property Office, which has current information on how to patent new ideas and protect existing patents.

Imagine you are head of R&D at a small fabrication company that has developed a revolutionary new type of aluminium can for drinks. Write an email to the chairperson of the board of directors explaining how and why the company should patent the new design.

Supply and demand

Supply refers to a company's ability to produce goods or services for sale. **Demand** refers to consumer desire to purchase those goods or services.

> ### Key terms
>
> **Supply:** a company's ability to produce goods or services for sale.
>
> **Demand:** consumer desire to purchase goods and services.

Supply may be affected by several factors, for example material or component availability and cost, complexity of the manufacturing process, worker productivity and technological advances.

Demand may vary depending on whether:
- a product is new and 'fashionable' (for example a new smart phone or new trainers)
- a product is seasonal (for example Christmas decorations or disposable barbeques)
- the market is already saturated (i.e. all the customers that might want a product have already bought it).

A well-run business will adapt its products as demand changes.

If supply is greater than demand, a company might end up with products left in stock. It may be necessary to sell these off cheaply or even throw them away.

If demand is greater than supply, a company will have to step up production or run the risk of letting customers down. Competition might react more quickly to fill the gap in the market.

> ### Research
>
> At the height of the 2020 COVID-19 pandemic, there were problems with supply and demand of PPE for health workers. Find examples of how this problem was solved. How long did it take?

13.2 Business and commercial practices

Depending on the circumstances you find yourself in, you might change the way you talk and act. For example, the way you talk and act in college may be very different to how you talk and act when out socialising with your friends. When you start working, you will learn a whole new vocabulary that is specific to your chosen industry.

This section focuses on the use of legal language, where words have specific meanings that cannot be misunderstood. It involves a finely tuned and accurate form of words, 'engineered' language in a way.

> ### Industry tip
>
> When you are working in a company, you will be expected to use the correct tone and language when talking to management, co-workers, customers and suppliers. This is more formal language than you may have used during your education and with your friends.

Legal practices

In business, most communications, **contracts** and conversations are between a **customer** and a **supplier**.

> ### Key terms
>
> **Contracts:** formal and legally binding agreements between two or more parties that detail what each party must or must not do.
>
> **Customer:** someone who buys a product or service.
>
> **Supplier:** someone who sells a product or service.

Documents such as contracts are written in legal language. This is very different from normal spoken English and uses specialised vocabulary with precise meanings. Legal language ensures that important information is communicated clearly and without any ambiguity, thereby avoiding confusion or arguments between the parties.

Contract clauses

Contracts and other long legal documents are divided into **clauses** of various types, which makes it easier to find important pieces of information.

Key term

Clauses: clearly defined sections of a contract or other legal document.

Warranties

When you purchase a high-value item, it may come with a 12-month or 24-month warranty.

A warranty is a written promise by a company to repair or replace a product that develops a fault within a specified timeframe. Where offered, it is considered a term of the contract, and outlines the conditions that might give rise to a claim. It might also explain the circumstances that will not be covered, for example due to improper use of a product. The document is written in legal language so that there are no disputes.

▲ Figure 13.9 A warranty will not usually cover improper use of an item, such as jumping on a sofa

Force majeure clauses

A force majeure clause is included in a contract to excuse one or both parties from fulfilling their obligations when there is an unforeseeable event, beyond their control, that prevents them from doing so. Typical events that may invoke a force majeure clause include war, terrorism, earthquakes and pandemics.

Indemnity clauses

An indemnity clause states that if one party causes loss or damage through their own fault and actions, the other contractual party will not have to pay.

For example, if a company supplies the wrong type of brick and a building falls down, it is seen as that company's responsibility to put it right.

Liabilities

A liability is a debt or obligation. A contract might state that a customer is liable for £10,000 as part of the agreement with the supplier. That is a liability for the customer, who will be expected to pay the amount within a given timeframe. Often that is immediately or within 30 or 90 days.

There could also be a liability for the supplier to deliver the correct products by a certain date or pay some of the money back. A contract will describe exactly what it is that the supplier will supply, including costs, maintenance requirements, when it will arrive and anything else that the customer might reasonably expect to know.

A company larger than one person will have liabilities for the owner. If the company goes bust, the owner may be liable for prosecution and damages for any employee that is left without money. Any creditors that the company owes money to will want paying back, even if that means the owner having to sell their house and belongings.

Other clauses

A confidentiality clause, also known as a non-disclosure agreement, requires one or more parties to keep certain information secret for a specified period.

A termination clause sets out how a contract can be ended (terminated) by either side without breaching the contract.

An intellectual property clause serves to protect unique information developed by a company. In most cases, intellectual property created by an employee in the course of their employment automatically belongs to the employer, and this type of clause explains this.

Test yourself

Imagine you are working in the legal department of a large engineering company. A new supplier has approached the company, wanting to sell it a range of items.

What sort of information would you include in a new contract for this supplier?

Improve your English

In suitable language, write a formal letter applying for an apprenticeship at a large engineering company of your choice.

Management

When a company is formed (sometimes called a start-up), there may only be a few staff and an informal management structure. Everyone knows the boss and information flows freely between the different members of staff. It is unlikely that they will need regular meetings and everyone will just go to the person they need when there is something to discuss. The manager/owner will probably know what is going on at all times.

Some companies can work well like this for years, but if a company expands and takes on new work, it will need more people to cope with the extra workload. Consequently, there will be a need to add some structure to the company.

▲ Figure 13.10 In a start-up, there may only be a few staff and an informal management structure

As a company grows, the staff might be split into different departments, each with its own departmental manager. These departments should be able to control themselves, thus freeing up the executive management to plan, lead and control the business, and meaning

they only need to contact the heads of the departments and not each individual member of staff. This ensures decision-making processes are quicker and easier and the business is more efficient at responding to change.

The larger a company becomes, the more management personel will be required, and often, more layers will need to be added to the company structure.

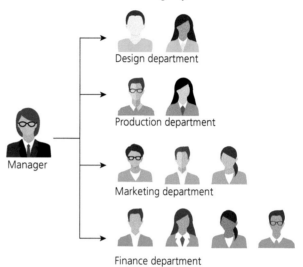

Design department

Production department

Manager

Marketing department

Finance department

▲ Figure 13.11 As a company expands, the staff might be split into separate departments

Company growth will also require more capital to invest in new equipment or premises. There are many ways of obtaining the extra money (discussed later in this chapter), but one way often used is to issue **shares**. The company is valued and that value is split into shares, which are then sold. Shareholders buy shares in the hope that the value of the company will go up, allowing them to sell their shareholding for more than they bought it. This process means that the company is owned by the shareholders and not the original owner, although the owner often retains some shares and keeps a **controlling interest**. The company becomes a private limited company, with the letters 'Ltd' at the end of its name. If the company sells its shares on the stock market and anyone can buy them, it becomes a public limited company or PLC.

Key terms

Shares: units of ownership of a company that can be bought and sold; their value varies according to the overall value of the company.

Controlling interest: where a shareholder owns more than half the shares in a company, meaning they can make decisions without needing the support of the rest of the shareholders.

The word 'limited' refers to the owner's liability if the company fails or is sued. The amount of money for which the owner is liable is limited to the face value of their share in the business.

Once a company becomes limited, it has to have a board of directors, who usually hold shares in the company. There will be a managing director (sometimes called a chief executive officer or CEO) and other directors who may have specific job roles, for example sales director or production director. The board makes all the important organisational decisions.

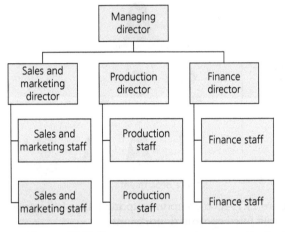

▲ Figure 13.12 Once a company becomes limited, it has to have a board of directors

If a company is not limited, the owner usually chairs the board and the members are called managers.

A board typically has representation from each department, although some directors/managers may take responsbility for more than one department. For example, a sales director/manager might also oversee marketing and purchasing.

While the directors/managers may have individual responsibilities for their own departments, the board comes together to make plans for the whole company. In fact, planning is one of the most important functions for management.

Each department has a critical function in ensuring the smooth running of the company:
- The sales and marketing department makes sure the company has new business and a regular source of income.
- The design department and R&D department look at new products and provide information to the rest of the company so that the products can be made.

- The production department plans and executes all the manufacturing of the products and might also be resonsible for buying in all the materials, tools and equipment needed.
- The shipping department packs all the products and sends them to the customers.
- The human resources department handles recruitment, payroll, employment policies and benefits. It also checks on the staff to make sure everyone is content and that there aren't any problems between members of staff that stop them working effectively.
- The finance department is responsible for controlling and documenting incoming and outgoing cash flows for the company.

If everyone knows what their responsibilities are, a company will work more efficiently and there will be less friction between departments.

Health and safety

In the early days of heavy industry, there was little responsibility placed on management for the health and safety of the workforce. Over time, the requirement for safe working practices has grown, and this is now one of the major responsibilities of any company. A company can be shut down if it does not keep its employees safe.

Planning

We have already mentioned that a production department needs to plan the manufacturing of products. However, the rest of the company needs to plan as well, for example:
- the IT department will need to plan the best time to roll out software updates to company computers in order to avoid disruption
- the transport department will need to plan vehicle services so that they are on schedule and not all vehicles are taken out of use at the same time
- the human resources department might need to plan for the employment of temporary staff ready for an increase in demand.

Resource allocation

Resource allocation is about assigning the best people, materials, tools and equipment to tasks and projects. When planning, the company has to make sure it has everything and everybody it needs to secure its future. A good plan is clear and easy to understand, with everyone's roles explained and all required resources clearly flagged.

The design department will supply design drawings with materials labelled. It is the production department's job to look at the drawings and decide which machine or process will be most efficient and what material **stock forms** will allow the best use of time and money.

Key terms

Resource: something that can be used to complete a task.

Stock forms: standard shapes and sizes in which materials are available.

Staffing

The people who work for a company might be called staff, employees or colleagues. They are the human element of a company.

Management has to decide how many people it needs for each department and what skills they require. If a company increases its production, it will need to take on more staff by advertising the new positions. However, if a company is struggling or has redesigned its way of working, it might also have to make some people redundant.

It is important to get the number of employees right, as issues such as absence due to ill health, maternity leave or sudden resignations will mean that the remaining staff have to adjust to cover the gaps. This can have an effect on projects, as staff may need retraining if they take on a different role and someone will still need to do their original job. Staff from other departments might be moved on a temporary basis to help with large production runs or in emergencies. The COVID-19 pandemic meant that a lot of people worked from home, although some still needed to be in the workplace.

Training and development

Staff may choose to do the same job year after year until they retire. Most people, however, want promotion, challenge and variety in their career.

Apprenticeships are designed around the gradual learning of skills, to take someone from being a school-leaver to a fully trained member of staff, ready to perform a valuable function in the company.

Existing staff may ask for training, or be encouraged to take training, in order to upskill and be able to undertake more responsible or complex tasks. Sometimes, an experienced member of staff may actually deliver the training.

People may need to be trained following the introduction of a new piece of equipment or software.

It can be useful to send a group to another site for training, or a trainer could be invited on site to address the group. This always needs careful planning, as staff will not be available for production while they are training.

While training can make an employee feel valued, they might also want to progress in the management structure. A company could manage this by advertising posts to internal staff first, before going through the external recruitment process.

Business models

Our discussion so far has concerned traditional management structures with fixed roles and responsibilities, where projects are run by a senior member of staff and more junior staff follow instruction. However, modern business models allow for more flexible structures and give employees more opportunities.

Agile business models

Project teams

When a new project is announced, a team of personnel is chosen from different areas of the business to look at the project from all angles. The project leader will be selected according to their relevant skills and experience, rather than their seniority. Therefore, a junior member of staff may end up running a project with much more senior members of staff, but their knowledge of the project makes them the ideal person to push the plan forward.

Response to change

Changes happen in every business. A customer might change their mind about the details of a project, and the company needs to be able to respond to that change swiftly and constructively. For example, a customer might initially define the length of a cable for a telephone network as 20 m and then decide that it needs to be 25 m. This change would require amendments to drawings, **procurement**, production plans and possibly even to the shipping requirements.

Use of software

The COVID-19 pandemic and subsequent lockdown showed that there is more than one way of working. Changes to working patterns were already evident before the pandemic, but the crisis accelerated the need for remote working.

Meetings do not need to take place in a conference room with all relevant personnel attending in person. It can be just as effective using an online collaboration platform such as Microsoft Teams or other dedicated software packages.

▲ Figure 13.13 The COVID-19 pandemic resulted in increased levels of remote working

Customer collaboration

Online collaboration software can also be used for communicating with customers. Documents can be changed by multiple stakeholders, at the same time, to allow everyone's voices to be heard until **consensus** is reached.

Policies and procedures

When a company is small, all decisions can be made by the owner. However, once a company has grown to a size where this is unworkable, it can introduce policy and procedure documents that give direction on what to do in different situations. Some actions are mandatory and required by law, for example concerning discrimination, health and safety and sick pay, but these documents can also cover other topics that management consider important.

A policy is a set of rules or guidelines for the organisation. For example, it could be a policy that all staff meet with the human resources (HR) department once a year to review progress and decide on training or progression. Another policy could be that the fire alarm is tested on the first Friday of each month.

Procedures detail the steps required to ensure policies are accomplished. Referring back to the example above, each member of staff might be sent an automatic email on the anniversary of their joining the company, inviting them to a meeting with a member of HR. The corresponding HR member gets a reminder email also, and a meeting is set up. Minutes are taken and any decisions are recorded.

A fire alarm procedure would explain who the fire marshals are, how to test the alarm, where the staff are to evacuate to and when to call the fire brigade.

13.3 Financial and economic concepts

Financial responsibility

A company needs to keep clear and accurate financial records. It could employ an **accountant** to do this as a **contractor** or take on a full-time member of staff.

Company accounts must be updated regularly and some of the information has to be reported to Companies House, a government agency that oversees UK businesses. This is known as 'filing accounts'. There is a lot of information on the Companies House website about this subject.

It is a requirement for all companies, as well as individuals, to make sure they are paying the correct amount of tax. This is regulated by HM Revenue and Customs (HMRC). A specialist accountant is often employed to manage this for a company.

Company employees also have to pay tax, and the company will take the right amount from its employees' pay and send it to the government. This is known as PAYE or 'pay as you earn'. It will also deduct national insurance and pension contributions.

The use of computers has made keeping accounts much more straightforward, as before that everything had to be written down. The huge number of **transactions** taking place in any company made this a laborious and time-consuming task, and there was a much greater chance of making mistakes.

> **Key terms**
>
> **Accountant:** a qualified person who prepares and audits a company's financial records.
>
> **Contractor:** someone who is self-employed but agrees to complete work for a company for a set fee.
>
> **Transactions:** exchanges of money when selling or buying goods or services.

Sources of finance

As mentioned earlier in this chapter, a company can sell shares when it needs money. However, there are other sources of finance.

Loans

A loan is when a company borrows money from either a bank, another company or a venture capital organisation.

Whatever the source of the loan, it has to be repaid, usually with interest. The lender will define a rate of interest for the loan, depending on the perceived level of risk. If there is a significant chance of the company going out of business, the lender might impose a high rate of interest, but if they anticipate the business growing well, they can offer a lower rate. The rate of interest and the duration of the loan are known as payment terms.

Capital

Capital is money that is already saved in the business. It is normally held in a bank account where it can be accessed quickly when needed, but where it can accumulate interest in the meantime.

> **Improve your maths**
>
> A company takes advantage of an interest-free bank loan and borrows £10,000.
>
> How long will it take the company to pay back the loan if its monthly payments are £200?

Budgets and budgeting

Whenever a company plans something new, it should establish a budget to cover the cost of any new machinery, equipment, materials, personnel, training and software. Without this planning, it will not know how much money to borrow from a bank or move out of capital reserves.

The company should also look at the type of costs involved:

▷ Direct costs include things such as the equipment and raw materials needed to make something. They are items that are easy to point at, for example a casting furnace.

▷ Indirect costs are more difficult to estimate, for example maintenance of the new equipment. The value of equipment will also **depreciate**, and that is considered an indirect cost.

▷ Overheads are costs related to the day-to-day running of a business, even when machinery is not in use. These could include rent on a building, heating and lighting, running a cafeteria and maintaining the company's IT systems.

> **Key term**
>
> **Depreciate:** reduce in value over time.

Useful financial words and phrases

Revenue is the total amount of income generated by a company. It does not take any costs into account and is sometimes known as the 'top line'.

A creditor is someone to whom money is owed. A debtor is someone who owes money. If a bank lends money to a company, the bank is the creditor and the company is the debtor.

Cash flow is the amount of money coming into and going out of a business at any given time.

A profit and loss (P&L) statement is a special set of accounts that shows clearly the income and expenses for a business in a given period of time.

Break-even (or break-even point) is when income and expenditure are the same and there is neither profit nor loss; they are said to balance.

Assessment practice

1   A company decides to make a new plasma cutter so it asks several local companies what features they would find useful on such a machine. What is this called?

2   An engineering company makes a small electric vehicle for a client. The total costs are £39,000, comprising wages for the manufacture at £10,000, materials at £5,000 and design work at £24,000. It then sells the vehicle for £45,000. What is the difference of £6,000 called?

3   A company needs to buy a supply of brass rivets for a new project. Is this an indirect cost, a direct cost or an overhead?

4   What does the M in the TIMWOODS acronym stand for?

5   Give two reasons why a company might ask for a product to be redesigned.

6   What happens if supply is greater than demand?

7   What is a product warranty?

8   Why do shareholders invest in companies?

9   How are project team members selected in an agile business model?

10 Define what is meant by a company policy and give an example.

Project practice

You are an apprentice engineer working at a large manufacturer of aluminium drinks cans. Prepare a short presentation (no more than 15 minutes) that you could present to the managers of your company on how to minimise waste in the design and manufacture of the product.

14 Professional responsibilities, attitudes and behaviours

Introduction

It is vital that engineers understand the wider context within which their work falls, and that they always act appropriately in the workplace. This chapter looks at the main responsibilities of both employers and employees in the engineering workplace and identifies different job roles and to whom they are accountable. You will also learn about the moral and ethical responsibilities of engineers and the importance of professional standards in the engineering sector.

Learning outcomes

By the end of this chapter, you will understand:
1 the purpose, function and content of job descriptions
2 the behaviours and personal conduct expected in an engineering workplace
3 the responsibilities and accountabilities of different roles within an engineering company
4 the main duties of an engineering organisation regarding equality, diversity, accessibility and inclusion
5 the purpose and importance of continuing professional development (CPD) and professional recognition
6 the human factors that affect the work that takes place in an engineering workplace.

14.1 Professional conduct and responsibilities in the workplace

Job descriptions

It is important that people who work in an engineering environment have a clear understanding of their roles and responsibilities. A **job description** is a written document that details exactly what an employee is expected to do in a particular job role. Job descriptions are also often used when a company is advertising for new staff, so that applicants understand what is required of them to be successful.

A job description usually includes:
▶ the job title, for example maintenance engineer or health and safety manager
▶ information about the purpose of the job
▶ an outline of the main roles and responsibilities
▶ details of any qualifications and prior experience required.

Workplace behaviours

In addition to fulfilling the specific technical roles detailed in their job description, engineers are expected to behave in an appropriate manner at work. This includes:
▶ ensuring good punctuality and attendance
▶ demonstrating a positive attitude to work, both as an individual and within a team
▶ treating others, including colleagues, clients and customers, with respect
▶ ensuring a professional appearance that is appropriate for the job role being performed.

If these behaviours are not followed, it can have a negative impact on the working environment, for example:
▶ If someone is consistently late for work or regularly misses important meetings, this can cause tensions with colleagues who have a good attendance record.
▶ An untidy appearance can give a negative impression of the company's professionalism to potential customers.

On the flip side, someone who demonstrates a positive 'can do' attitude is more likely to be able to work to the best of their abilities. This positivity often rubs off on others, who will raise their standards to match.

Seeking advice and guidance

Sometimes it is necessary for engineers to seek advice and guidance from others within the workplace. This could be regarding a specific engineering task, such as how to complete a particular maintenance activity, or about a wider workplace issue, such as how to resolve a conflict with another colleague or settle a pay dispute.

Usually, the first point of contact for a formal query or dispute is the employee's immediate **line manager** or team leader, who will either deal with the issue personally or escalate it to a more senior or appropriately qualified colleague if necessary. For example, a human resources manager could provide support related to the worker's contract of employment. A query regarding a safety concern may be referred to a health and safety manager.

Less experienced engineers, or those new to the job, will often seek informal help and support from more experienced colleagues. This helps them to develop their knowledge and skills. Some companies operate coaching, mentoring and/or 'buddy' schemes that allow colleagues with different skills and experiences to observe, speak with and learn from one another.

▲ Figure 14.1 Engineers will often seek the advice and guidance of others in the workplace

Engineers should always ask for help if they feel unable to successfully or safely complete a task by themselves. Any potential health and safety issues should be reported immediately.

Key terms

Job description: a written document that outlines the main roles and responsibilities associated with a job.

Line manager: someone who has direct responsibility for managing frontline workers, such as production staff or machine operators.

Test yourself

Think about different scenarios where you may need to ask for support and guidance in the workplace, for example when unsure how to operate a piece of equipment.
1   Who would you seek advice from?
2   How would you approach them?
3   What questions would you ask?

Personal conduct

Reputation

The **reputation** of an individual or organisation is how they are perceived by the public based on how they behave or operate. This could be in terms of their competence and/or the quality of their work, their ability to complete jobs on time or how they treat colleagues and customers.

A reputation can be positive or negative. It can also change over time. A good reputation is important because it gives confidence to potential customers and clients that a job will be done well. It is often hard to repair a reputation once it has been damaged.

Ethical responsibilities

Ethics are the moral principles that govern how an individual or organisation behaves. Engineers have a personal responsibility to ensure they act in an ethical manner.

Most companies have their own set of ethical standards that must be followed by all employees. In addition, the Engineering Council has specified a set of ethical principles that its members are expected to follow. To comply, engineers should always:
▶ act with integrity and respect, for example by being aware of how their actions could affect others and acting in a trustworthy and reliable manner
▶ show respect for life, the law and the overall public good, for example by making sure health and safety issues are taken seriously and considering sustainability issues in their use of materials and resources
▶ perform their work with accuracy and rigour, for example by keeping their knowledge and skills up to date and only performing work that they are suitably trained or qualified to complete
▶ promote high standards of leadership and communication, for example by encouraging equality, diversity, accessibility and inclusion.

Case study

Volkswagen emissions scandal

The Volkswagen emissions scandal is an example of how an engineering company's reputation can be affected by the actions of its employees. In 2015, regulators in various countries began investigating one of the largest vehicle manufacturing companies in the world, the Volkswagen Group, for potential discrepancies between engine emissions test results and their actual road-use outputs. It was suggested that certain diesel-powered vehicles had been fitted with 'defeat device software' that could tell when a car was being tested, so that they looked as though they met the required standards to enable them to be sold.

The scandal caused the company's share price to drop dramatically as market confidence in the company's products fell. The group's chief executive at the time, Martin Winterkorn, said the company had 'broken the trust of our customers and the public'.

▲ Figure 14.2 Volkswagen was accused of fitting software to vehicles which could affect the results of exhaust emissions tests

Questions

1   Research the additional impacts of the emissions scandal, both on the company involved and on the wider vehicles market.
2   Investigate other examples of where engineering companies have experienced either a positive or negative impact on their reputation.

Key terms

Reputation: how an individual or organisation is perceived by the public based on how they behave or operate.

Ethics: the moral principles that guide how an individual or organisation behaves.

Discuss, using examples, the importance of acting in an ethical manner when working in an engineering environment.

Research

Visit the Engineering Council website: **www.engc.org.uk**.

Find, read and check that you understand the ethical standards for engineers set by the Engineering Council.

Levels of accountability within organisational structures

Responsibilities of different roles within an organisation

There are a number of different job roles within an organisation, each with their own specific responsibilities. Failing to meet these responsibilities can have an impact on other people within the organisation. For example, disorganised management of a production line could mean that machine operators do not have the required materials and consumables available to manufacture a product or part. If directors do not have a clear strategic vision for the organisation,

managers could be left confused about what they are trying to achieve with their teams. It is therefore important that everybody within the organisation understands the responsibilities of their specific roles and executes them to the best of their abilities.

Accountability is important within an organisational structure because it ensures performance is of an appropriate standard and allows for potential improvements to be identified. Each worker has a specific person to whom they report and are directly accountable, and who is responsible for managing their performance. Usually this is the next most senior person. For example, a machine operator is usually immediately accountable to the production or manufacturing manager. If an issue cannot be dealt with by their immediate line manager, for example a serious disciplinary matter, it may be sent further up the organisation to be resolved by a more senior colleague.

Key term

Accountability: having responsibility for an action, task or job.

Test yourself

Explain why accountability is important in the engineering workplace.

▼ Table 14.1 Responsibilities and accountabilities of different roles within an engineering company

Role within organisation	Main responsibilities	Accountable to
Apprentice	Developing their knowledge and skills while 'on the job' Completing the relevant qualifications for the job role that they are working towards, e.g. machine operator	Line manager Coach/mentor Apprenticeship/training provider
Operator	Safely operating tools, equipment and machinery used to manufacture, test or maintain products	Line/department manager Team leader
Management	Leading and organising teams or groups of employees Setting targets, e.g. for production Reviewing performance of individual team members	Senior managers Directors
Director	Overall management and oversight of the company and its operations Strategic decision-making	Chief executive officer (CEO) Shareholders External regulators

Main duties of an organisation regarding equality, diversity, accessibility and inclusion

All organisations have duties regarding **equality**, **diversity**, **accessibility** and **inclusion**.

▶ Equality is about ensuring nobody suffers prejudice based on age, disability, gender reassignment, marriage/civil partnership, pregnancy/maternity, race, religion or belief, sex or sexual orientation. The Equality Act sets out these **protected characteristics** that must be observed by law.

▶ Diversity involves employing, accepting and including employees from all backgrounds. Diverse workplaces can bring different perspectives on problems and solutions, cultural insight and localised market knowledge.

▶ Accessibility is concerned with the removal of barriers that prevent workers performing their roles, and ensuring equal access to workplace facilities and equipment for all. For example, people with disabilities should be able to perform the same tasks as those who are non-disabled.

▶ Inclusion is the practice of creating a workplace culture where the different skills, talents and perspectives of all workers are accepted, appreciated, supported and valued.

Test yourself

Outline the main protected characteristics that are set out in workplace equality legislation.

14.2 Continuous professional development (CPD) and professional recognition

What is continuous professional development (CPD)?

Continuous (or **continuing**) **professional development** (CPD) is about ensuring that an individual can continue to develop their skills, abilities and experience after formally qualifying for their job role. It can take many different forms, such as face-to-face or remote training courses, observation of colleagues, and coaching, mentoring or buddying schemes. It can also include taking further qualifications relevant to current or future potential job opportunities within a company or sector. For example, a manufacturing engineer may wish to take a management qualification to put them in a good position when applying for a production manager role.

▲ Figure 14.3 Training sessions can be face-to-face or can be run remotely

Key terms

Equality: ensuring nobody suffers prejudice based on the protected characteristics set out in the Equality Act 2010.

Diversity: employing, accepting and including employees from all backgrounds.

Accessibility: ensuring equal access to workplace facilities and equipment for all and removing barriers to achieve this.

Inclusion: the practice of creating a workplace culture where the different skills, talents and perspectives of all workers are accepted, appreciated, supported and valued.

Protected characteristics: nine aspects of a person's identity set out in the Equality Act. The Act legally protects people from being discriminated against on the basis of these characteristics.

Continuous professional development (CPD): the learning activities that workers engage in to improve the knowledge and skills needed to perform their current or future job roles.

Many aspects of CPD require the engineer to take a level of personal responsibility for their needs and requirements. For example, it is reasonable to expect a professional engineer to keep up to date with current safety standards related to their role.

How CPD motivates staff and improves performance

New processes, technologies and ways of working are constantly being developed within the engineering sector, so it is vital that engineers continually update their knowledge and skills. For example, training on a range of new machines increases the skillset of the operator and allows more flexibility of staffing for the company. This in turn can lead to greater efficiency and productivity.

Effective CPD is also essential for maintaining workforce morale and motivating staff to perform as well as possible. For example, sessions focusing on the wider ethos and goals of the company increase staff's sense of belonging and contribute to increased confidence, enthusiasm, discipline and loyalty.

CPD allows workers to keep up to date with changes to legislation, quality standards and company standard operating procedures (SOPs). For example, if changes are made to wiring regulations, all engineers working on electrical equipment need to be retrained so they can follow the new procedures safely and effectively. Failure to do this could leave the company or individual engineer open to prosecution.

> **Test yourself**
>
> Explain the purpose and benefits of effective CPD in the workplace.

Professional standards and recognition

Professional bodies set the **professional standards** that they expect their members to follow.
- ▶ The Engineering Council acts as the main regulatory authority for chartered and incorporated engineers.

- ▶ The Institution of Engineering and Technology (IET) sets and reviews various professional and technical standards, most notably the wiring regulations for electrical engineers (BS 7671).
- ▶ The Institution of Mechanical Engineers (IMechE) mainly sets and reviews standards relating to the field of mechanical engineering.

Appropriately qualified and experienced engineers can apply for chartered status. This is a peer-reviewed process that is managed by the Engineering Council. To achieve this level of recognition, applicants must demonstrate extremely high levels of professional competence in their specialist field. Chartered engineers are highly sought after by companies and can command high salaries.

Outstanding performance within the engineering industry can also be recognised and rewarded through professional awards. For example, the Engineering & Manufacturing Awards (EMA) are held annually. Categories include Design Team of the Year, Digital Innovation and Manufacturing Technology Innovation.

> **Key terms**
>
> **Professional bodies:** organisations that represent, support and set the standards of work and behaviour for a profession.
>
> **Professional standards:** agreed standards, set by professional bodies, that describe the skills, knowledge and behaviours that characterise excellent practice and support within a profession.

> **Test yourself**
>
> Explain, using an example, the benefits of professional recognition within the engineering sector.

> **Research**
>
> Investigate the requirements for and benefits of Chartered Engineer status: www.engc.org.uk/ceng.

14.3 Human factors within engineering and manufacturing contexts

Human characteristics

Any workplace is only as successful as the people who work in it. Therefore, **human characteristics** must be considered and taken into account at all times.

Physical characteristics are defining features of a person's body, for example their strength or appearance. **Mental characteristics** are to do with how a person thinks, for example their ability to accept change in the working environment or their confidence in dealing with new situations.

The characteristics, capabilities and limitations of people should be taken into account when setting expectations, scheduling work and defining job roles. For example, machine operators can become tired over the course of long shifts. Therefore mistakes or accidents are more likely to occur if they do not take enough breaks or holiday time to recuperate. A higher level of errors results in reduced production efficiency and increased waste.

A company can also develop a poor reputation for its treatment of workers if it does not take their physical and mental needs into consideration, potentially leading to difficulties in recruiting staff. Where workers have physical or mental disabilities, it is the responsibility of the employer to make reasonable adjustments to ensure they can perform their role.

Test yourself

Explain the difference between physical and mental characteristics.

Health and safety

Workers suffering from fatigue are more likely to make mistakes that can endanger both themselves and others working around them. Engineers should never operate machinery, tools or equipment when they are tired.

Workplace design

Effective **workplace design** allows tasks to be performed safely and efficiently. It also allows management to have good oversight of the work being carried out.

The design considerations and assessment criteria depend on the types of task carried out in the workplace. For example, in a production facility, machines should be located so as to minimise the amount of movement required for the different product parts being manufactured. Other considerations could include:

▶ whether the layout is open plan, closed or a mixture of the two
▶ the proximity of different departments and teams for maximum efficiency
▶ the spacing of machinery to ensure safety and reduce disruption from noise
▶ how and where tools and equipment are stored
▶ how the design fits with the company ethos and working style
▶ where meetings and presentations can be held
▶ access to power and computer equipment
▶ areas for workers to take breaks and eat
▶ seating and furniture for comfort and ergonomics.

▲ Figure 14.4 Design must be carefully considered in an open-plan manufacturing workplace

Key terms

Human characteristics: the physical and mental features of people.

Physical characteristics: defining features of a person's body.

Mental characteristics: characteristics relating to how a person thinks.

Workplace design: how the workplace is arranged to ensure maximum safety and efficiency.

The workplace should also be designed so that all employees can both access and make use of the facilities and equipment. For example, reduced-height CAD/CAM workstations allow ease of operation by wheelchair users.

Human error

Mistakes made by people can result in production errors and safety issues in an engineering and manufacturing environment. Although the likelihood of **human error** can never be fully eliminated, it can be reduced.

▼ Table 14.2 Causes of human error and how they can be avoided

Cause of human error	How it can be avoided
Insufficient training	Ensure all workers are suitably qualified for their role
	Ensure workers are trained in the safe operation of all tools and equipment that they need to use
	Use standard operating procedures
	Provide and engage in ongoing CPD
Fatigue	Ensure appropriate limits on working hours
	Ensure workers take regular breaks
	Ensure workers do not operate tools or machinery while tired
Excessive workload	Plan effectively to ensure workers have enough time to complete tasks allocated to them
	Allocate tasks, tools, equipment and resources effectively
	Ensure good teamworking
Stress	Encourage mental wellbeing within the workplace
	Make use of flexible hours and remote working
	Provide opportunities for counselling and mental health support

15 Stock and asset management

Introduction

When anything tangible is made or processed, storage is required for materials, tools, parts, work in progress and products waiting to be shipped. This is known as stock.

It is important to know what stock there is, how much there is, where it is and its value. The movement of stock must be planned so that it is in the right place at the right time and in the right amounts. Some types of stock have special storage requirements.

Learning outcomes

By the end of this chapter, you will understand:
1. the purpose of effective stock and inventory management and control
2. the key issues, risks, advantages and disadvantages associated with different practices
3. the purpose and methodology of effective asset management
4. the advantages and disadvantages associated with methods of capacity management.

15.1 Stock and inventory management principles and practices

To understand the importance of stock control, you first need to understand what stock is. Stock (sometimes called inventory) is made up of the things owned by a company that it needs to carry out its business.

If a company makes cast aluminium pulleys, stock would include aluminium as a raw material, moulds, lathe tools, drill bills and casting sand, as well as any half-finished or completed pulleys.

If a company makes automatic gearboxes, the stock will be considerably larger and more complex, but the stock control principles will be the same.

Stock has a value. This could be the amount of money that it:
▶ cost to buy
▶ would cost to replace
▶ could be sold for.

These figures may not be the same because values of materials change over time (see 'depreciation' later in this chapter).

Stock needs to be tracked. It might arrive in a van or truck along with lots of other purchased items, so it will need to be sorted and added to a **catalogue**.

> **Key term**
>
> **Catalogue:** a detailed list that can be easily searched, usually by use of an index.

▲ Figure 15.1 Incoming stock needs to be sorted and added to a catalogue

Types of stock

Stock could be many things:
▶ Raw materials: these are the basic materials from which products are made, for example sand, cement and water are the raw materials for making concrete blocks.
▶ Sub-assemblies: these are units that are assembled separately but will eventually be incorporated with other units into a larger finished assembly. They may be manufactured in another department, factory or even by a different company. As they may not be needed immediately, they have to be stored as stock.
▶ Components: these are individual parts of a product, for example electronic components include LEDs, resistors and sensors.
▶ Consumables: these are materials that are used during manufacturing but are not part of the finished product, for example machine cutting oil, printer paper and grease.
▶ Work in progress: this means partially finished goods awaiting completion and sale. Sometimes production may have halted while waiting for parts or due to a break. These unfinished goods need to be stored as stock.
▶ Fasteners/fixings: these are hardware used to join objects together. As they are needed so often during manufacturing, it is common for a large quantity to be kept in stock. The staff that monitor stock levels need to keep track of the stock and order more if there is a possibility that the company might run out. Of course, ordering too much will waste money.

At any time, the company might need to know the value of its stock, so this will need to be checked and recorded regularly.

Taking care of stock

The conditions in a storage area or **warehouse** must be appropriate for both the people working there and the type of stock being stored. At a minimum, the environment needs to be controlled to ensure adequate ventilation, lighting and temperature.

> **Key term**
>
> **Warehouse:** a building where stock is kept prior to use or sale.

However, some stock requires special storage conditions to ensure it does not get damaged. For example, natural materials such as leather, paper and timber will deteriorate unless temperature and humidity levels are carefully controlled. If perishable food is being stored, refrigerated facilities are required.

Some types of stock require careful handling and storage, for example glassware, electronic components and delicate metal parts. Stacking something too high might result in instability and damage. Fragile items need to be stored with packing materials to keep them from breaking.

High-value stock will need security. Formula 1 tyres have 24-hour security at the Pirelli headquarters, as well as careful environmental care to ensure they are kept at 20–25 °C.

▲ Figure 15.2 Formula 1 tyres require careful and secure storage

If something has to be stored for a long time, it should be kept out of the way of stock that has a high turnover. Stock items that are needed often should be kept in an accessible position.

> ### Test yourself
>
> Imagine you are working in the purchasing department of a soup manufacturer. Your company buys carrots to add to various soup recipes. Suggest what precautions should be taken to keep the carrots from going bad during transport, handling and storage.

> ### Improve your English
>
> Imagine you are working at a manufacturer of bicycle bells. Write an email to the head of the stores department, suggesting that the steel for the bells should be stored in a dry area so it does not come to the production line rusty. The head of the stores department is a senior member of the company and should be addressed in a formal manner.

Stock and money

As well as purchase costs, stock also costs money to:
- store
- keep
- move
- dispose of.

Costs of storing stock

If we think about the Formula 1 tyres mentioned earlier, there are costs associated with the need for continual environmental control in the warehouse, i.e. the installation, running and maintenance of heating and air-conditioning systems. It also costs money to install and maintain CCTV and other security equipment, as well as to pay security personnel.

Even stock that does not require this type of care still needs to be stored in a warehouse, with the costs that this entails, for example:
- rent
- building maintenance
- staff wages
- facilities for staff (heat, light and eating/toilet facilities).

Even mostly robotic warehouses need some human staff. The robots themselves need environmental control, electricity and maintenance.

Every day that something sits in the warehouse, it costs the company money.

> ### Health and safety
>
> In heavily automated warehouses, it can sometimes be dangerous for humans to work in the same area as robotic equipment. However, modern robotics can be designed to work with humans and be aware of where workers are at any given point, stopping if there is anyone nearby. This is called 'cobotics' and it increases safety and efficiency in automated workplaces.

Costs of keeping stock

When stock is bought in, it should be used straight away or it can become out of date. Short-life food stuffs are a good example here, but chemicals such as grease, paint and oil can also have a use-by date. Sometimes stock may even become **obsolete**, for example when imperial bolts do not fit metric fittings.

Stock may also depreciate, which means it decreases in value over time. This reduces the overall value of the company.

> **Key term**
>
> **Obsolete:** no longer used.

Costs of moving stock

Even if stock is located in the same building as production, it will need moving.

Under the principles of just in time (JIT, see below), this needs to be done quickly and efficiently, so that what is needed arrives where and when it is needed. Staff may be able to carry small stock items, but usually forklift trucks and robotic carriers are required to move items.

When stock moves between factories, across the country or between countries, it needs to be transported by road, rail or air, all of which will cost the company money.

Costs of disposing of stock

If a company finds it has stock that it no longer needs, or which has become obsolete, it will need to dispose of it so it does not take up valuable space. This may be as simple as putting the stock in the waste, but this is rare. Most companies recycle wherever possible, with parts and materials sorted and sent to relevant recycling facilities. While this takes time and effort, which in turn costs money, there are regulations in place that companies must follow.

Just in time (JIT)

Toyota introduced the principle of just in time (JIT) to their car-manufacturing facilities in the early 1970s. It defined JIT as making (or buying) 'what is needed, when it is needed, and in the amount needed'. This principle was so successful that it was taken up by many other manufacturers in many different fields, and it is regarded as standard practice today.

JIT means that there is never extra stock in the warehouse, as only what is needed is ordered. The tools, materials, fasteners, chemicals and fixings needed for production are delivered to the employee at the point of production, meaning that materials travel the shortest distance possible and no time is wasted going back and forth to the stores during production.

Some companies resist the implementation of JIT but, where used, it has helped modern manufacturing become more streamlined and has helped increase efficiency and productivity. However, in some circumstances, for instance when making prototypes or one-offs, it can stifle innovation and slow down a skilled practitioner.

Some companies have taken the ideas and technologies of JIT and developed them to fit their own ways of working. For example, Airbus in Broughton uses JIT to control the specific tools needed for production. Only the tools required are delivered to each workstation, rather than a generalised toolbox, to avoid the operative wasting time by having to choose a tool for the job.

Companies need to be careful, however. It is easy to be complacent and think that once JIT has been set up for a process it will always be the optimum way to work. It is important to regularly review practices, to ensure they are as efficient as possible.

Made to order and made to stock

Made to order

In a simple business, a customer orders something, the factory makes it and it is sent to the customer. This is the basis of 'made to order' and it has the advantage that there is no need for finished product to be sitting around in the factory or stores. It is cost-efficient and straightforward.

An example of this would be a custom-made yacht, where the customer describes the features and fittings they want, as well as the size, type of engine and style. It would be impossible to keep an example of every type of yacht in stock, so the yacht is made to order.

▲ Figure 15.3 A custom-made yacht is an example of a made-to-order product

Unfortunately, this approach can mean that employees are standing around waiting for orders, or a large volume of orders can come in suddenly when staff do not have time to make the products.

These disadvantages are due to it being a reactive, rather than proactive, way of working, as well as reliance on orders coming in, meaning that a lack of orders will cause problems with turnover. This is a risk for a company of any size, and it should be a key company objective to actively find new business – the main role of a marketing department.

Made to stock

If a company knows there is a good chance of a product selling in large quantities in the near future, for instance plant pots in the spring, it can plan to upscale its production so that there is sufficient stock when the large orders come through. This calls for expert marketing and forecasting of future requirements. If a company makes too much of something that is then not saleable, it will have a large quantity of 'dead' stock which will need to be sold off at a discount or possibly a loss.

Ideally, there should be just enough product in stock for it to be shipped out to customers at the same rate as it is made.

Material requirements planning

Products may be made up of raw materials, components or sub-assemblies. These all need to be available at the appropriate time and in the appropriate quantities, following the principles of JIT.

This is the responsibility of the purchasing department, which should work to ensure efficiency. For example, a company making injection-moulded bicycle parts might require polymer for the pedals. However, there may also be a parallel production run of mudguards using the same polymer. The purchasing department needs to be aware of this, as often there is a lower price for buying materials in larger quantities.

If a mistake is made in the order quantity or quality, it may impact manufacturing, so thorough checking processes are required.

Careful planning of material requirements will reduce the amount of stock held by the company, which is cost-efficient. However, it does reduce the flexibility of the R&D department, which will need to order in everything it needs when redesigning a product or looking at new approaches to manufacturing.

Supply chain

It is rare for all the materials and parts of a product to be made in the same factory. More often, they are bought in from external suppliers, even if some sub-assemblies are made in house. These suppliers may in turn have bought from another company, and this is known as a supply chain – a network of companies or individuals involved in creating a product and delivering it to market. Any problem in the chain can cause difficulties for several manufacturers.

A recent example of a broken supply chain was the shortage of semi-conductors to the automotive industry in 2019. With the worldwide demand for personal computers, smart phones and tablets skyrocketing, the semi-conductor industry shifted its focus to fulfil that demand. This meant that the supply of semi-conductors (microprocessors and sensors in particular) to the automotive industry was thin and prices went up. The pandemic made the situation worse, and some car companies had to halt production. Most have now made deals with semi-conductor manufacturers to secure supplies for the future, and production has steadily increased since then.

▲ Figure 15.4 There were problems supplying semi-conductors to the automotive industry in 2019

Packaging and storage

Not many types of product can be stacked on a shelf or pallet without some sort of protection from damage.

▲ Figure 15.5 Items stored in warehouses require some sort of protection from damage

Since space can be short in warehouses, stock needs to be stored in such a way that it will not get damaged. It might be wrapped in protective material such as polystyrene, cardboard, wood pulp, bubble plastic or packaging bubbles, or it may be in containers or 'bins'.

If items are to be moved, they will need even more protection. While pallets need to hold a significant amount of stock to make them efficient to store and move, they need to be loaded carefully to ensure both stability and protection of the items at the bottom, which may be crushed if the pallet is overloaded.

The bumps and jolts of road traffic will make matters worse, and this is why most products come in some sort of box with packing that holds the contents in place.

Valuable items will have specially designed packaging, and some items will be supplied disassembled in order to help avoid damage. Flat-pack furniture is a good example here, as not only does it take up less space but it is also less likely to get damaged in transit.

Some products require environmental control during transportation as well as in the warehouse. Perishable food stuffs need refrigeration and paper goods need to be kept dry.

The use of technology in stock control and movement

It is important to know how much stock a company owns, as it is part of the company's value.

In the early days of industry, stock would arrive at a warehouse in plenty of time and stores staff would place it on shelves using their own numbering system. They would enter into a large ledger (book) what stock had arrived and its location in stores. They would regularly go round the shelves and count everything there – called a 'stock take'. Anything running low was mentioned to the purchasing department, who would order more. When an item was needed, the stores staff would look it up in the ledger and go to fetch it for the production department.

Since the introduction of computers, stock control is more automated, with new stock arriving just before it is needed and in the right quantity. New items might be stored briefly but will normally move to the production department for use soon after delivery. Instead of using ledgers, stock levels and numbers are stored in an electronic database where the information can be easily accessed. All this means that the value of stock is easy to measure, new stock can be ordered automatically once a minimum stock level is reached, and the locations in stores can be used to control stock robotics.

▲ Figure 15.6 Automated stock control can support the use of robotics in warehouses

The use of computer databases instead of ledgers means that stock levels can be checked by anyone with access. When shopping online, there is often an indication of how many items remain in stock, so that a customer can decide how many to buy. This number is automatically updated once purchases are made.

In a company, the purchasing department can get an automatic alert if stock is getting low. This could even be intercepted by an automated ordering system, which sends an order to the supplier for new stock.

If it is possible to know stock levels instantly from a database, there is no need to carry out a formal stock take, with staff going around the warehouse and counting the numbers of items. A sensible company might do spot checks, particularly on high-turnover items, to make sure the system is up to date, but computers are usually accurate as long as there hasn't been any human error.

Human error tends to occur with repetitive tasks, such as counting and typing numbers or remembering where things are. Technology can help to eliminate these errors.

Barcodes are a way of showing a multi-digit number in a form that can be read by a scanner and entered directly onto a computer system. They can be used to simplify the entry of stock into a catalogue or database, avoiding the need for a human to enter the numbers manually and making the process quicker and more accurate.

If the stock control system requires more information than a simple number, the company might use QR codes, as they can hold large amounts of data and even direct staff to a relevant page of information on an item.

▲ Figure 15.7 QR codes can hold large amounts of data and are therefore useful in stock control

In large warehouses, the storing and recovering of items might be performed by robotic units. They can fetch the stock that is needed from where it is stored and add new stock without human intervention, updating the database as they go. Human staff can be used to perform more delicate tasks, such as assembly or packaging.

The use of technology in production

Technology is used to support many activities in a production environment, for example CNC machines, computer-controlled conveyor belts and robotic welding or painting equipment. However, some tasks still require human intervention, especially if they are complicated, varied or need human dexterity.

While most printed circuit boards are made using pick and place machines, and everything is automated including the soldering, sometimes there are components that cannot be handled easily by machines. A human operator might solder the components but they could be helped by a place-to-light system, which shows them where the component should be soldered, reducing the likelihood of mistakes.

When collecting components for an assembly task, some companies might use a pick-to-light system. This comprises a set of boxes (or 'bins') that contain all the required components, with a light over each box. When the parts are collected, the light shows the operator which box to retrieve the components from. These can then be packed up and sent to the appropriate manufacturing area. The system is often used for fastenings and fixings, such as nuts and bolts. Supplying the correct number to the operator can avoid the problem of some being missed out.

> ### Research
>
> Research the use of pick-to-light systems and write a short report on how they could be used in an industry of your choice.

15.2 Asset management and control principles

Capacity management

If there is too little or too much stock, a company is not planning efficiently:
- Too little stock means there will be a halt in production while more stock is brought in, causing the company to lose time and probably money. This is called a bottleneck, because it slows the throughput and delays all subsequent stages of production.

- Too much stock could mean spare stock left in the stores after production. This can become obsolete or go out of date, wasting money and requiring disposal.

The amount of stock required for production needs to be calculated at the planning stage. Design drawings should include accurate numbers for each part and, at each stage, these numbers should be reviewed.

Sometimes, new storage might be needed when a new product goes into production, for example if that product includes a greater number of components. The company will need to decide whether to the store the extra stock on site, or ask suppliers to deliver it when needed.

Planners should also check whether a new product uses components or materials that haven't been used before in that company. This might mean finding a new supplier who can satisfy the demand and then adding the new items to the stock catalogue before the production department can request them.

Asset management and budgetary control

Not everything that a company owns is straightforward stock. It is also important to keep track of items such as hand tools, machines, computers, desks, chairs and many other pieces of equipment used around the company. These items are called assets and they need counting and valuing. They might be kept in the stores or in and around the desks of employees, but it should still be possible to identify where they are, how old they are, when they need maintenance and how much they are worth.

Key stages

Planning
During the planning phase, planners should know what assets the company has that could be used for a project and their availability.

Acquisition
The purchasing department has a duty to buy new assets at a good price. It should look for a reliable supplier who can deliver the requested goods on time and in good condition.

Operation
Whether assets are kept near to where they will be used or in stores, the location must be controlled

so that employees can access them. If a tool needs calibrating (for example measuring equipment), it should be carried out at a time when it will not interfere with production.

Maintenance

During production, assets such as machines need regular, planned maintenance, and this should be included in the plan so it does not interfere with the workflow. Maintenance is often completed by specialist teams.

Disposal

Disposing of assets used to be a case of simply putting them in a skip. Nowadays, each item needs careful disposal:

▶ Electrical and electronic waste must be managed in line with the Waste Electrical and Electronic Equipment (WEEE) Directive. Polluting PCB materials need to be disposed of so as not to contaminate the environment. Some valuable components, such as gold and silver, may be retrieved and reused.

▶ Recyclable items, such as metal, paper and most plastics, have to be sent to recycling facilities. Sometimes the company may even get paid for the waste.

▶ Non-recyclable items go to landfill, so the use of these should be reduced as much as possible during the early planning phase.

The waste that a company produces might be checked by the authorities and a company can be fined heavily if it does not dispose of waste correctly.

Research

Research the types of materials used for packaging. Which are the best from a cost point of view and how does that compare with the most recyclable materials? How can a company improve its use of recyclable packaging materials?

Methods of capacity management

Capacity management is so important in keeping a company working efficiently that there are often members of staff whose entire roles are dedicated to it. They will work closely with the planning department.

There are three basic methods of controlling capacity in production:

▶ Product capacity planning: this is where the amount of stock is calculated to provide just enough to deliver against orders, for example ordering enough LEDs, wires and power supplies to make 10,000 sets of specially designed fairy lights for a chain of fast-food restaurants.

▶ Workforce capacity planning: this focuses on how much product can be made with the people and hours available. It shows whether temporary staff need to be employed to fulfil orders and how soon those workers would need to be recruited. For example, a small engineering company has been asked to make railings for a local theme park. The publicity for doing this is worth the extra effort, but there are not enough welders to complete the job in time. The company can take on some short-term contractors to support the existing welders.

▶ Tool capacity planning: this looks at how much product can be made with the tools, equipment and workshop facilities available. It may be necessary to order more tools and equipment. If they are only required for a short time when there is an unusually large order going through, they could be hired. For example, a building company has been asked to build a new car park in the town centre. It has all the equipment needed to make the concrete, lay the stone and build the flowerbeds, but it does not have a roller to flatten the surface. It could therefore hire a roller to complete this project.

Budgetary control practices

As the value of stock is such a large proportion of the value of a company, correct budgetary control is vital. For discussion purposes, this aspect of manufacturing management can be split into three sections: life cycle, whole-life approach and depreciation.

Life cycle

▶ Design: the life cycle of stock starts at the design stage. The design engineer should take into account the cost of any stock that needs to be purchased. If, for example, a design requires an extruded strip, it should be a standard size because ordering extrusions in unusual sizes will cost more.

▶ Planning: the planning team will look at the whole process and might even be involved before the design stage.

- Ordering: it is important to order stock in cost-efficient quantities. For example, it may be more economical to buy a bulk pack of ten motors when only nine are needed, and the extra one could be kept for spares. Ordering from the cheapest supplier will only be useful if it can supply the materials or parts in time for production. If a member of the purchasing team spots that a change in the design would lower the cost of materials, they could go back to the design engineer to discuss this.
- Goods in: this department unpacks incoming goods and puts them in the warehouse. It will check orders carefully to make sure all the goods have arrived and are in good condition. Any issues need to be dealt with quickly to ensure production is not affected.
- Production: as already mentioned, the production department will use a JIT approach to make sure everything runs perfectly.
- Packing and dispatch: in order to save money, the planners might ask the packing department to put orders together so that they can be sent in fewer vans and lorries. This will save on fuel and driver salaries. Reducing the amount of packaging needed would also help to reduce costs.
- Disposal: waste materials, returned goods and obsolete stock should be disposed of correctly.

Whole-life approach

The life cycle can, and should, be examined in the context of the whole life of the product, from the raw materials used to make it all the way to the final destination of any waste produced.

Extending this even further might mean analysing what happens to used tools and equipment, how waste heat from manufacturing might be used to generate electricity, or how water that has been used for cooling or washing is disposed of.

Depreciation

Anything that has a cost will gradually reduce in value. For example, a new car loses around five to ten per cent of its value the moment it is driven off the forecourt, and will continue to lose value over time. This is called depreciation. Some exceptions include antiques, vintage items, fine art, collectable or unique products that might increase in value due to changes in fashion.

▲ Figure 15.8 A new car depreciates in value the moment it is driven off the forecourt

A company's value includes the value of its stock and assets, so these must all be taken into account when considering depreciation.

Depreciation is sometimes measured in percentage per annum (how many per cent a year). For example, a robot that cost £120,000 when new could depreciate at 30 per cent a year as follows:

- Original price = £120,000
- End of year 1: $120,000 - 30/100 \times 120,000 = £84,000$
- End of year 2: $84,000 - 30/100 \times 120,000 = £48,000$
- End of year 3: $48,000 - 30/100 \times 120,000 = £12,000$
- End of year 4: $12,000 - 30/100 \times 120,000 = -£24,000$

This means that the robot has a negative value by the end of year 4. At this point, it is 'written off' and does not need to be included in the value of the company. It may still be useful though and could last for several more years. The value calculated in this way only relates to the value of the robot if it were to be sold. The value to the company is harder to calculate but it is not worthless.

Industry tip

Not all engineering takes place in a workshop or sitting at a computer. Sometimes, engineers have to complete tasks that at first do not seem to be directly related to engineering. For example, they may need to work with the purchasing department, the sales department or possibly the human resources department. It can be useful to have an engineer involved in the decision-making processes in these other areas. When working in the engineering industry, it is important to be flexible, as much can be learned from other specialists.

Assessment practice

1 What is another word for stock?
2 Some stock requires environmental control during storage and transportation. What conditions might be required to keep perishable food stuffs fresh?
3 Which type of stock can be described as 'the basic materials from which products are made'?
4 As well as the amount to purchase, what other four costs are associated with stock?
5 Explain how the use of computers has helped with stock control.
6 What are the advantages of using barcodes on stock items?
7 What is the name of the system where an operator is directed to the correct location for a component by a light?
8 A company is disposing of a set of electronic scales. What legislation should it follow when it does this?
9 A small engineering firm has received an order that requires extra people to complete all the stages. Which part of capacity management will it need to consider?
10 If a forklift truck cost £50,000 when new and its value depreciates at 15 per cent a year, how many years will it be before it can be 'written off'?

Project practice

You are working in the production department of a large electronics manufacturer which has to assemble a prototype printed circuit board.

Make a list of all the tools and equipment you might need to solder three boards, together with possible purchase costs. Outline any health and safety measures and training that might be required in order to carry out the work.

16 Continuous improvement

Introduction

In an ideal world, every manufacturing activity would be carried out with perfect effectiveness at the minimum possible cost. In the real world, however, this almost never happens, and there are many practical reasons why not. There are usually opportunities to improve performance, in order to increase profits or be more competitive in the marketplace.

In this chapter, we will explore the principles and practice of how manufacturing companies can continuously improve their performance.

Learning outcomes

By the end of this chapter, you will understand:

1. methods of gathering feedback and evidence about performance, including types of KPIs and how these can be used to evaluate continuous improvement activities
2. how the eight wastes affect the performance of engineering activities
3. that continuous improvement involves incremental change
4. the purpose, methodology, benefits and limitations of a variety of continuous improvement techniques.

16.1 Continuous improvement principles and practices

When manufacturing products, activities can be classified as either value-added or non-value-added:

▶ Value-added activities change the form, structure, properties or appearance of a product in some way to increase its worth to the customer, for example changing the shape of a piece of material by machining.
▶ Non-value-added activities do not change the characteristics of a product, for example moving parts around on a forklift truck. They consume resources but do not increase the worth of a product.

In production, any non-value-added activity can be classified as a **waste**. Eliminating or reducing waste activities improves the performance of a company.

▲ Figure 16.1 Moving parts around on a forklift truck is a non-value-added activity

The eight wastes

Seven types of waste (*Muda* in Japanese) were originally identified by Toyota. More recently, popular use has added an eighth waste: failing to make full use of the skills and knowledge of the workforce. The mnemonic TIMWOODS shown in Table 16.1 is a useful way of remembering the eight wastes (see also Chapter 13).

▼ Table 16.1 The eight wastes

Type of waste	What it means
Transportation	Unnecessary movement of materials and products
Inventory	Having materials and products that are not currently being processed
Motion	Wasted time or effort related to unnecessary movements of people or products
Waiting	Time spent waiting for the next process to be carried out
Over-production	Making more products than customers have asked for (but to the quality required)
Over-processing	Making products to a higher specification than needed (but in the quantity required)
Defects	Products or services failing to meet customer expectations
Skills (unused talent)	Not using the skills and knowledge of the workforce

How the wastes affect performance

Toyota's original seven wastes affect performance by increasing costs in two ways:

▶ Increasing the cost of making products:
  – Transportation, motion and over-processing all require labour time, which must be paid for.
  – Transportation requires equipment costs for resources such as forklift trucks, cranes and conveyor belts.
  – Defects increase costs, either due to the time required to rectify them or the time and materials required to manufacture replacement parts.
▶ Increasing the cost of operating the business:
  – Inventory and waiting increase the space required for materials. This extra space costs money in terms of rent and rates.

– Inventory, over-production and waiting increase the amount of materials and work in progress held by the company. Companies often borrow money to buy materials, either from banks or shareholders, and they have to pay this back with interest. More materials in stores or between processes means more money borrowed, more interest to pay and therefore more cost overall.

Reducing the amount of waste in a business therefore reduces the costs of making products and of operating the business. The ideal scenario, where a company manufactures products with zero waste, is an aspirational goal referred to as **lean manufacturing**.

Research

Using the internet, search for an example of how each type of waste has affected a different engineering company. Identify the steps taken by each company to reduce the waste.

Industry tip

Using KPIs allows employees to 'speak with data'. These objective measures mean that subjective opinions can be avoided, reducing disagreements between staff.

Key performance indicators (KPIs)

An important first step towards eliminating waste and improving performance is to evaluate current performance. This helps to diagnose where waste is occurring and identify areas with the most potential for improvement. This is essential to ensure the correct approaches are applied for improving performance. One way of carrying out this evaluation is to use **key performance indicators (KPIs)**. These are measures of how well a company is operating, and they vary depending on the type of business. Table 16.2 gives some examples.

Key terms

Lean manufacturing: a production method that involves manufacturing with the minimum waste.

Key performance indicators (KPIs): quantifiable measures used to evaluate the success of an activity or organisation.

▼ Table 16.2 Examples of common KPIs

KPI	What it measures	Example
Throughput	Number of products manufactured per unit of time	1200 products/day
Changeover time	Time taken to switch processes between making different products	90 minutes
Total processing time	Sum of the actual machining time of the different processes needed to manufacture a product	46 minutes
Total cycle time	Time to manufacture a customer order from start to finish	28 days
Scrap	(Total number of products scrapped/total number of products produced) × 100	7%
Yield	(Total number of good products produced/total number of products produced) × 100	93%
Capacity utilisation	(Output achieved/potential output) × 100	68%
Operation effective efficiency (OEE)	(Total number of good products produced × ideal processing time/planned production time) × 100	60%
On-time delivery	(Total number of products delivered on time/total number of products produced) × 100	85%

Key term

Operation effective efficiency (OEE): a measure of how well a manufacturing process is used in comparison to its potential.

Test yourself

Without looking at Table 16.2, write down all the KPIs you can recall and what each one measures.

By reflecting on the processes and methods they use, companies can evaluate how well they are currently performing and identify where there is potential for improvement. Sometimes this involves completely reorganising their facilities and implementing new processes, but this can be time consuming and expensive and can also disrupt production. More often, companies will ensure continuous improvement by making small, incremental positive changes; the Japanese refer to this approach as **Kaizen**.

Key term

Kaizen: a Japanese term meaning continuous improvement through small, incremental positive changes.

Plan, do, check, act

Once an opportunity for improvement has been identified, a company can focus its efforts on achieving the maximum benefit for the time, cost and resources it has to invest.

One approach to managing the improvement activity is to use the 'plan, do, check, act' (PDCA) cycle, as shown in Figure 16.2. It is also known as the Deming cycle, after the engineer who first created it.

The cycle involves four steps:
▶ plan – identify what improvement is required, how it can be achieved and how it will be measured
▶ do – implement the planned improvement activity

▶ check – compare performance against what it was hoped would be achieved
▶ act – identify where changes are needed and adjust the processes and practices used.

The PDCA cycle is carried out iteratively. This can mean going 'round the loop' several times, making small, incremental improvements each time, until the desired improvement in performance is achieved.

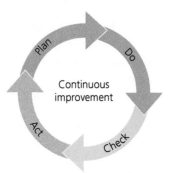

▲ Figure 16.2 PDCA cycle

Approaches to continuous improvement

A wide variety of tools and approaches can be used to improve performance. Each requires effort and resources and has different benefits and limitations.

If a company started randomly applying different approaches to improvement, it might get lucky and see some benefits, but it would more likely find the effort and cost of doing so exceeded the benefits. As an analogy, if someone goes to the doctor, the doctor diagnoses what is wrong and provides an appropriate treatment. They do not just provide a random treatment and hope for success. Likewise, companies need to select the most effective tool to achieve the improvement they want.

Value stream mapping

Value stream mapping (VSM) is a technique for analysing how materials and information flow through a company. It can be used to identify the eight wastes, or to compare actual flows to the ideal flow. The ideal flow is often called the future state map.

Key term

Value stream mapping (VSM): a technique for analysing how materials and information flow through a company.

VSM involves 'walking the process'. In the case of materials, this means actually following the route that they take from arriving at the factory to leaving as finished products. Whenever there is an activity involving the materials, it is recorded, often using a VSM form like the one shown in Table 16.3. Each activity is classified as either value-added (for example machining the required features) or as one of the eight wastes.

▼ Table 16.3 A VSM form

Step number	Activity	Value added?	Type of waste
1	Move pallet from stores to work area	No	Transportation
2	Pallet in work-in-progress area	No	Waiting
3	Move pallet to machine	No	Transportation
4	Load part into machine by hand	No	Transportation
5	Machine drill three holes	Yes	N/A

The benefit of VSM is that it requires relatively little effort to highlight areas for improvement. This means that any subsequent improvement activity can be focused to achieve the maximum return on the effort. It can also be used to create future state maps that have the minimum possible amount of waste, allowing for the complete reorganisation of a company.

The main limitation of this approach is that each value stream map follows a single product. If a company batch manufactures a wide variety of products, the flow of materials will be different for each. This may require multiple different value stream maps, to ensure changes that benefit one product do not adversely affect another.

Industry tip

A simple way to illustrate the amount of transportation in a company is to measure the distance travelled by all the parts in a product, from when they arrive at the company until the product leaves. This provides an easily understood value that can be used as a measure of success.

Visual management

Visual management is a management approach where information or operational status is communicated in a visual form, rather than using written or verbal instructions. It allows information to be easily and quickly understood.

Key term

Visual management: a management approach where information or operational status is communicated in a visual form.

There are a range of visual management tools, as shown in Table 16.4. Each has a different purpose and specific benefits.

▼ Table 16.4 Examples of common visual management tools

Visual management tools	Purpose	Benefits
Progress/countdown boards	To show, for example, the number of: • parts produced • products still to be made • days since an accident	Focus workers' attention on the KPIs presented on the board Motivate workers to achieve the KPIs
Shadow boards	To show that all the tools needed for a task are available and in the correct place	Stop time being wasted searching for tools that have been misplaced or borrowed by other workers
Standard operating procedures (SOPs)	To specify a consistent and uniform method for carrying out an activity	Help to maintain consistent quality when making repeat products
Traffic lights	To indicate when it is safe to enter an area	Prevent accidents and collisions
Health and safety signs	To indicate hazards and/or required control measures	Reduce the risk of injury
Floor markings around machines	To show areas that may be unsafe to enter when machines are operating	Prevent accidents and injuries

In general, visual management tools are most effective when the work requires standard methods or the same tools each time, and when there are clearly defined KPIs. The main limitations include the effort required to keep information up to date and the fact that workers need to actively engage with the tools. For example, a shadow board is of limited use if no one bothers to replace tools that are missing from it.

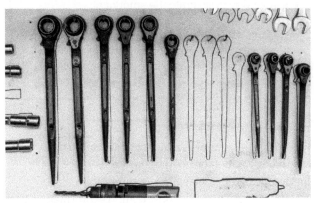

▲ Figure 16.3 A shadow board is a visual management tool

Test yourself

Give three examples of how visual management can be used to support health and safety in an engineering workplace.

6S

6S is an approach that is used to help create an organised, efficient and safe workplace. A 6S activity usually focuses on a single working area and involves making sure that:
▶ all items that are not required for the work activity are removed from the area
▶ all items that are required for the work activity are available and positioned as conveniently as possible.

Key term

6S: an approach that is used to help create an organised, efficient and safe workplace.

Each step of 6S is represented by a keyword and has different benefits, as outlined in Table 16.5.

In order to maintain the benefits, a 6S activity requires active participation by workers in the area and ongoing engagement. However, workers should be incentivised by the prospect of a cleaner, more organised workplace, and this should support engagement and make their daily activities easier to carry out.

The main limitation of 6S is the time needed to train workers in the approach and then to implement it – this can take one to two days for each process area.

▼ Table 16.5 The steps of 6S

Step	What it means	Example	Benefits
Sort	Identify items that are required in the workspace	Sort through all the items in a location and remove any that are not needed	Reduces the time taken to find required items Increases available space Eliminates obstacles
Set in order	Put items in the easiest place for when they are needed	Place all tooling so it is easy to reach from where it is needed Place the most frequently used items nearest to the workspace	Reduces the time taken to find required items
Shine	Clean the workplace	Regularly remove waste products Clean machines and equipment	Improves process efficiency and safety Makes maintenance issues visible
Standardise	Establish procedures to ensure the first three Ss are maintained	Develop visual controls and standard operating procedures	Ensures consistency of practice
Sustain	Ensure the good practices are maintained	Use visual controls and standard operating procedures	Maintains the benefits of the first four Ss
Safety	Manage and minimise hazards and risks	Carry out risk assessments and implement control measures	Ensures workers are safe and fewer accidents occur

 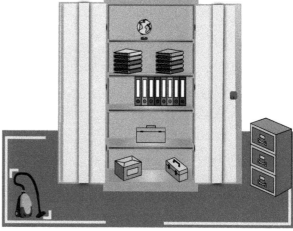

▲ Figure 16.4 Before and after 6S

Single minute exchange of die

In order to produce different products during batch manufacturing, it is necessary to make changes to processes. This may involve, for example, swapping over tools, repositioning jigs, resetting machine parameters and checking the first part produced. While these activities are taking place, no products are being made, so there is no value added, despite workers still being paid. Minimising the time to change over between products therefore reduces non-value-added costs.

Single minute exchange of die (SMED) is a term that was first used by Toyota. The goal of SMED is to complete every changeover between products in less than a minute. However, the term is now commonly used to describe most activities undertaken to reduce changeover times, even if they take a lot longer than a minute.

> ### Key term
>
> **Single minute exchange of die (SMED):** a system used to reduce changeover times when making different products.

The procedure for carrying out a SMED activity is as follows:

1 Identify all the steps in the current changeover activity. This may involve recording the process and documenting it in detail.
2 Remove all external elements from the changeover process. These are things that can be carried out while the process is running. For example, the tools and documentation needed for a changeover could be found while the previous batch is being processed, not when the process is being changed over.

3 Convert internal elements to external elements. Internal elements are things which currently have to be done when the process is stopped, for example sharpening a cutting tool when changing between different products. This could be changed from an internal to an external element by having a second cutting tool, which is sharpened while the previous batch is being processed.
4 Standardise the process. Document the new, fastest, most efficient method of carrying out the changeover, for example using a standard operating procedure.
5 Train staff. Ensure all workers follow the improved standardised process.

Faster changeover times result in less equipment downtime and lower manufacturing costs. Alternatively, more frequent changeovers can be carried out without increasing manufacturing costs. This means that batch sizes can be reduced, resulting in fewer finished products being held and greater customer flexibility to customise the design of batches.

There are some limitations of the SMED approach:
- ▶ SMED activities tend to be most effective in reducing changeover times when products are being manufactured in batches, with common, understood characteristics between different products.
- ▶ The implementing team needs to have a good understanding of the practical requirements of the changeover process.
- ▶ It may require investment in jigs and fixtures or additional tooling.
- ▶ It takes time to train workers in the approach and then to implement it.

▲ Figure 16.5 Faster changeover times result in less equipment downtime and lower manufacturing costs

Total productive maintenance

Total productive maintenance (TPM) is a strategy to keep equipment in effective working condition and available for operation by ensuring that the whole workforce is involved in its maintenance rather than just a maintenance team. Its main objective is to increase operation effective efficiency (OEE) – one of the KPIs outlined earlier in this chapter.

Key term

Total productive maintenance (TPM): a strategy to keep equipment in effective working condition and available for operation by ensuring that the whole workforce is involved in its maintenance rather than just a maintenance team.

Higher OEE means less unplanned downtime and fewer interruptions to production, leading to greater productivity. This in turn results in either increased income or the ability to lower prices (due to overhead costs being divided between more products) in order to gain a competitive advantage. TPM also helps to maintain a consistent product quality. The combination of more reliable production and a more consistent product can improve customer satisfaction and confidence in the manufacturer.

The only limitation of TPM is the time required to implement it.

Implementing TPM involves a number of activities, often referred to as the eight pillars of TPM, and these are described in Table 16.6. A key feature of the pillars is that they rely on the experience of operators, as they are typically the most aware of when a machine is working at its best (or not).

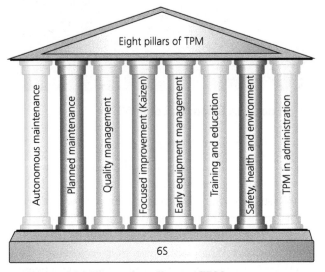

▲ Figure 16.6 The eight pillars of TPM

▼ Table 16.6 The eight pillars of TPM

Pillar	Description
Autonomous maintenance	Making operators responsible for routine maintenance activities, such as lubricating, cleaning and inspecting machines
Planned maintenance	Replacing components just before they are predicted to fail, to ensure continuity of performance
Quality management	Including error detection and prevention in the production process
Focused improvement (Kaizen)	Making continuous small changes to improve performance
Early equipment management	Using the experience of operators to improve the design of new equipment
Training and education	Ensuring all staff (operators and managers) understand the requirements of TPM
Safety, health and environment	Ensuring the working environment is safe
TPM in administration	Looking beyond the immediate activity to improve things such as order processing and scheduling, to get the best performance out of equipment

Kanban

When batch manufacturing products, there may be several processes involved, which may not necessarily be carried out in the same sequence. How these products flow through the manufacturing processes has a significant impact on the performance of the company.

Traditional batch-manufacturing companies may start the manufacture of a batch but allow the workers at each step of production to decide which batch to do next. This often means that there are several different batches of work in progress awaiting the next process step. The advantage of this approach is that workers can choose a job that requires the minimum amount of effort to change over. The disadvantages are that they may choose an easy job rather than one needed by the customer, and the cost of the extra space needed to hold all the work in progress can be high.

Kanban is a system used to manage the flow of work in a manufacturing company. It is a type of pull system, meaning that products are only made when there is customer demand (rather than making stock for a warehouse). Typically, Kanban is used where companies are batch manufacturing products that involve several different processes or operations.

It is often implemented using a card system or Kanban board, where instructions are issued for each process, telling the worker which batch of products to work on. Workers must follow the Kanban sequence of manufacture, making the products in the order required. While it is straightforward for them to follow, a lot of work may be required in the background for the production scheduling and to create the instruction cards.

> ### Key term
>
> **Kanban:** a system used to manage the flow of work in a manufacturing company by regulating the supply of materials between processes through the use of instruction cards or boards.

The advantage of Kanban is that, compared to conventional approaches, it normally reduces the amount of inventory. As mentioned above, this in turn reduces the space required for work in progress and the amount of money that the business needs to operate. Furthermore, if a quality defect occurs in a batch, for example due to setting up a process incorrectly, it can be quickly detected, thereby minimising scrap and waste.

The main limitation is the time needed to schedule each process, which may increase labour time and the cost of the scheduling activity.

▲ Figure 16.7 A Kanban board is used to issue instructions for each process, telling the worker which batch of products to work on

Assessment practice

1. What is meant by a waste activity in manufacturing?
2. Explain why 'waiting' is a waste activity.
3. Explain the difference between over-production and over-processing.
4. Explain two ways in which the eight wastes can affect the financial performance of a manufacturing company.
5. Give two reasons why companies use key performance indicators (KPIs).
6. A lathe is used to manufacture shafts. The ideal processing time for each shaft is ten minutes. In a typical eight-hour shift, the lathe manufactures 33 shafts. Calculate the operation effective efficiency (OEE) of the lathe.
7. Other than OEE, give two KPIs that are commonly used in manufacturing.
8. Describe how a value stream mapping activity is carried out.
9. Give two examples of visual management that are commonly used in an engineering workplace.
10. Explain one benefit and one limitation of using the Kanban system.

Project practice

1 Identify a working area in a workshop you have access to and undertake a 6S activity. Before you start, carry out a risk assessment and be sure to implement any identified control measures.

2 Identify a single process or machine that is being used to manufacture different products. This could be, for example, a lathe that is being used to manufacture parts for different student projects. Time and observe the changeover process, identifying the internal and external elements. Carry out a SMED activity, then time the changeover process again.

3 Write a report of your SMED activity. Summarise how the activity has modified the changeover process.

17 Project and programme management

Introduction

Having spent the previous chapters exploring the different aspects of a project, we can now bring together areas such as health and safety, finance, professional behaviour, stock control and continuous improvement, and look at how to plan a whole project.

In order to control a project, it is essential to have a clear, accurate and feasible plan. This plan must be understandable not only to those closely involved with the technical aspects of the project but also to those who do not have an engineering background, for example finance, sales and marketing personnel.

In this chapter, we will identify different ways of planning and explore how various techniques can pull together all the required information.

Learning outcomes

By the end of this chapter, you will understand:
1. how projects are defined and structured
2. the management practices, processes and documentation needed at each stage of the project
3. types of risk and how these are managed throughout the life of the project, including the role of research and development
4. the benefits and limitations of collaborative working
5. the responsibilities of different roles and how they contribute to a project

6. how to identify the resources required to carry out a project
7. the benefits and limitations of the different planning methods
8. how to plan projects using the different methods
9. how to monitor and evaluate the progress of projects
10. the reasons for reviewing and evaluating projects in order to improve subsequent projects.

17.1 Principles of project management

What is a project brief?

Whenever there is more than one person involved in a project, there needs to be a clear and approved direction for the team to follow. This starts with a **project brief** that outlines the scope and objectives of a project.

> **Key term**
>
> **Project brief:** a short description of the scope and objectives of a project.

A project brief could read as follows:

> 'Design and make a fitting to attach to a set of railings that allows safe securing of bicycles.'

The brief is straightforward but lacking in detail, and the project team may have to go back to the client to ask for more information. An improvement would be:

> 'Design and make a fitting for the local council to attach to a set of railings that allows safe securing of bicycles. The railings are of the vertical, wrought-iron type and the fitting must be weather-resistant.'

This brief has more detail but the project is not contextualised. Let's improve on that:

> 'The theft of bicycles is increasing, especially with the use of e-bikes becoming more widespread. To combat this, the local council is going to install fittings to the railings around the council buildings so that employees can secure their bicycles to the railings and prevent theft. Design and make a fitting to attach to a set of railings that allows safe securing of bicycles. The railings are of the vertical, wrought-iron type and the fitting must be weather-resistant.'

The brief is improving with each piece of information added, but the team still does not know other important details such as timescales, quantities and safety requirements. Let's try again:

> 'The theft of bicycles is increasing, especially with the use of e-bikes becoming more widespread. To combat this, the local council is going to install

fittings to the railings around the council buildings so that employees can secure their bicycles to the railings and prevent theft. Design and make a batch of 20 fittings to attach to a set of railings that allow safe securing of bicycles. These should be available to be launched at the town spring fair on 21 May 2024. The railings are of the vertical, wrought-iron type and the fitting must be weather-resistant and made of steel with a powder-coated finish.'

This brief is much better, and the team could probably start the project with this level of information. The client could make it even clearer by supplying a drawing with dimensions.

In industry, projects can be vastly more complicated than this, and the discussions between the various stakeholders can be lengthy. For large projects, it can take months or even years before a project brief is agreed.

> **Health and safety**
>
> When undertaking any project, it is important to consider health and safety:
> - Does the product itself present any health and safety issues (toxic materials or other hazards to health)?
> - Will the project include any dangerous activities (hazardous processes or materials)?
> - Are there any product uses that might be a health and safety issue?
>
> If any issues are found, they should be clearly noted in the project plans and documentation, so that control measures can be implemented.

How to identify a good project brief

A good project brief will have clear information about:
- timescales
- quantities
- quality
- materials
- safety requirements
- consumer profiles (consumers are not necessarily the same as the client or customer).

There should be indications of what part of the project will be needed by what date. For instance, the bicycle fittings mentioned above might need to be at the prototype phase by early January so that the client can agree on the design. There could be a range of sizes specified, and these would need checking before the project could be completed. These are known as **project goals** or success criteria.

It is common to have points in a large project where the client can check that everything is going to plan. Internal projects, where the client is the company itself, would be checked by the **management**.

Test yourself

Think of a product that is well known to you and try to write a project brief for it. Do not include its common name, for example you cannot say 'design a hammer'. Imagine it is a new product and keep it simple, such as a pencil or a chair.

Check your brief against the information on identifying a good project brief, then show it to a fellow learner or your tutor. Can they work out what the product is? Don't forget, you cannot use its common name!

Project life cycle

Initiation

A typical project will start when one of three things happens:

▶ the opportunity for a new product is identified by a company (internal client), often the research and development department (technology push)
▶ the marketing department (internal client) identifies that the company needs to design a new product in order to keep up with the current market (market pull)
▶ an external client makes an enquiry (market pull).

If the company is interested in pursuing the project, management will nominate a project leader. They will investigate the idea and help create the project brief with the client (whether internal or external).

Planning

Once the project leader has a good project brief to work with, and if the project is complicated enough to require the involvement of many departments, they will need to form a project team. The project team will create a plan for the company to follow, with timescales and resources (staff, tools, equipment and materials) clearly identified.

Implementation

This is about putting the plan into action. All the relevant people will be given tasks to follow, with the project leader checking on progress.

Monitoring and reporting

Monitoring and reporting involve checking project progress against the plan to identify any possible changes that may affect the outcome.

At defined points in the project, the team will meet to discuss progress and make any changes required. If there have been any major problems, the project may have to be restarted.

A large project may be divided into several smaller projects that require more complicated control and checking.

Evaluation

Once the project is complete and the client has agreed that the outcome is what they needed, the project will be closed. It is useful at this stage for the company to look at any lessons that could be learned from the process in order to help future projects.

Constraints

Quite often, constraints might be put on the project team that limit what it can do and that could affect the project's success:

▶ Available time: often a project needs to be finished within a short timescale and the project team will need to plan around this.
▶ Costs of the project: a client will usually put a limit on the cost of the final product and this can influence quality.
▶ Customer or client requirements: as an example, religious beliefs may rule out the use of animal products such as leather. Some oil seals are leather, so a suitable polymer replacement would need sourcing.
▶ Available staffing/skills: for example, while a project might require the use of a complex composite material (such as carbon fibre), there may not be anyone available in the company with the correct skill-set to handle it.

▶ Available machinery: for example, while the project team might want to use a robotic welding machine, the company may not have one available and another way of welding joints would need to be found.

▶ Benefit to the business: even though an opportunity has been identified, the company may decide that it is not worth pursuing as there will be no benefit to the business in terms of future developments or contracts.

▲ Figure 17.1 The availability of machinery may act as a constraint on the project team

How to manage risks in a project

Risks need to be taken into account at all stages of a project:

▶ Risks to costs and budget: if a project spends more than was budgeted, often because it takes too much time, the company might end up not making any money. This means that the project leader and their team need to keep track of costs at all times and let the management know if they foresee any problems.

▶ Risks to quality: if projects are rushed, quality can suffer. If the quality of the product is not good enough, the client might refuse to accept it and the project will fail. It is always a case of balancing time, costs and quality.

▶ Risks to safety: employees need to be kept safe when designing and making a product. If a project includes new processes or materials, a company will need to ensure it includes updated precautions in its working practices. For example, it is the company's responsibility to ensure that all staff understand the dangers of working with hazardous substances and how to avoid injury. The product could be corrosive or explosive under certain conditions, and everyone needs to be kept safe. If the end product carries any safety risk, the client needs to be informed in order to take the correct precautions.

▶ Risks to the reputation of the company: a company's reputation will be negatively affected if it fails to take safety seriously or produces low-quality items.

▶ Risks to the project:
  – Poor communication can be a problem. If a project brief is not clear and accurate, or the information passed to the production department is incomplete, the project will fail when progress is checked. It is essential to communicate clearly with the client at the outset of the project.
  – A shortage of resources (staff, materials, equipment, tools) will also lead to project failure.
  – If the requirements agreed in the project brief change, the whole project may need to start again. Small changes can sometimes be accommodated, but the later they are introduced, the harder it is to change the course of the project.

Most risks can be foreseen with good planning. A process known as FMEA can help to identify them.

▲ Figure 17.2 All staff must understand the dangers of working with hazardous substances and how to avoid injury

Using FMEA as a risk management tool

Failure mode and effects analysis (FMEA) is a process that allows all stakeholders to be involved at the early planning stage to try to identify any possible risks to the project from several points of view.

Representatives for all stakeholders are nominated to attend the project meeting and identify any risks. All points raised must be considered relevant and should be explored to see if there is any way of avoiding the risk. Sometimes this could be as simple as not using certain materials or tools, or changing a small detail of the design; more serious issues could be taken back to the client to discuss changes to the brief or an increase in time/costs.

Examples:

▶ A member of the production department might notice that the existing tooling cannot cope with the depth of cut required by the client. Possible solutions could include investing in new tooling or changing the design (in conversation with the client) to be achievable with the current tooling.

▶ A member of the sales department might know the client well and identify a cultural requirement that is not clear in the brief. A possible solution could be to contact the client for clarification.

▶ Packing and dispatch staff might point out that a product's final dimensions or weight might be too large to fit in one of the current trucks. Possible solutions might include investing in new transport, hiring a truck (if it is a one-off project), shipping the product in component parts or contacting the client to reduce the final product size.

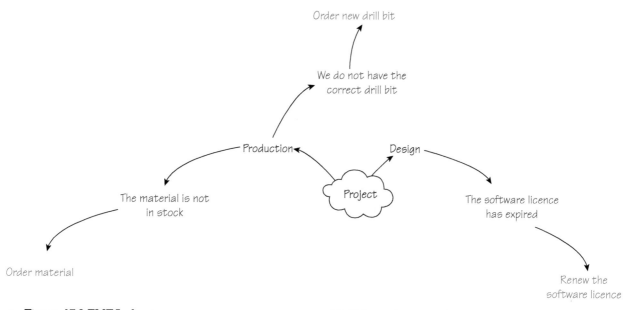

▲ **Figure 17.3 FMEA chart**

FMEA can be carried out at the design stage, known as DFMEA, or at production, known as PFMEA, but the process is the same.

How to work together

As we have already seen, modern project management involves effective teamwork. Including all stakeholders at all stages of the project allows for a wide-ranging input of knowledge and opinions, as well as generating a feeling of investment in the outcome. Employees who are engaged with the process are more likely to work carefully.

In a project team, there should be representatives of several departments, to ensure a broad base of knowledge, experience and expertise. As management may have oversight of several ongoing projects, and cannot lead each one, they will usually appoint someone from within the team as project leader. This may not be the most senior employee.

Working within a multi-level, multi-specialist team is sometimes called matrix working, because reporting is not linear (from bottom to top) and people with different levels of seniority work together. Rather than holding **meetings** at long tables in conference rooms, which can tend to feel more formal with a head-of-the-table effect, they tend to be held in designated areas of the shop floor, using noticeboards, screens and whiteboards, with no tables or chairs. This not only keeps people focused but also makes the meetings shorter and more task-oriented

Team communication

Working as a team means members communicating with one another. Regular meetings, whether face to face or remote, will help to avoid misunderstandings that could cause mistakes and delays. Communication can also involve emails, texts or meeting **minutes**.

Minutes record what is discussed during a meeting and any actions. An action is a task assigned to a member of the group, to be completed before the next scheduled meeting. The minutes are circulated to all members after the meeting is closed, so that everyone has a clear record and knows what they are supposed to be doing. Using the formal minute-taking process can give structure to team communications.

Meetings typically run as follows:
- An agenda is sent out in advance of the meeting, often by email, which lists everything that needs to be talked about.
- The meeting usually starts by recording attendance and apologies for absence.

- The minutes of the last meeting are discussed. If there were any actions, these might be looked at (matters arising).
- The last section is any other business (AOB), where new items not on the agenda are discussed.

Collaborative technologies

Email has made a huge difference to the way companies communicate. From the late 1980s onwards, it steadily took over from letters, fax and telephone to become the main form of communication.

▲ Figure 17.4 Email has made a huge difference to the way companies communicate

Email became even more important in the early twenty-first century, with the ability to organise meetings and events and pass **documents** and images to collaborators across the world.

The use of **video calls** started in the late 1990s, but these mostly took place using expensive systems that worked over the standard telephone network (slowly) or via satellite. However, as internet bandwidth increased, using the internet to communicate face to face became commonplace.

Key terms

Meetings: planned occasions when several members of a team come together to discuss a project.

Minutes: the record of what is discussed during a meeting and any actions.

Documents: written, printed or electronic records of data and/or text.

Video calls: telephone calls that use video to allow participants to see each other in real time.

The global COVID-19 pandemic in 2020 accelerated the use of video conferencing and it became widespread. Easy-to-use applications such as Zoom or Microsoft Teams meant that meetings could be held without special equipment, cameras or microphones and using just a simple laptop. Working from home and social distancing changed the way people worked and studied, possibly forever.

Cloud technology allows documents to be stored online and accessed simultaneously by different members of the same team, facilitating **collaboration**. The changes made by each person can be tracked, saved, changed, reversed and annotated.

> **Key term**
>
> **Collaboration:** working together towards a common goal.

Possible disadvantages of collaboration

While collaboration has many advantages, some members of the team might be hesitant to offer new ideas, deferring to more experienced members and their opinions. This can result in a lack of fresh thinking.

Unless the process is controlled by the project leader, collaboration can also result in team members having either too much or too little to do, or an unclear idea of what is required of them personally. It is a management responsibility to monitor the work of the team to ensure everything runs smoothly.

> **Research**
>
> Research how school/college work was delivered before, during and after the COVID-19 pandemic. Produce a small presentation to compare how things changed.

17.2 Roles and responsibilities in projects

Defined roles in a project team

Except for very small projects, where one person can complete all the tasks on their own, a project team should comprise members with different roles and areas of knowledge, experience and expertise. Without a range of viewpoints, it would be impossible to discuss an issue.

▲ Figure 17.5 A project team should comprise members with different roles and areas of knowledge, experience and expertise

There are many roles in a project that must be considered.

External stakeholders

Clients will initiate a project and agree a project brief with the team. Small changes brought about by issues in the design or production processes can be overseen by the project leader, but the opinion of the client may need to be sought for larger problems. Sometimes customers or consumers have the final say on a project, as they are the ones who are going to buy a product or service.

Some areas of work have regulations that need to be followed. This can sometimes be overseen by the project team or a special department in the company, but often there are regulators who check the company's work, for example:

▷ the Environment Agency can fine an individual or a company if they are not working in an environmentally friendly way, such as letting waste liquids flow into a local river

▷ the Food Standards Agency can stop production of food for human consumption if it thinks there is a danger to health

▷ the Health and Safety Executive enforces workplace health, safety and welfare.

Management

Someone from the management team will be involved with the project. If it is a fairly routine, low-value or small project, they may only check on progress occasionally. If the project is important or high value, they might attend project meetings or ask to be copied into emails and documents.

For projects of significant importance that might directly affect the future of the company, a manager might be put in charge of the whole project and control it personally.

Project team members

The project leader controls the project, making sure everyone knows what they should be doing and by when. If something goes wrong and the project slips, the project leader will try to get it back on track by allocating extra resources or by negotiating with the client over the end date. It is also the responsibility of the project leader to keep management informed of progress.

Some team members might be given specific responsibilities, such as:
▶ health and safety
▶ liaising with regulators
▶ keeping documentation up to date
▶ ensuring clear communication of information
▶ overseeing financial aspects of the project
▶ reporting back to management.

> **Industry tip**
>
> When you are working in industry, every project you are part of will have a plan. Always look at the whole plan and work out how and where your tasks fit in.

17.3 Project planning and control

What do we need to know to plan?

Imagine you are made project leader of a new project. There are things you need to know before work begins.

First, you need to check who is available to join your team. It is important to choose:
▶ employees with relevant experience (perhaps not on an identical project but on something similar or that used similar skills); if they have had experience leading a project before, they might be able to support you in your leadership role
▶ representatives from different departments who can give a range of viewpoints and help to develop the project plan.

Next, it will be important to work with the client to produce the project brief that outlines the scope and objectives of the project, including information on design, constraints and timescales. All this is necessary to be able to make a plan.

Finally, you need to understand any software or processes that the company uses in its planning. It is unlikely that you will be able to use a planning technique of your choice; the company will probably have policies, procedures and processes in place for all projects.

In general, you will need information on the following:
▶ Time: how long have you got? What is the maximum timeframe? Is there a possible extension?
▶ Budget: how much money have you got to implement the project? This must include any employee hours spent on the planning itself and also any time spent in meetings, as well as wages for the manufacturing time.
▶ Human resources: who is available to complete the work?
▶ Training needs: do the people available have enough experience or do they need further training? (This also needs to be included in the budget.)
▶ Communication needs: what information do people need and how is it best communicated? Is it best to use email, noticeboards (online or physical) or face-to-face communication?
▶ Production facilities: do you have the equipment available to make what is needed in the required quantities and within the time available, or do you need to budget for more?

All these points need resolving during the planning stages of the project.

> **Improve your English**
>
> On small pieces of paper, write out the planning points:
> ▶ time
> ▶ budget
> ▶ human resources
> ▶ training needs
> ▶ communication needs
> ▶ production facilities.
>
> On separate pieces of paper, write short descriptions for each planning point in your own words.
>
> Ask a fellow learner or your tutor to try to match the keywords and descriptions.

Planning techniques

Gantt charts

A **Gantt chart** is a two-dimensional, graphical representation of tasks against time. It provides information on how long each task is expected to take and what resources are required.

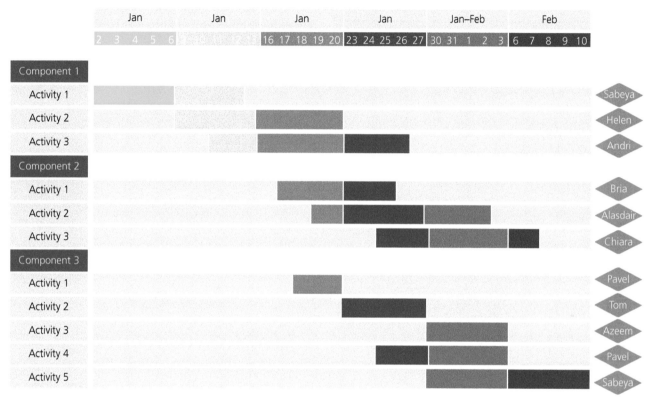

▲ Figure 17.6 A Gantt chart

The project is split into tasks and each team member is given a set of tasks to complete. The scale of time at the top could be hours, days or even months, depending on the size of the project. A large project will be split into several Gantt charts, to avoid the main one becoming too cluttered.

Advantages of Gantt charts include:
▶ they are easy for non-specialists to read
▶ they are widely used in more or less the same format
▶ they allow for resources (such as people and equipment) to be allocated to each task
▶ they show a good estimate of the length of the project.

Disadvantages include:
▶ they can quickly become extremely complicated and physically large if printed out
▶ if tasks are not completed all in one go but are instead spread out over the weeks, this can become hard to show on the chart.

PERT charts

A **program evaluation and review technique (PERT) chart** focuses on the tasks themselves. Each task is shown as a shape and is annotated with the estimated time to complete it. Ideally, the estimated times are added by someone with the correct experience, but sometimes different sources have different opinions, so they are calculated using a simple formula:

E (estimated time) = (O + 4M + P)/6

Where:
▶ O = optimistic time: if a member of staff is ready to work hard or does not understand the scope of the project, they might give an optimistic estimate
▶ M = most likely time: if a member of staff has broad experience and their opinion can be relied upon, they might give a most likely time as long as there are no problems or delays

▶ P = pessimistic time: if a member of staff has had a previous bad experience of this sort of task, they may give a pessimistic estimate.

The tasks are linked together in dependencies; this means that if task x has to finish before task y starts, then task y is dependent on task x. This could be because both tasks are to be completed by the same person or because both tasks need the same equipment.

Once the dependencies are sorted, the planner can see how the plan might flow from one task to the next. With careful examination, the planner can see where the longest linked list of tasks is, and this will form the critical path.

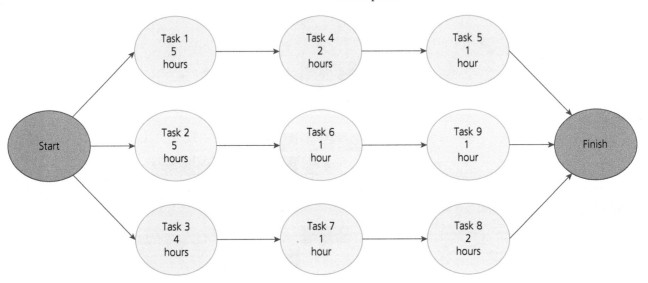

▲ Figure 17.7 A PERT chart

Critical path analysis (CPA)

The **critical path** is the list of tasks that will need the most careful control if timescales are not to slip. Paths that are not critical might have time available to take up any slack and accommodate tasks that take longer than planned. The critical path does not allow for **contingency**, so other ways of coping with change have to be found.

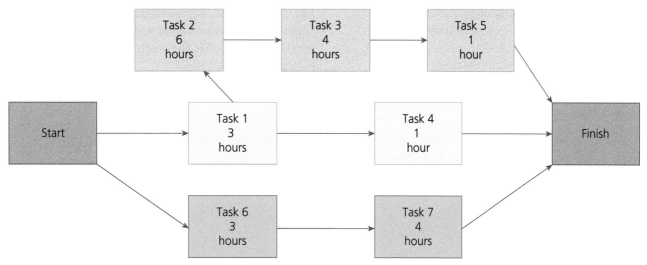

▲ Figure 17.8 Critical path analysis

In the example shown in Figure 17.8, there are seven tasks that need to be completed before the project is finished. Task 1 needs to be complete before task 2 can start, and this is shown by the arrow. Tasks 2, 3 and 4 can run on after each other, and tasks 6 and 7 can run separately.

What can be seen here is that tasks 1, 2, 3 and 5 add up to 14 hours; tasks 1 and 4 add up to four hours; and tasks 6 and 7 add up to seven hours.

Tasks 1, 2, 3 and 5 will take the longest, so they form the critical path. If something goes wrong at task 6, and it takes four hours instead of three, the whole project should still finish on time. However, if task 3 runs late, the whole project will be shifted.

Contingency planning

Once the critical path is recognised, the project team should put contingency plans into effect. This could include having extra personnel ready to help if needed or implementing backup plans if equipment breaks down at a crucial time. This helps to protect the critical path from failure.

Planning technologies

There are many computer applications that can be used to support planning activities. Some are based on Gantt or PERT charts, while others may use their own graphic layouts to represent project flow. Simple applications will only allow entry of tasks and lay out a provisional plan chart; the more complex (and usually expensive) variants will generate and control documentation, including meeting minutes, task control records and reports. Some packages include financial control (for example wages and tax) and the hosting of documents in the cloud for security. They

can help identify a critical path and notify the project leader if there is any slippage in the project.

Project control

One of the key management tasks is tracking how projects are progressing. Once a project is running, it may be tempting for management to keep asking how it is going, to the extent that it interrupts the flow of work and causes the project to overrun.

A more efficient way of working is for the project team to provide management with regular updates. This relies on accurate information being reported, so it will usually follow a formal reporting structure defined in the company's procedures. If the responsibility for this communication is delegated to a single team member, this will avoid disrupting the whole team.

Of course, it is also the responsibility of high-level management to monitor the project, and they might occasionally check the documentation and progress. They will ask for reports to be generated so that they can easily access the required information.

Monitoring reports might involve looking at the budget and costs, the quality of the product and the time the project is taking. The 'manage by stages' approach could be implemented, where management checks on the project at defined points in the plan. Everyone knows where these monitoring points will be and therefore has the information ready for checking.

Another process is 'manage by exception'. This requires greater expertise from the manager, who will work with the project leader to identify unusual stages in the plan where new processes are used or where the critical path is particularly tight. They will ask for reports at only those stages, so that they can help resolve any issues without the overhead of regular meetings and excess documentation.

The conclusion of a successful project will offer the opportunity to review any lessons learned and to make sure all useful knowledge is carried forward for use in the next project.

Assessment practice

1 At the beginning of a project, the project team and the client produce a short description of the project's scope and objectives. What is this called?

2 Suggest three ways in which a typical project might be initiated.

3 Suggest three constraints on a project team that might limit what it can do and in turn affect the project's success.

4 Explain why training staff will affect the project budget.

5 How does FMEA help with risk management?

6 What is the purpose of a meeting agenda?

7 Describe some of the effects that the 2020 COVID-19 pandemic has had on working practices.

8 How is estimated time calculated in a PERT chart?

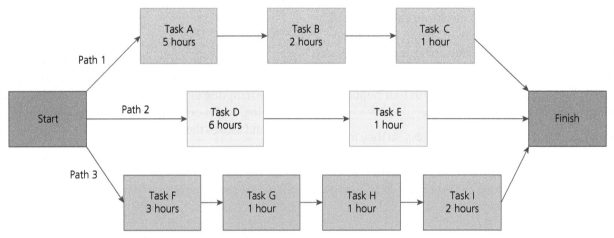

▲ Figure 17.9

9 Examine the PERT chart in Figure 17.9. Which is the critical path?

10 In Figure 17.9, which task is dependent on task G?

Project practice

Consider a project you are involved in and use what you have learned in this chapter to produce a considered plan.

This might involve calling a meeting of everyone concerned, performing an FMEA, constructing a PERT chart and Gantt chart and looking for the critical path.

Present this information in an electronic format for delivery to a group of stakeholders. The presentation should be no longer than 15 minutes, so be concise and clear.

Glossary

6S An approach that is used to help create an organised, efficient and safe workplace.

Absolute pressure Sum of gauge pressure and atmospheric pressure.

Absolute temperature Temperature of an object measured on the Kelvin scale, where zero is absolute zero.

Acceleration Rate at which velocity changes.

Accessibility Ensuring equal access to workplace facilities and equipment for all and removing barriers to achieve this.

Accidents Unplanned, undesired occurrences which may result in harm.

Accountability Having responsibility for an action, task or job that is to be completed.

Accountant A qualified person who prepares and audits a company's financial records.

Accuracy Degree of closeness between a measured value/dimension and its true value/dimension, often expressed as a percentage.

Actuators Devices that convert electrical, electronic or mechanical signals into physical movement.

Additive manufacturing Creation of complex components using 3D computer models; material is added in a series of thin layers which are gradually built up into the required shape.

Aerodynamics Branch of fluid dynamics that is concerned with how air moves around objects.

Alloys Materials comprising two or more metals in solid solution with one another.

Analogue signals Continuous signals, usually represented as sine waves.

Anions Negatively charged ions.

Anisotropic Having different mechanical properties when measured in different directions.

Anode Positive terminal of an electrolytic cell.

Anthropometric data Research-based data that provides guidance on typical measurements for different parts of the human body.

Arc Partial length of the circumference of a circle.

Arithmetic progression Sequence of numbers where the next term is found by adding or subtracting a common difference (d) to or from the previous term.

Artificial intelligence (AI) Ability of machines to gather information, perform analysis and make autonomous decisions.

Assurance Certainty about something.

Atom Smallest unit of a chemical element that exhibits all the physical properties of that element.

Augmented reality (AR) Technology that allows users to enhance their real-world experience with real-time, computer-generated augmentations or overlays.

Autonomous systems Systems that integrate AI, vision systems, sensor technology and robotics to achieve independent and self-directed operation.

Auxiliary view View used when a standard orthographic view is unable to show the required detail for a component; commonly necessary where features such as holes are present on sloping or inclined surfaces.

Balanced forces Forces that are equal in size and act in opposite directions.

Base units Fundamental units of the SI system that are defined arbitrarily and can be used to derive other units of measurement.

Bell-crank linkage Type of linkage where the direction of motion is converted between horizontal and vertical; the output force can be increased or decreased compared to the input force, depending on the positioning of a fixed pivot.

Bending moment Measure of the bending effect that occurs when a force is applied to a beam or other structural component.

Bessemer process Method of producing steel by burning off carbon and other impurities in pig iron by blasting air through the molten metal.

BIDMAS An acronym for remembering the set order of priority for the sequence in which operations are carried out: Brackets, Indices, Division, Multiplication, Addition, Subtraction.

Block diagrams Diagrams that use blocks and arrows to provide a top-down overview of a complex system in terms of its inputs, processes and outputs, and how they connect together.

Bonuses Extra money over and above the salary or wages of an employee.

Boyle's law Law stating that for a fixed amount of gas at a constant temperature, pressure and volume have an inverse relationship.

Braking force Force needed to slow down an object or bring it to a stop.

Buoyancy An upward force exerted by a fluid that opposes the weight of an immersed object.

Calibration Process by which measuring equipment is checked to ensure measurements taken are accurate and repeatable.

Cam mechanism Mechanism that uses a cam and a follower to turn rotary motion into linear or reciprocating motion.

Capacitance Ability of a circuit or component to store electrical charge.

Capital Money or valuable items that a business can use to generate income.

Cartesian coordinates A pair of numbers or three numbers that specify the distances from the coordinate axes.

Casting A shaping process which involves heating a material above its melting point and pouring the resulting liquid into a mould, where it then cools and solidifies into the required shape.

Catalogue A detailed list that can be easily searched, usually by use of an index.

Cathode Negative terminal of an electrolytic cell.

Cathode ray tubes (CRTs) Vacuum tubes that direct a beam of electrons onto a fluorescent surface to produce images.

Cations Positively charged ions.

Characteristic gas equation Equation showing that the product of pressure and volume is directly proportional to the temperature and the amount of gas.

Charles' law Law stating that for a constant amount of gas and pressure, volume and temperature have a direct relationship.

Chemical etching Process that uses the reaction of acids and alkalis to create markings on a metal surface.

Circuit diagrams Diagrams drawn up by electronic design engineers to show how components (such as resistors, capacitors and integrated circuits) are connected in an electronic circuit.

Circular economy Model for sustainable production that aims to make products last as long as possible, consume only renewable energy and create zero waste at the end of product life.

Circumference Outside border of a circle.

Clauses Clearly defined sections of a contract or other legal document.

Closed-loop system System with one or more feedback loops.

Closed system Thermodynamic system where only energy transfer takes place and mass remains constant.

Cloud computing Where users remotely access computer services such as data storage, software and networking through the internet.

Coefficient of heat transfer Coefficient that indicates how well heat will be transferred through a series of resistant mediums.

Collaboration Working together towards a common goal.

Commissioning Process of getting a newly installed machine or piece of equipment up and running, ready for handover to the customer.

Competence The ability to do something effectively, with sufficient knowledge, judgement and skill.

Compliance Being in accordance with commands, rules or requests.

Composites Materials made up of a mixture or combination of distinctly different materials that work together to provide improved mechanical properties.

Compound Substance made from two or more elements where the elements are bonded together and difficult to separate.

Computer-aided design (CAD) Using computer software to create 3D models of engineering components that can be manipulated to create isometric views, exploded views and orthographic projections.

Concurrent forces Forces where the lines of action all meet at the same point.

Condition-based maintenance Maintenance enabled by modern sensor and communication technology that allows real-time monitoring of the condition of complex equipment.

Conduction (thermal) Transfer of heat through direct contact between atoms or molecules.

Conformity Compliance with standards, rules or laws.

Consensus A point where everyone agrees with a design or document.

Contingency Extra time and resources reserved for any possible change in the plan.

Continuous professional development (CPD) The learning activities that workers engage in to improve the knowledge and skills needed to complete their current or future job roles.

Contractor Someone who is self-employed but agrees to complete work for a company for a set fee.

Contracts Formal and legally binding agreements between two or more parties that detail what each party must or must not do.

Controlling interest Where a shareholder owns more than half the shares in a company, meaning they can make decisions without needing the support of the rest of the shareholders.

Convection Transfer of heat through the movement of heated particles in liquids and gases.

Co-planar forces Forces that are all acting in the same plane.

Copyright A legal right to be identified as the creator of a literary or creative work and to control its reproduction.

Cosine rule $a^2 = b^2 + c^2 - 2bc \cos A$

Costs Money required to make, store, sell or market a product.

Crack propagation Process of crack enlargement and movement through a material that will eventually lead to material failure.

Creep Process where materials subject to constant loading well below their yield point experience gradual elongation due to plastic deformation and structural degradation.

Critical path The most time-critical set of tasks in a plan.

Crystalline Where atoms, ions or molecules come together to form solids with well-ordered, repeating lattice structures.

Cumulative frequency Sum of all frequencies so far in a frequency distribution.

Current Flow of electrons through a conductive material.

Customer Someone who buys a product or service.

Cyber-physical systems Systems that integrate computerisation, networking, and physical processes, equipment or machinery.

Datum Common reference point, surface, plane or axis from which dimensions are measured.

Demand Consumer desire to purchase goods and services.

Depreciate Reduce in value over time.

Derived units Units of the SI system that are derived by multiplying or dividing base units in specified combinations.

Design brief A document which defines the purpose of a new product and outlines all the needs that must be met.

Design for assembly Design approach that ensures assemblies can be put together easily.

Design for manufacture Design approach that ensures products incorporate features that support the use of a particular process in their manufacture.

Diameter Straight line passing from side to side through the centre of a circle.

Differentiation Process used to calculate the instantaneous rate of change of a function.

Digitalisation Process of converting physical objects such as documents, paper-based systems or other artefacts into digital data that can be read, displayed, processed and distributed by computer.

Digital signals Discrete signals, usually represented as square waves.

Dimensions Measurements that define the size of the features shown on an engineering drawing.

Displacement Change in position of an object; its length is the shortest distance between the object's start and end positions.

Dissociation Process of breaking up a compound into simpler components that could recombine under other conditions.

Distance How far an object moves.

Distributed energy systems Small-scale, localised electricity generation systems that usually use renewable energy sources.

Diversity Employing, accepting and including employees of all backgrounds.

DMAIC A data-driven, problem-solving method used to help identify and fix problems within a process and improve future output.

Documents Written, printed or electronic records of data and/or text.

Drag Force that acts in opposition to the movement of an object through air; also referred to as air resistance.

Drones Unpiloted aerial vehicles.

Duty holders People with a legal responsibility.

Electrical control systems Control systems that make use of electrical power supplies, components and signals.

Electrical power sources Power sources that generate energy in the form of electricity.

Electrolysis Process that causes chemical decomposition through the use of electricity.

Electrolyte Liquid that conducts electricity.

Electrons Negatively charged sub-atomic particles that orbit the nucleus of an atom.

Electroplating Metal finishing process that uses electrolysis to deposit a metal on the surface of a part or component.

Element Substance that cannot be broken down into smaller constituents and consists of only a single type of atom.

Employees People who work for a company in return for wages or a salary.

Employer Someone who sources the pay of an employee and controls the work they do.

Energy (electrical) Capacity for an electrical circuit to do work.

Entropy The measure of a system's thermal energy per unit temperature.

Equality Ensuring that nobody suffers prejudice based on the protected characteristics set out in the Equality Act 2010.

Equilibrium When the magnitude and direction of the forces acting on an object are balanced.

Ergonomic design Design approach that ensures a product is safe and comfortable to use by taking into consideration both human behaviour and the size, strength and limitations of the body.

Ethics The moral principles that guide how a person or organisation behaves.

Expansivity Amount by which a material expands or contracts due to a change in temperature of one degree.

Exploded views Drawings that bring together a series of isometric component drawings, positioned on the page to illustrate how they fit together into an assembly.

Fatigue Material degradation process that can cause components to fail when they undergo repeated cycles of loading and unloading.

Ferrous metals Iron and alloys containing iron.

Finishing Adding a coating to the outer surface of a component that protects it from corrosion. It can also improve aesthetics.

Fit Amount of clearance between two mating parts.

Fit for purpose Where a product fulfils the purpose for which it was designed.

Flowcharts Diagrams used to visualise the steps in a process.

Fluid power control systems Control systems that use a fluid, such as liquid or a compressed gas, as the power transmission medium.

Force Push or pull on an object that causes it to change its speed, direction or shape.

Force (electrical) Attractive or repulsive interaction between any two charged objects.

Forming Application of force to change the shape of a material.

Freehand sketches Drawings produced by hand without the use of drafting aids such as rulers, set squares or compasses.

Friction Force that acts in opposition to an object sliding along a surface.

Functional requirements What a product must be able to do in order to fulfil its basic purpose.

Galvanic protection Coating a base metal with another more reactive metal to protect it from corrosion.

Gantt chart A two-dimensional, graphical representation of tasks against time, which clearly shows how long each task is expected to take and what resources are required.

Gears Toothed wheels that are linked together to transmit drive.

General gas equation Equation showing that the product of pressure and volume is directly proportional to temperature for a given amount of gas.

Geometric dimensioning and tolerancing (GD&T) System that allows tolerance limits to be applied to shape (or form), as well as orientation and location relative to a reference or datum surface, plane or axis.

Geometric progression Sequence of numbers where the next term is found by multiplying the previous term by a fixed (non-zero) number called the common ratio.

Gravitational force Force that attracts all objects with mass towards each other.

Hazards Anything with the potential to cause harm.

Histogram A graphical chart that displays data using bars of different heights; it groups the data into ranges and the height of each bar shows the amount of data in each range.

Human characteristics The characteristics related to the physical and mental features of people.

Human error The possibility of mistakes made by people.

Hybrid technology systems Systems that combine two or more different technologies in a single product or system to optimise utility and/or efficiency.

Hydraulic pressure Pressure exerted by water or liquid on a surface.

Hydraulic systems Systems that use a liquid as the power transmission medium.

Hydrostatic pressure Pressure applied at a given point by fluid at equilibrium due to gravity.

Hypothesis Testable statement of the expected outcome of a study following experimentation or any other verification process.

Ideal gas A gas composed of molecules that do not attract or repel each other.

Imperial system of units Traditional system of measurement introduced by the British Weights and Measures Act 1824 which includes units such as feet and inches, pints and gallons, and pounds and ounces.

Inclusion The practice of creating a workplace culture where the different skills, talents and perspectives of all workers are accepted, appreciated, supported and valued.

Inclusions Foreign matter or impurities trapped in the structure of a metal.

Inclusive design Design approach that focuses on making design solutions that are accessible to the widest possible proportion of society.

Income Money that comes into a company.

Indices Small superscript digits that appear after a number or letter to show many times it is multiplied by itself.

Ingots Rough rectangular castings used as the raw material for worked or wrought metals.

Input blocks System blocks that take signals from the real-world environment and change them into signals that process blocks can understand.

Integration The opposite of differentiation.

Interface devices Modules that ensure accurate communication takes place between PLCs and input/output devices.

Internal combustion engine Engine where fuel is burned inside the engine itself.

International System of Units Globally agreed set of base units and derived units used for standardised measurement in science and engineering.

Internet of things (IOT) A network of physical machines, equipment and devices that are embedded with computerised sensors and networking technology to make them capable of interconnection through the internet.

Ions Elements that have an electric charge due to losing or gaining electrons.

Isolated system Thermodynamic system where neither energy nor mass is transferred.

Isometric drawings Drawings used to achieve a basic three-dimensional (3D) appearance by using only vertical lines and lines receding at $30°$ from the horizontal.

Isotropic Having the same mechanical properties when measured in any direction.

Iterative design methodology An iterative approach to the design process, with structured opportunities to improve the design solution over several design cycles.

Job description A document that outlines the main roles and responsibilities associated with a job.

Joining A process used to fix two or more components together into a larger assembly.

Kaizen A Japanese term meaning continuous improvement through small, incremental positive changes.

Kanban A system used to manage the flow of work in a manufacturing company by regulating the supply of materials between processes through the use of instruction cards or boards.

Key performance indicators (KPIs) Quantifiable measures used to evaluate the success of an activity or organisation.

Kinetic energy Energy that an object possesses because of its motion.

Laminar flow Flow pattern where fluid particles move smoothly in parallel layers with little or no mixing.

Latent heat Heat added to or removed from a substance that results in a change of phase without causing a change in temperature.

Lean manufacturing A production method that involves manufacturing with the minimum waste.

Legislation A law or set of laws passed by Parliament.

Lever Rigid rod that turns on a pivot or fulcrum.

Lift Upward force generated in response to an object moving through the air that acts against gravity.

Lifting operation An operation concerned with the lifting or lowering of a load.

Light-emitting diodes (LEDs) Electronic devices that produce light when current flows through them.

Linear design methodology A right-first-time approach to the design process, with no structured opportunities to improve the initial design solution.

Line manager Someone who has direct responsibility for managing frontline workers, such as production staff or machine operators.

Lines of action Geometric representations of how forces are applied.

Linkages Systems made up of levers or rods connected by pivots.

Logarithms Mathematical operations that determine how many times a base number is multiplied by itself to produce another number; they are the inverse functions to exponentials.

Logic gates Process devices that receive digital input signals (1s and 0s) and produce digital outputs depending on those inputs.

Magnetic field Area surrounding a magnet where magnetic forces are observable.

Magnetic flux Total magnetic field that passes through a given area.

Magnetic flux density Amount of magnetic flux that passes through a given area at right angles to the magnetic field.

Maintenance operations Practical actions carried out by maintenance staff and machine operators in order to keep equipment and machinery in good working condition.

Management Collective term for experienced and qualified members of staff who are in charge of a company and/or its projects.

Market research Collecting information about customers' wants and needs to help a business make decisions about design, production and marketing.

Mass Quantitative measure of an object's inertia.

Mean average, $\bar{x}$ Sum of all the terms (x_i) divided by the quantity of terms, n: $\bar{x} = \dfrac{\sum x_i}{n}$.

Measurement Action performed to obtain an accurate and precise value of a physical variable in accordance with national and international standards.

Measurement gauge An instrument used to check that measuring equipment is still recording precise measurements.

Mechanical power sources Power sources that generate energy in the form of electricity through mechanical motion, vibration or pressure.

Mechanical properties Characteristics that define the behaviour of a material in response to an applied stress.

Mechanical work Amount of energy transferred by a force, measured in joules (J).

Mechatronics The integration of mechanical and electronic systems and sub-systems in order to create fully functional systems.

Median Middle number.

Meetings Planned occasions when several members of a team come together to discuss a project.

Mental characteristics Characteristics that affect how a person thinks.

Methodology A system of methods or rules used for a particular activity.

Microcontrollers Small computers, contained in microchips, that respond to input devices and control output devices.

Microprocessors Powerful central processing units mainly used in computing applications.

Minutes The record of what is discussed during a meeting and any actions.

Mixtures Substances formed by combining two or more chemicals which can be separated, as they are not bonded permanently.

Modal average Number that occurs most frequently.

Molecule Group of two or more atoms held together by chemical bonds.

Moment Turning effect of a force.

Momentum Product of the mass and velocity of an object.

Motion Change in position of an object with respect to time.

Motor Drive device that turns electrical energy into rotary movement.

Motor driver IC Integrated circuit used to ensure that there is enough current for motors to operate correctly.

Multimeter Device used to take different electrical and electronic measurements, such as voltage, current, resistance and power.

Network (electrical) Arrangement of connected electrical or electronic components.

Neutrons Sub-atomic particles without an electric charge found in the nucleus of an atom.

Newton's first law of motion A body will remain at rest, or will continue moving at a constant speed in a straight line, unless a force is acting upon it.

Newton's laws of motion Fundamental laws that describe the relationship between the motion of a body and the forces that are acting on it.

Newton's second law of motion The acceleration of an object depends on the mass of the object and the net force acting upon it.

Newton's third law of motion For every action or force, there is an equal and opposite reaction.

Non-contact forces Forces acting between objects that do not physically touch each other.

Non-concurrent forces Forces where the lines of action do not meet at the same point.

Non-crystalline or amorphous Having randomly arranged molecules without a regular lattice structure.

Non-equilibrium structure Structure that is not arranged in the most efficient way of packing atoms together into a crystal lattice.

Non-ferrous metals Metals that do not contain iron.

Nucleus Centre of an atom made up of protons and neutrons.

Obsolete No longer used.

Open-loop system System without a feedback loop.

Open system Thermodynamic system where both mass and energy transfer take place.

Operational envelope Area of three-dimensional space representing the maximum extent or reach of a robotic arm.

Operation effective efficiency (OEE) A measure of how well a manufacturing process is used in comparison to its potential.

Organic light-emitting diodes (OLEDs) Special types of LED that contain an organic material layer that produces light when current flows through it.

Orthographic projections Drawings that use multiple views of a product or engineering component and are used in formal engineering drawings.

Output blocks System blocks that take signals from process blocks and turn them back into real-world environmental signals.

Overcurrent When the safe load current for a circuit is exceeded.

Parallel circuit Circuit where the components are arranged in loops or branches.

Parallel-motion linkage Type of linkage where two or more parts of the linkage move in the same direction but parallel to each other.

Patent A legal document that grants an owner exclusive rights over an invention.

Personal protective equipment (PPE) Clothing or equipment designed to protect a user from workplace hazards.

Phasor diagram Diagram used to show the phase relationships between two or more sine waves with the same frequency.

Physical characteristics The defining features of a person's body.

Physical properties Material characteristics that can be observed or measured without changing the nature of the material.

Pneumatic systems Systems that use compressed air to transmit power.

Point load Force applied at a single point along a beam.

Polar coordinate system Coordinate system used to determine the location of a point on a plane. Can be two- or three-dimensional.

Potential divider Circuit arrangement that uses two resistors connected in series to divide the initial supply voltage, resulting in a smaller output voltage.

Potential energy Energy that is stored by an object due to its position.

Power Rate at which energy is transferred or converted.

Power (electrical) Rate at which electrical energy is transferred.

Power series Infinite sequence of numbers which are power functions of variable x.

Precision Degree of closeness between measurements taken repeatedly under the same conditions.

Pressure Force applied perpendicular to the surface of an object per unit area over which that force is distributed.

Preventative maintenance Maintenance that uses manufacturer-recommended fixed service schedules to regularly repair or replace components prone to wear, before they fail.

Primary chemical cell Single-use chemical cell.

Probability Measurement of the likelihood that an event will occur, presented as a fraction (in its simplest form), a decimal or a percentage.

Process blocks System blocks that respond to signals they receive from input blocks and alter them in some way, before sending them to the output blocks.

Procurement The process of identifying and buying materials, tools and equipment for use by a company.

Product design specification (PDS) A list of specific and measurable criteria that define how a product will meet the requirements of a design brief.

Professional bodies Bodies that represent, support and set the standards of work and behaviour for a profession.

Professional standards Agreed standards, set by professional bodies, that describe the skills, knowledge and behaviours that characterise excellent practice and support within the profession.

Profit What is left when costs are subtracted from income.

Program evaluation and review technique (PERT) chart A representation of a set of tasks that shows the expected time needed to complete them and their dependencies.

Programmable logic controllers (PLCs) Programmable devices used in manufacturing and industrial control systems.

Project brief A short description of the scope and objectives of a project.

Project goals The desired outcomes of a project as agreed by the client and company.

Protected characteristics Nine aspects of a person's identity set out in the Equality Act 2010. The Act legally protects people from being discriminated against on the basis of these characteristics.

Protons Positively charged sub-atomic particles found in the nucleus of an atom.

Provision of services Supply and connection of electricity, gas, ventilation, compressed air or water to equipment or machinery.

Pulley Mechanism that uses wheels, an axle and a rope to reduce the effort force needed to lift a load.

Pulse amplitude modulation (PAM) The amplitude of the pulsed carrier signal is varied according to the amplitude of the message signal.

Pulse width modulation (PWM) The width of the pulsed carrier signal is varied according to the amplitude of the message signal.

Pure metals Metals that contain a single type of metal atom and are not mixed or alloyed with other elements.

Pythagoras' theorem For right-angled triangles, the square of the longest side (called the hypotenuse) is equal to the sum of the squares of the two shortest sides: $a^2 = b^2 + c^2$.

Quenching Rapid cooling of hot metals, usually achieved by submerging them in oil or water.

Radiation Transfer of heat via electromagnetic waves.

Radius Straight line from the centre of a circle to its circumference, equal to half the length of the diameter.

Range Difference between the highest value and the lowest value.

Reaction forces Forces that act in opposition to action forces, in line with Newton's third law of motion.

Reactive maintenance Maintenance carried out when there is an equipment failure.

Renewable power sources Natural power sources used to generate energy that are replenished at a higher rate than they are consumed, for example wind and solar.

Reputation How an organisation or individual is perceived by the public based on how they operate.

Resistance Opposition to the flow of electrical current.

Resolution Smallest increment or decrement that can be displayed as a result of measurement.

Resource Something that can be used to complete a task.

Reverse-motion linkage Type of linkage where the output direction of motion is the reverse of the input direction of motion.

Risk The likelihood that a hazard will cause harm.

Risk assessment Systematic process of identifying and evaluating hazards in order to determine whether sufficient control measures are in place to prevent harm.

Robotics Design, development and manufacture of robots that are able to perform physical tasks that would previously have been carried out by humans.

RSS feed (Really Simple Syndication feed) A web-based application that notifies users of any updates on a selected website.

Sacrificial anode Sacrificial metal forming the anode in electrolytic protection; this corrodes in preference to the metal forming the cathode, which is therefore protected from corrosion.

Sampling The process of selecting batches of products for testing.

Scalar Value with magnitude but not direction.

Schematic diagrams Diagrams that show a system in terms of its individual components, represented by standard symbols.

Schematics Simplified visual representations of the connections between components in an electrical, a pneumatic or a hydraulic system.

Scientific method Systematic and objective approach used to acquire knowledge that includes observation, questioning, formulation of hypotheses, testing and experimentation, data analysis and the formation of conclusions based on empirical evidence.

Secondary chemical cell Rechargeable chemical cell.

Section view View used to reveal a cross-section of a component when sliced along a specified cutting plane.

Sector Pie-shaped part of a circle, comprising two radii at an angle of θ joined by an arc.

Semiconductors Materials that conduct current better than insulators but not as well as conductors.

Sensible heat Heat added to or removed from a substance that results in a change in its temperature without causing a change of phase.

Sensors Electronic components that detect changes in the physical environment around them.

Separation point The point where the boundary layer detaches from the surface and forms a wake, or turbulent flow.

Series circuit Circuit where the components are arranged in a single line.

Shaping Pouring or injecting liquid material into a mould where it solidifies into the required shape.

Shareholders Individuals or entities that have invested money in a company in exchange for shares in the company's stock.

Shares Units of ownership of a company that can be bought and sold; their value varies according to the overall value of the company.

Shear forces Forces that act in a parallel direction to the surface of an object.

SI Officially recognised and internationally accepted measurement system used in science, technology and engineering, as well as everyday commerce.

Signal conditioning Modifying the signal from a sensor or other device so that it meets the requirements for processing by a PLC.

Significant figures The quantity of digits in a number that are required to present it to the correct approximation, not including zeros used to locate the decimal point.

Simple chemical cell Two electrodes placed in an electrolyte and connected by a conductor.

Simply supported beam Beam that rests on two supports and is free to move horizontally.

Sine rule $\dfrac{a}{\sin A} = \dfrac{b}{\sin B} = \dfrac{c}{\sin C}$

Single minute exchange of die (SMED) A system used to reduce changeover times when making different products.

Six Sigma A set of quality management techniques that aim to improve processes within an organisation and in turn reduce the amount of manufacturing errors and product defects.

Smart materials Advanced modern materials characterised by their ability to react in response to an external stimulus.

Solubility Measurement of the extent to which a chemical can dissolve in water.

Solutions Homogenous mixtures formed by the combination of two or more substances.

Specification limits Boundaries that define where a product works and does not work; they assess how capable a process is of meeting customer requirements.

Speed Rate at which an object moves over a certain distance.

Standard deviation How much the data varies from the mean value.

Standard operating procedures (SOPs) Step-by-step instructions for routine operations. These are put together by an organisation and all workers must follow them.

Start-up A company that has been set up recently, often by someone with a new idea or product.

Stock forms Standard shapes and sizes in which materials are available.

Summing point The part of a system that produces the algebraic sum of the reference signal and the feedback signal.

Supervisory control and data acquisition (SCADA) software Software used to supervise, monitor and control industrial engineering processes.

Supplier Someone who sells a product or service.

Supply A company's ability to produce goods or services for sale.

Suspensions Heterogenous mixtures formed when solid particles are mixed in liquid without them dissolving.

Sustainability Ability to fulfil our current needs without compromising the needs of future generations.

Sustainable design Design approach that ensures environmental impact is minimised through reduced use of energy, natural resources and materials.

Technical specifications Documents that define the requirements for projects, products or systems.

Temperature Amount of kinetic energy produced by particles.

Testing The process of assessing product sample batches to check they are within customer specifications for the final product; this determines whether corrective actions are needed in the manufacturing process.

Thermionic valve Electronic device that amplifies electrical signals by controlling the flow of electrons in a vacuum tube.

Thermodynamics Branch of physics that is concerned with the relationship between heat, energy and work.

Thermoplastic polymers Polymers that soften when heated and can be moulded before being allowed to cool and reharden.

Thermosetting polymers Polymers that set permanently during manufacture and do not soften when heated.

Thrust Force that makes an object move forward by overcoming drag.

Title block Element of an engineering drawing that provides essential information about the drawing, what it shows and the identification number under which it is filed.

Tolerance limits Range of acceptable values for a dimension, defined by a maximum and a minimum value.

Tolerances Allowable amounts of variation for a specified quantity, especially in the dimensions of a product or part.

Torque Force that causes an object to rotate around an axis or another point in an angular direction.

Total productive maintenance (TPM) A strategy to keep equipment in effective working condition and available for operation by ensuring that the whole workforce is involved in its maintenance rather than just a maintenance team.

Tractive force Force needed to overcome the resistance caused by friction.

Transactions Exchanges of money when selling or buying goods or services.

Transducer Umbrella term used to describe both electrical/electronic sensors (input devices) and actuators (output devices) that convert one form of energy, or signal, into another.

Transfer function Algebraic representation of the contents of a system block.

Trigonometric functions Set of functions – sine (sin), cosine (cos) and tangent (tan) – that can be used to calculate the other dimensions of a triangle when only the length of one side is known.

Turbulent flow Flow pattern where fluid particles move in such a way that they mix in a zig-zag pattern.

Turnover Total amount of money received by a company from the sale of goods or services over a certain period of time.

Unbalanced forces When the force acting in one direction is greater than the force acting in the other direction.

Uncertainty Margin of doubt about a measurement, quantified by the width of the margin and a percentage confidence level.

Uniformly distributed load Load applied evenly over the entire area or length of a beam.

User-centred design Iterative design approach that focuses on feedback from the end user of a product during its design and development.

Valence Ability of an atom to combine with other atoms.

Value stream mapping (VSM) A technique for analysing how materials and information flow through a company.

Vector Value with both magnitude and direction.

Velocity Speed at which an object moves in a particular direction.

Video calls Telephone calls that use video to allow participants to see each other in real time.

Virtual reality (VR) Immersive technology that allows a user to experience a digitally generated scene as if they inhabited the virtual space.

Viscosity Measurement of a fluid's resistance to flow in centipoise or Nsm^{-2}.

Visual management A management approach where information or operational status is communicated in a visual form.

Voltage Difference in electrical potential between two points in a circuit.

Volume Amount of three-dimensional space occupied by something.

Vortices Regions of a fluid where the flow revolves about an axis line.

Warehouse A building where stock is kept prior to use or sale.

Waste Any production activity which does not add value to a product.

Wasting Removal of material from a workpiece to produce a component of the required size and shape.

Weight Downwards force that opposes lift.

Wiring diagrams Diagrams that show a system in terms of its physical wiring layout, with components shown as pictorial representations.

Worked or wrought metals Metals that have undergone plastic deformation processes.

Work harden Increase in hardness and reduction in ductility due to a build-up of dislocations and grain distortions caused by plastic deformation.

Workplace design How the workplace is arranged to ensure maximum safety and efficiency.

$y = mx + c$ Formula representing straight-line graphs, where m is the gradient of the line and c is the value where the relationship crosses the y-axis, known as the intercept.

Answers

Chapter 1

Test yourself

Page 2
- Additive manufacturing
- Joining
- Wasting

Page 4 (left)
Client brief, product design specification (PDS), investigation, ideas, development, manufacturing, testing, evaluation

Page 4 (right)
Linear design relies on getting the design right first time, with no structured opportunities to evaluate and improve the initial design solution. In iterative design, there are structured opportunities to evaluate and improve the design solution over several design cycles until it fulfils all the requirements of the PDS and client brief.

Page 6
Anthropometric data can provide guidance on the seated height of adults that can be used to determine the required height of the desk.

Page 7
Jargon and complex technical information

Page 8 (left)
Effective maintenance:
- maximises the service life of machinery and equipment so that it lasts longer and needs to be replaced less frequently
- minimises downtime caused by unplanned breakdowns, thereby avoiding costly delays in production
- ensures equipment runs correctly and efficiently
- reduces impact on the environment by ensuring waste materials are managed safely and disposed of responsibly.

Page 8 (right)
Reactive maintenance is carried out when there is an equipment failure and therefore involves unplanned downtime. Preventative maintenance uses manufacturer-recommended fixed service schedules to regularly repair or replace components prone to wear before they fail, thereby avoiding unplanned downtime.

Page 9
Working on a live system would be dangerous and could result in serious injury or even death. Shutdown involves taking a machine or piece of equipment out of service so that it can be worked on. To ensure the safety of maintenance engineers, all services connected to the equipment must be isolated and locked off, and the system must be de-energised. Steps must also be taken to ensure equipment is not accidentally activated while it is being worked on, for example by using a lock out/tag out system.

Page 11
Mass production is used to manufacture large amounts of the same product.

The set-up costs are high because of the specialist equipment and high levels of automation involved, but manufacturing operations are performed as efficiently as possible and the manufacturing time and cost per product are low.

It is not flexible and so very little variation is possible. Tasks that are not automated are done by a low-skilled workforce, trained to perform the same operation on each product.

Page 13
Unexploded munitions or explosive devices are unpredictable and dangerous. They can detonate without notice, causing severe injury or death. It is not safe for a human operator to approach a suspect device. Using a robotic vehicle with an arm-mounted camera and manipulator allows a human operator to get a close look at a device and often make it safe without putting themselves at risk.

Page 14 (left)

The later in the production process that a quality issue is identified, the more difficult and expensive it is to put right. Faulty goods that reach the customer are particularly problematic, because they can damage the reputation of the company and involve additional postage or freight costs to collect faulty items and deliver replacements.

Page 14 (right)

Check that:

▶ there are no scratches or damage
▶ all component parts are fitted correctly
▶ there are no leaks
▶ electrical systems are safely installed
▶ the machine functions as expected.

Assessment practice

1 Powder coating will protect the steel frame from corrosion and improve its aesthetic appeal.

2 Linear design relies on getting the design right first time, with no structured opportunities to evaluate and improve the initial design solution. In iterative design, a prototype is made and tested. Following this, an evaluation stage generates suggestions for improvement that are fed back into the investigation stage. This cycle is repeated until the final design fulfils all the requirements of the PDS and the client brief and further improvement is unnecessary. This approach enables and encourages a culture of continuous improvement, which ultimately leads to better products and higher sales.

3 The use of self-jigging features, such as locating slots, allows components to be positioned easily and accurately prior to welding. This speeds up the process of assembly, as no measuring or marking out is required.

4 Asymmetrical mounting holes on symmetrical components would ensure they could only be fitted in the correct orientation.

5
- **Reduce:** reduce the amount of material, energy and waste involved in manufacturing and using a product.
- **Refuse:** refuse to use harmful or polluting materials or processes.
- **Rethink:** rethink conventional, unsustainable approaches, for example the choice of materials or production processes.
- **Repair:** design products that enable straightforward repair and refurbishment to extend their useful life.
- **Reuse:** design products that can be reused and avoid single-use disposable products. Repurpose existing components, products or waste materials.
- **Recycle:** use recycled materials and/or ensure materials are recyclable. The design should allow easy separation of different material types when the product reaches the end of its useful life.

6 Well-maintained machinery and equipment operate efficiently and so minimise energy usage and any associated emissions that may be harmful to the environment. Maintenance also helps to prevent leaks of fluids like oil or hydraulic fluid that might otherwise cause local contamination around the equipment.

7 Automated processes involve low numbers of low-skilled workers while providing high production rates, low per-item costs and consistent quality. This approach reduces overall manufacturing costs, which in turn allows companies to increase profit margins.

8 Using conventional manufacturing processes to make a one-off prototype component is highly skilled, time consuming and expensive. A CAD model can be used to manufacture the component prototype directly on a 3D printer. This is a less expensive and quicker approach that will produce a prototype that exactly matches the CAD model.

9 It is vital that bought-in components and raw materials comply with their purchasing specifications, are fit for purpose and can be used in the manufacture of products. Use of faulty or unsuitable components will cause quality and/or performance issues. At best, this leads to rework, delays and additional expense for the manufacturer. At worst, faulty products reaching the customer will have to be replaced, which is costly financially but also in terms of reputation.

10 QA is an overall quality strategy that incorporates a wide range of activities to monitor, control and reduce defects in manufacturing processes. This includes statistical process control (SPC), process standardisation, process documentation, identification of defects, process improvement etc. QC is the part of a QA system that deals with inspection and the identification of defects.

Chapter 2

Test yourself

Page 19

Electrical and electronic control

Page 21

Tractors and farm machinery using internal combustion engines replaced horses and steam-driven traction engines to power agriculture, leading to greater efficiency and increases in food production.

Page 22

Thermionic valves were the forerunner to modern transistors and were used to amplify electrical signals in early radios and televisions by controlling the flow of electrons in a vacuum tube.

Page 24

The invention of the silicon chip and microprocessor in the 1970s

Page 25

While VR describes a self-contained virtual world, AR allows users to enhance their real-world experience with real-time computer-generated augmentations or overlays.

Assessment practice

1 Chemical engineering
2 Research, development and design of systems to optimise the processes and activities involved in supply-chain management, such as purchasing, storage, warehousing and distribution
3 The invention of the Bessemer process in the 1850s enabled the mass production of steel, making it cheap and plentiful. Steel is much stronger and more versatile than cast or wrought iron, which it largely replaced. It can be cast, forged or rolled into large beams or thin sheets and joined by riveting, bolting together or welding. Steel or steel-reinforced structures allowed new designs for buildings, bridges and dams, transforming our built environment. Steel was widely used to build the tools, machinery and factories that led to the mass production of consumer goods. Many of these, including cars, washing machines and refrigerators, also relied on components made from steel. Even today, with the wide range of other material at our disposal, steel is still a cornerstone of industry and the most widely used metal in engineering. It has contributed more than any other metal to the development of modern society.

4 Any two from the following suggested answers:
   - Cars made possible by the internal combustion engine gave people unprecedented freedom to travel for work. This opened up the labour market, making it easier for companies to recruit the workers they needed to grow.
   - Cars also allowed people to travel to tourist destinations, helping the development of the tourist industry and contributing to economic growth.
   - Tractors and farm machinery using internal combustion engines replaced horses and steam-driven traction engines to power agriculture, leading to greater efficiency and increases in food production.
   - An enormous industry emerged to satisfy the demand for cars, which itself helped to grow the economy and provided well-paid jobs for thousands of factory workers.
   - The popularity of the internal combustion engine caused a rapid increase in demand for oil, which soon became an important source of income for oil-producing countries and supported the growth of their economies.
   - Demand increased for steel, rubber and other raw materials essential in car manufacturing, and so did the size of the industries that supplied them.

5 Any two reasons from the following:
   - They are more efficient and so cheaper to run.
   - They have fewer moving parts and so are easier to maintain and repair.
   - They provide instant power, unlike a steam engine which must be allowed to work up steam.
   - They cause less local pollution, as there are no emissions from the motors themselves.

6 Interchangeable parts allow worn or broken components to be replaced quickly and easily by ensuring any new components are available off the shelf and fit perfectly. This vastly reduces the time it takes to effect a repair and so minimises downtime when the machine is not available for use.

7 Early televisions used a cathode ray tube to direct electrons onto a fluorescent coating on the inside of the screen. When electrons hit the fluorescent coating, it glowed and the pattern of dark and light spots on the screen created the picture. The first colour television sets worked in a similar way but used advanced coatings that fluoresced in different colours to make up the image.

LCD or LED televisions no longer used a cathode ray tube. Instead, they used light-emitting diodes (LEDs) to backlight the screen and manipulated a liquid crystal display (LCD) layer to allow light through in certain areas to create the image. OLED televisions use individual organic light-emitting diodes to directly generate the colour of light needed to illuminate each pixel. These allow bright, high-contrast images to be displayed and offer extremely high image resolution due to the small size of individual OLEDs.

8 Online meetings, voice and video calls, access to shared resources via the cloud and email make remote working a practical and popular alternative to travelling into the office.
The convenience and low cost of online retailing has revolutionised the way we shop and means that some bricks and mortar high-street retailers are in decline.
Social media has become hugely influential and an extremely popular way of watching and sharing user-generated content.

9 Workers carrying out complex wiring tasks on large assemblies like aircraft must follow wiring diagrams carefully. Augmented reality can be used to overlay labels, diagrams or instructions on to what the worker can see, to guide them as they work.

10 Cyber-physical systems in agriculture use real-time sensor data, satellite positioning technology and drone imagery to monitor and control irrigation and the application of fertiliser and herbicide. Resources are directed to areas where they are needed and away from areas where they are not. This optimises efficient use of resources while maintaining ideal growing conditions, which minimises costs and maximises yield.

Chapter 3

Test yourself

Page 33
▶ Freehand sketch
▶ Exploded view
▶ Orthographic projection

Page 34

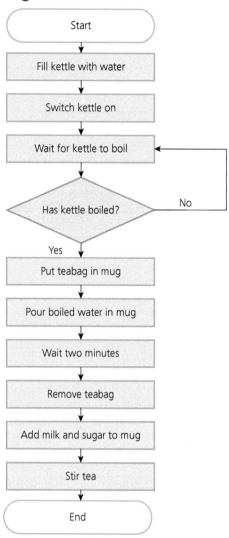

Page 37
CAD enables quick and easy editing, copying, storage and sharing of engineering drawings.

Page 40

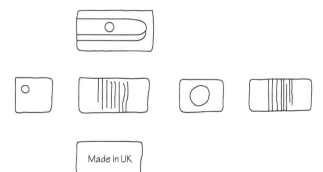

Page 41

▶ Chamfer
▶ Pitch circle diameter
▶ Material

Page 42

Component outline	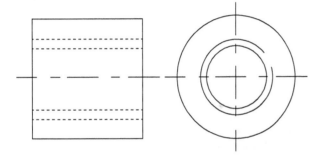
Centre line	
Hidden line	
Dimension leader line	

Page 43

Internal screw thread

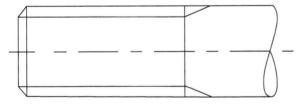

External screw thread

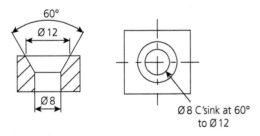

Countersunk hole

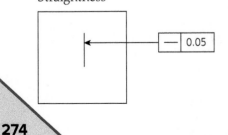

Ø 8 C'sink at 60° to Ø 12

Page 48

Straightness

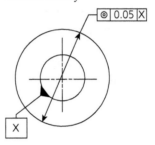

Perpendicularity

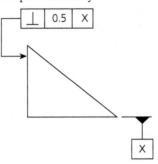

Concentricity

Improve your maths

Page 38

▶ 53 mm
▶ 10.6 mm
▶ 106 mm

Page 44

▶ 32.25–32.75
▶ 44.95–45.00

Page 45

Hole H7: 35 +0.025/-0, max. hole size 35.025, min. hole size 35

Shaft g6: 35 -0.009/-0.025, max. shaft size 34.991, min. shaft size 34.975

Assessment practice

1 A block diagram is used to simply represent a complex system and provide a basic overview of how its parts connect together. Labelled boxes or blocks are used to represent all the sub-systems or major components in the system. The inputs and outputs from each block are labelled, and connections between blocks are indicated using arrowed lines.

2 Variable resistor

3 Fuse

4 First angle orthographic projection

5 Undercut

6 —— — —— — — — —— —

7 External screw thread
8 Section view of a countersunk hole
9 From the ISO fits shaft-basis tables:
   – ring gear (hole) S7 tolerance limits = 320 -0.169/-0.226
   – flywheel (shaft) h6 tolerance limits = 320 +0.000/-0.036
10 Concentricity

Chapter 4

Assessment practice

1 $\frac{1}{2} \times 1.8 = 0.9\,\text{L}$

   $\frac{1}{3} \times 0.9 = 0.3\,\text{L}$

   Total used: $0.9 + 0.3 = 1.2\,\text{L}$

   Remainder: $1.8 - 1.2 = 0.6\,\text{L}$

2 $85 \times 5 = 425\,\text{mm}$

3 Rearranging, $F = \sigma A$

   $45 \times 225 = 10125\,\text{N mm}^{-2}$.

4 Convert to simultaneous equations:

   Equation (a) $4x + 3y = 26$

   Equation (b) $2x - 7y = -38$

   Multiply equation (b) by 2: $4x - 14y = -76$

   Subtract this from equation (a): $17y = 102$

   Rearrange to determine value of y: $y = \frac{102}{17} = 6$

   Insert value for y into equation (a): $4x + 18 = 26$

   Rearrange to determine value of x: $4x = 8$

   $x = \frac{8}{4} = 2$

5 Using geometric progression:

   $15 = 30 \times 0.9^{n-1}$

   $0.5 = 0.9^{n-1}$

   $\log 0.5 = n \log 0.9 - \log 0.9$

   $n = \frac{\log 0.5 + \log 0.9}{\log 0.9} = 7.57$

   Therefore it takes 8 hours of practice to reach 15 minutes.

6 $\frac{dy}{dx} = 4t - 3$, equating to zero and rearranging $t = 0.75$

   Inserting this into the original equation:

   $s = 2t^2 - 3t + 4 = 2.875\,\text{cm}$

7 $A = \sqrt{700^2 + 820^2 - (2 \times 700 \times 820 \times \cos 40)} = 532\,\text{mm}$

   $b = \sin^{-1}\left(\frac{\sin 40}{532} \times 820\right) = 82.2°$

   $c = 180 - 40 - 82.2 = 57.8°$

8 $\mathbf{a}.\mathbf{b} = (6 \times 4) + (8 \times 0) = 24$

   $|\mathbf{a}| = \sqrt{6^2 + 8^2} = 10$

   $|\mathbf{b}| = 4$

   $\theta = \cos^{-1}\frac{24}{10(4)} = 53.1°$

9 Mean = 27.055, median = 27.05, mode = 27.0, standard deviation = 0.143

10 6A3 = 1699 = 110 10100011

Chapter 5

Test yourself

Page 76

Metre, kilogram, second, ampere, kelvin, mole and candela

Page 77

$\frac{13.5}{0.45} = 30\,\text{lb}$

Page 78 (left)

1 c, a, b: 0.005 kilograms (= 5 grams), 20 grams, 50 grams
2 kilobyte, megabyte, gigabyte, terabyte
3 0.108 terametres

Page 78 (right)

Any two from: speed, distance, mass, volume, time, energy and power

Page 79

A quantity that has both magnitude and direction

Page 81 (left)

The location of a point on a plane determined by the distance from a reference point and the angle from a reference direction

Page 81 (right)

(4, 7.4)

Page 87

Digital calliper to measure flange diameter.

Micrometre to measure the inner diameter of the hole.

Page 88

1 The smallest unit of a chemical element that exhibits all the physical properties of that element
2 Negative
3 Electron

Page 89

A compound is a substance made from two or more elements that are bonded together and therefore difficult to separate; a mixture is formed by combining two or more chemical substances which can be separated, as they are not bonded permanently.

Page 92

Electrolysis uses electricity to break down ionic compounds. A positive electrode called an anode and a negative electrode called a cathode are connected to an electricity supply and placed in an electrolyte.

Negatively charged ions (anions) are attracted towards the positive anode; positively charged ions (cations) are attracted towards the negative cathode.

For example, electrolysis is used for electroplating, where a metal object is coated with a thin layer of another metal to prevent it from rusting.

Page 93

$$p = \frac{F}{A} = \frac{15}{6} = 2.5 \text{ Pa}$$

Page 95

$100 \text{ mm} = 0.1 \text{ m}$

$M = F \times d$

$M = 12 \times 0.1 = \text{Nm}$

Page 96

A coplanar force system is a system where all forces are acting in the same plane.

Page 97

Reciprocating motion is motion that moves back and forth along a straight line, whereas oscillating motion is motion that moves back and forth about an equilibrium position as a result of torque.

Page 100 (left)

Laminar flow is when fluid particles move smoothly in parallel layers with little or no mixing, whereas turbulent flow is when the fluid particles move in such a way that they mix in a zig-zag pattern.

Page 100 (right)

The function of thrust in an aircraft is to overcome the resistance of air or drag to the aircraft passage.

Page 101

1 Convection makes bottom floors cooler than those higher up. Warmer air is lighter and rises up, while colder air is heavier and sinks down.
2 When the stove's element is turned off, convection causes heat loss to the air and the soup cools down.

Improve your maths

1 $80 + 28 = 108 \text{ kg}$
2 $220 - 100 = 120 \text{ km}$

Assessment practice

1 Base units are defined arbitrarily and can be used to derive other units of measurement.
   Derived units are derived by multiplying or dividing base units in specified combinations.
2 Speed, distance, mass, volume, time, energy, power
3 It is a vector quantity because it has both magnitude and direction.
4 A hypothesis is a testable statement of the expected outcome of a study following experimentation or any other verification process.
5 A vernier calliper
6 Calibration ensures that the hole depth measurements are accurate. If the measurement equipment is not calibrated then holes of incorrect depth could be measured as correct, and/or vice versa, leading to issues with fitting parts at the next stage of the manufacturing process.
7 The zinc used for the coating is attached to the anode and the machinery parts to be plated are attached to the cathode. These are then placed in a bath containing an acid solution. As the electric current flows through the circuit, the zinc coating is deposited on the machinery parts.
8 $150 \text{ mm} = 0.15 \text{ m}$
   Moment = $10 \times 0.15 = 1.5 \text{ Nm}$, clockwise (as the nut is being tightened)

9 P = 700 × 9.81 × 10 = 68 670 Pa

= 68.7 kPa to 3 s.f.

10 Conduction is the transfer of heat through direct contact between atoms or molecules, whereas convection is the transfer of heat through the movement of heated particles in liquids and gases.

Chapter 6

Test yourself

Page 109

The physical properties of a material include a wide range of physical, thermal and electrical characteristics that can be observed or measured without changing the nature of the material. Mechanical properties are characteristics that define the behaviour of a material in response to an applied stress.

Page 110

Pure metals contain a single type of metal atom. Alloys contain two or more metals in solid solution with one another.

Page 111

Cutting tools, springs

Page 114

Elastomers

Page 115

In GRP, the matrix is polyester resin and the reinforcement is glass-fibre strands or woven matting.

Page 116

Photochromic

Page 117

Ionic bonding

Page 118

Plastic deformation

Page 119

Worked or wrought

Page 120

Hot working

Page 121 (left)

Cold-working processes

Page 121 (right)

Heat-affected zone

Page 122

Brazing is carried out at a lower temperature than welding and avoids the issue of thermal distortion in thin-walled components.

Page 123 (left)

Martensite

Page 123 (right)

Normalising refines the grain structure of the steel by encouraging the formation of numerous small grains of uniform size and shape.

Page 125

Any three from the following:
▶ exposure to ultraviolet light (sunlight)
▶ exposure to atmospheric pollutants
▶ weathering
▶ humidity
▶ chemical attack
▶ accumulated wear
▶ accumulated impact damage
▶ thermal cycling.

Page 126 (left)

Any two from the following:
▶ voids
▶ inclusions
▶ surface damage
▶ tooling marks
▶ sharp corners
▶ sudden changes in cross-section.

Page 126 (right)

Primary, secondary and tertiary

Page 130

The amount of energy absorbed by the material test specimen when it fractures

Page 132

The maximum number of loading cycles the material can undergo at that load before fatigue failure

Improve your maths

Page 108

$V = 0.03 \times 0.04 \times 0.08 = 9.6 \times 10^{-5} \text{ m}^3$

$$\rho = \frac{m}{V} = \frac{0.756}{9.6 \times 10^{-5}} = 7875 \text{ kgm}^{-3}$$

Page 129

1 $\sigma = \dfrac{f}{A} = \dfrac{1200}{0.0113} = 106194 \text{ Pa}$

2 800 MPa

Assessment practice

1 $\rho = \dfrac{m}{V} = \dfrac{0.339}{1.25 \times 10^{-4}} = 2712 \text{ kgm}^{-3}$

2 An advanced modern material characterised by its ability to react in response to an external stimulus

3 Crystalline solids have well-ordered and regularly arranged lattice structures of atoms within a well-defined grain structure. Amorphous solids do not have a regularly arranged structure and do not form grains.

4 Any two from the following:
  - porosity caused by gas bubbles or thermal shrinkage
  - the formation of shrinkage cracks
  - changes to the grain structure of the metal with the formation of large and irregular grains
  - the introduction of impurities that form inclusions in the metal.

5 Low-carbon steel cannot be hardened by heat treatment because it contains insufficient carbon. Heating the mild steel in a carbon-rich atmosphere causes carbon atoms to diffuse into the surface layer of the steel. This increases the amount of carbon in solid solution with the iron to the point where it can be hardened by quenching from above the critical temperature of the steel. This gives a hardened jacket or case on a steel component but retains the toughness of the mild-steel core.

6 Painting and galvanising both form a physical barrier to exclude air and water from the surface of the steel that would otherwise cause corrosion. However, if the paint coating is damaged, any exposed steel will no longer be protected and will corrode. In contrast, if the zinc layer is damaged on a galvanised component, then the exposed steel will continue to be protected. The zinc and exposed steel will form an electrochemical cell with a zinc anode and steel cathode. The zinc anode will corrode in preference to the steel cathode, which will therefore continue to be protected.

7 Necking is the reduction in cross-sectional area caused by localised straining in a ductile tensile test sample just before it fractures.

8 $E = \dfrac{\sigma}{\varepsilon}$

$\sigma = E\varepsilon = 205 \times 10^9 \times 0.002 = 410 \times 10^6 \text{ Pa} = 410 \text{ MPa}$

9 A Rockwell hardness (HRC) test must be carried out on a flat, clean material sample with a smooth surface finish. The test apparatus measures the depth that a diamond indenter penetrates the surface of the material under a set load. This depth is displayed on the Rockwell test apparatus as a hardness number. The higher the number, the greater the hardness.

10 $\sigma = \dfrac{1}{RA} = \dfrac{0.10}{2.21 \times 10^{-3} \times 7.9 \times 10^{-7}}$
$= 57.3 \times 10^6 \text{ Sm}^{-1} = 57.3 \text{ MSm}^{-1}$

Chapter 7

Test yourself

Page 135 (left)

When designing space rockets, it is necessary to calculate the thrust required to escape Earth's gravity.

Maglev trains use the repelling forces of electromagnets to lift them up and push them forward.

Page 135 (right)

With concurrent forces, the lines of action all meet at the same point. By contrast, with non-concurrent forces, the lines of action do not meet at the same point.

Page 137

▶ Conservation of momentum:
$m_1 u_1 + m_2 u_2 = m_1 v_1 + m_2 v_2$

▶ Conservation of energy: $K_1 + U_1 = K_2 + U_2$

▶ D'Alembert's principle: $F - ma = 0$

Improve your maths

$F = ma$, so $a = \dfrac{F}{m}$

$a = \dfrac{4500}{1320} = 3.41 \text{ ms}^{-2}$

Assessment practice

1 A body will remain at rest, or will continue moving at a constant speed in a straight line, unless a force is acting upon it

2 Forces that are all acting in the same plane

3 Magnetic, electrostatic and gravitational forces

4 A beam that rests on two supports and is free to move horizontally

5 Point load, uniformly distributed load

6 Shear forces push the top part of the beam in one direction and the bottom part in the opposite direction. This can result in deformation and cracks appearing in the beam.

7 Force plus the negative of the mass multiplied by the acceleration is equal to zero: $F - ma = 0$

8 Potential energy is the energy that is stored by an object due to its position, whereas kinetic energy is the energy that an object possesses because of its motion.

9 Piezoelectricity uses crystals to convert mechanical energy into electrical energy. Generators take kinetic energy in the form of movement and convert it into an electrical current.

10 It is a sustainable/renewable source of power, as energy from the sun will be available for billions of years. It produces zero carbon emissions and so does not contribute to global warming.

Chapter 8

Test yourself

Page 142 (top)

1 Particles of an atom which have a negative electrical charge

2 When electrons flow through a conductive material

Page 142 (bottom)

Electrical power is the rate at which electrical energy is transferred, represented mathematically as P = E/t.

Page 144

Suggested answers:

▶ Maglev trains are able to levitate and move without any friction.

▶ Electromagnetic door locks create a magnetic field when energised, and this allows an electromagnet and an armature plate to hold together and stop the door opening.

▶ Generators can produce electricity when rotated.

Page 146

Kirchhoff's laws are required for more complex circuits, where Ohm's law cannot be used to model their behaviour.

Kirchhoff's current law is concerned with the conservation of charge within a circuit and can be used to describe current in a node.

Kirchhoff's voltage law is concerned with the conservation of energy within a circuit and can be used to describe voltage around a loop.

Page 149 (left)

1 In N-type semiconductors, electrons are the majority charge carriers. In P-type semiconductors, they are the minority carriers.

2 In forward bias mode, a positive voltage is applied to the P-type side of the junction, so current can only flow from the anode to the cathode. In reverse bias mode, a positive voltage is applied to the N-type side of the junction, so no current can flow until the electric field intensity is so high that the diode breaks down.

Page 149 (right)

Analogue signals are continuous and are usually represented using sinusoidal waves, whereas digital signals are discrete and are usually represented using square waves.

Page 150

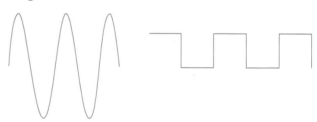

▲ Sinusoidal and square waveforms

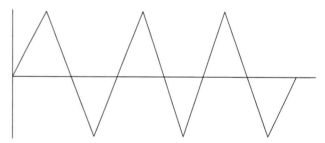

▲ Triangular waveform

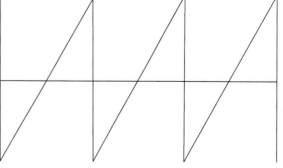

▲ Sawtooth waveform

Improve your maths

Page 146

1 $P = I \times V$

$I = \dfrac{P}{V}$

$= \dfrac{16}{12}$

$= 1.33 \text{ A}$

2 $P = I \times V$

$= 1.2 \times 12$

$= 14.4 \text{ W}$

Page 148

1 $C_{tot} = C_1 + C_2$

$= 47 + 47$

$= 94 \text{ } \mu F$

2 $(1/R_{tot}) = (1/R_1) + (1/R_2)$

$(1/R_{tot}) = 1/680 + 1/1000$

$(1/R_{tot}) = 0.0014 + 0.001$

$R_{tot} = 404.76 \text{ } \Omega$

Assessment practice

1 An electron has a negative electrical charge, whereas a proton has a positive electrical charge.
2 To represent the ideal properties and behaviours of their physical counterparts, in order to analyse and estimate how a circuit will work
3 The total magnetic field that passes through a given area
4 Choose two from:
   – maglev trains
   – electromagnetic locks
   – transformers
   – generators
   – or any other appropriate answers.
5 $R_{tot} = 330 + 330 = 660 \text{ } \Omega$
   $V = I \times R_{tot}$
   Therefore $V = 0.015 \times 660 = 9.9 \text{ V}$
6 A potential divider is a circuit arrangement that uses the way voltage drops across two resistors connected in series in order to divide/reduce the initial supply voltage. The values of each resistor determine the output voltage.
7 Fuse, circuit breaker
8 A positive voltage is applied to the N-type side of the junction. Current does not flow until the until the electric field intensity is so high that the diode breaks down.

9 To prepare and manipulate signals so that they meet the requirements of the next stage of processing
10 Fan in refers to the number of inputs that a logic gate is capable of handling safely, whereas fan out is the number of logic gate inputs that are driven by the output of another logic gate.

Chapter 9

Test yourself

Page 153 (top)

To create movement of two or more parallel levers in the same direction

Page 153 (bottom)

A lever where the load is between the fulcrum and the effort, for example a wheelbarrow

Page 154

A thermistor has a resistance that decreases as the temperature level increases, and vice versa.

Page 156 (left)

Standard motor, servo motor, stepper motor

Page 156 (right)

Unitary PLCs contain all the different parts and components within a single housing. Modular PLCs have different parts, or modules, that are connected together to form a customisable device.

Page 158

A cylinder, which forces a piston to move in a certain direction

Improve your maths

Gear ratio = 3:1

Assessment practice

1 A lever where the fulcrum is in the middle of the load and the effort, for example a seesaw
2 To transmit drive and to reduce the effort force needed when lifting loads
3 Light-dependent resistor (or any other appropriate sensor)

4 To convert electrical, electronic or mechanical signals into physical movement, for example motors and solenoids

5 A small computer on a microchip, used to respond to input devices and control output devices

6 Power supply, central processing unit (CPU), programming device and ports for connecting the input and output devices

7 Any two from:
 – analogue to digital conversion – changing a continuous analogue signal into a discrete digital signal
 – amplification – increasing the size, or gain, of a signal
 – attenuation – reducing the strength of a signal
 – filtering – removing unwanted features from a signal, such as certain frequencies.

8 L293D (or any other appropriate motor driver IC)

9 When higher speed is required, when less power is needed, when there is potential for contamination

10 Compressed air forces a piston to move in a certain direction, achieved by a rod moving back and forth in a barrel.

Chapter 10

Test yourself

Page 162 (left)

Input blocks take signals from the real-world environment and change them into signals that process blocks can understand.

Page 162 (right)

▶ Timers take an input signal and keep it high or low for a set time period. This can often be adjusted using different component values, for example by changing resistor or capacitor values, or by using variable resistors and capacitors.

▶ Comparators compare two different signals, often a measured signal and a reference signal, and indicate which is the highest.

▶ Pulse units produce a continuous sequence of digital pulses. As with timers, the duration of time that the signals are high and/or low can be changed by altering component values or by using variable resistors.

▶ Counters add up the number of digital signals or pulses received. This figure can then be outputted to a display.

▶ Latches take an input signal and keep it high or low until it is reset.

Page 163 (left)

▼ Truth table for an AND gate

Input A	Input B	Output
0	0	0
0	1	0
1	0	0
1	1	1

▼ Truth table for an OR gate

Input A	Input B	Output
0	0	0
0	1	1
1	0	1
1	1	1

▼ Truth table for a NOT gate

Input	Output
0	1
1	0

Page 163 (right)

Output blocks take signals from process blocks and turn them back into real-world environmental signals.

Page 164

An open-loop control system does not have a feedback loop, whereas a closed-loop control system has one or more feedback loops.

Improve your maths

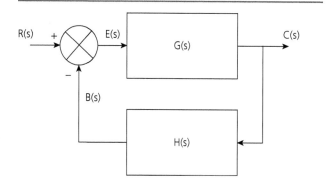

Assessment practice

1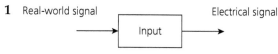

2 To keep a signal high or low until it is reset

3 NOT gates act as inverters, that is, the output signal is always the inverse, or opposite, of the input signal. They produce a high (1) output signal when the input signal is low (0), and vice versa.

4 To help with system modelling, analysis and fault finding, as it shows the relationships between the different parts that make up a system

5 To show how all the components and wires in a system will be physically laid out

6 The power transmission medium used – in electrical systems this is electrical power, in hydraulic systems it is a liquid such as oil or water, and in pneumatic systems it is a compressed gas

7 A type of sensor that sends a signal into the environment and measures the responses that it gets back

8 Motor, solenoid

9 An actuator that responds to and/or outputs continuous signals

10 To improve accuracy, improve efficiency, reduce human error, cut labour costs, reduce waste and increase quality

Chapter 11

Test yourself

Page 170

The European Union (EU)

Page 173

ISO 9001

Page 176

Either 100% inspection or sampling

Page 177

Six Sigma

Page 181

A hierarchical SOP is used for complex procedures with a greater number of steps.

Improve your maths

Page 177

2000 components will be needed for testing.

Assessment practice

1 The Welding Institute (TWI)

2 Quality assurance is a management methodology used by organisations to prevent defects in manufactured products. Quality control is a corrective tool used to ensure manufactured products meet accepted standards. It implements the inspection and testing requirements of quality assurance in order to detect and rectify defects.

3 Plan, do, check, act

4 ISO 9001 is an international standard that specifies criteria for a quality management system within an organisation.

5 Traceability means being able to identify and track all the processes within an organisation, from the purchase of raw materials, through to manufacture of the product and finally to disposal of the product.

6 Three purposes of document version control are:
   – to keep track of amendments
   – to record key decisions
   – to prevent the use of out-of-date documents.

7 Failure mode and effects analysis

8 Two advantages of quality circles are:
   – They result in increased productivity and product quality.
   – They boost employee morale and improve teamwork skills.

9 The Pareto principle states that, for many incidents, roughly 80 per cent of consequences come from 20 per cent of the underlying causes.

10 Step-by-step SOPs break down processes into numbered lists of detailed steps. The steps should be clear enough so that someone can follow them without supervision.

Chapter 12

Test yourself

Page 188

1 Health and Safety at Work etc. Act 1974
2 Examples include:
▶ Management of Health and Safety at Work Regulations (MHSAWR) 1999
▶ Provision and Use of Work Equipment Regulations (PUWER) 1998
▶ Personal Protective Equipment (PPE) Regulations 1992
▶ Control of Noise at Work Regulations 2005
▶ Manual Handling Operations Regulations (MHOR) 1992 (amended 2002)
▶ Lifting Operations and Lifting Equipment Regulations (LOLER) 1998

▶ Work at Height Regulations (WAHR) 2005 (amended 2007)
▶ Electricity at Work Regulations 1989
▶ Control of Electromagnetic Fields at Work (CEMFAW) Regulations 2016
▶ Reporting of Injuries, Diseases and Dangerous Occurrences Regulations (RIDDOR) 2013
▶ Control of Substances Hazardous to Health (COSHH) Regulations 2002 (amended 2004).

Page 191

Health and safety inspectors have the legal right to enter a workplace without giving notice, although notice may be given where the inspector considers it appropriate.

HSE inspectors can:
▶ inspect engineering sites
▶ give guidance and advice on work or processes
▶ take photos and samples
▶ talk about and discuss situations and/or problems
▶ seize evidence, equipment or materials.

Page 192

Employers must:
▶ carry out risk assessments of all the company's work activities
▶ identify and implement adequate control measures
▶ inform all employees of the risk assessments and associated control measures
▶ review the risk assessments at regular intervals
▶ make a record of the risk assessments if five or more operatives are employed
▶ ensure the safety of visitors, contractors and members of the public.

Employees must:
▶ take reasonable care of the health and safety of themselves and others
▶ comply with the employer's health and safety policy
▶ not recklessly interfere with anything that may affect health and safety.

Page 195

A risk assessment defines which workplace hazards are likely to cause harm to employees and visitors.

Page 196

Where proposed work is identified as having a high risk, strict controls are required. Examples of when a permit-to-work is required include:
▶ when completing 'hot works', e.g. welding/plasma cutting
▶ when working in confined spaces, e.g. in a tank

▶ when lone working, e.g. in a sewer/tank
▶ when maintaining machinery and equipment, e.g. maintenance of a disc cutter, etc.

Page 197

The risks of using oxygen in the workplace are:
▶ Oxygen is a very reactive gas that poses fire and explosion hazards.
▶ When pure oxygen is at high pressure, for example from a cylinder, it can react destructively with materials such as oil and grease.
▶ Fires also burn hotter and faster in the presence of oxygen.
▶ Some materials, such as oily rags, may catch fire spontaneously due to oxidation.

Page 202

▶ Landfill is cheaper than recycling processes.
▶ Reuse cuts down single use items and reduces waste.
▶ Recycling ensures that less waste goes to landfill, and recycling materials can require less energy/carbon than producing new materials.
▶ Controlled waste cuts down on toxic waste going into landfill.

Improve your maths

20% of £250,000 = 0.2 × 250,000 = 50,000

50,000 + 10,366 = 60,366

Total fine = £60,366

Assessment practice

1   The main objectives of the Health and Safety at Work etc. Act are to:
    – secure the health, safety and welfare of people at work
    – protect people other than those at work (for example, visitors or the general public) from risks to health or safety caused by work activities
    – control the possession and use of explosive, highly flammable or otherwise dangerous substances.
2   A health and safety culture in the workplace includes shared values, beliefs, expectations and attitudes about how to behave safely within an organisation.
3   HSE inspectors have powers to enforce health and safety legislation, for example:
    – carrying out workplace inspections
    – issuing informal cautions and formal legal notices to duty holders
    – initiating prosecutions.

4 Employees are required to:
   – use machinery, equipment, dangerous substances, transport equipment, means of production and safety devices in accordance with training and instruction
   – report work situations which pose a serious and imminent danger to health and safety
   – report shortcomings in their employer's protection arrangements for health and safety.
5 Typical injuries include:
   – cuts and bruises
   – broken bones
   – muscular sprains and strains.
6 The five stages of risk assessment are:
   1 Identify hazards.
   2 Decide who might be harmed and how.
   3 Evaluate the risks and decide on control measures.
   4 Record any findings (when there are five or more employees).
   5 Review the risk assessment and update if necessary.
7 A safe system of work is a set of procedures designed to eliminate or minimise the risks involved in specific operations.
8 Oxygen is a very reactive gas that poses fire and explosion hazards in the workplace.
9 Lock out, tag out is a safety procedure that ensures the isolation of hazardous energy sources before maintenance work is carried out.
10 Benefits:
   – Less waste goes to landfill.
   – Recycling materials can require less energy/carbon than producing new materials.
   Limitations:
   – Recycling waste can be expensive.
   – Some wastes cannot be recycled.
   – The separation of useful material from waste can be difficult.

Chapter 13

Test yourself

Your answer should include all aspects of a contract discussed in this chapter, for example:
▶ details of both parties
▶ subject of the contract
▶ any warranty
▶ a definition of force majeure
▶ liabilities if any product is found to be faulty
▶ duration of the contract and what happens when time runs out.

Improve your maths

Number of months = 10,000/200 = 50

50 months = 4 years and 2 months to pay back the loan

Assessment practice

1 Market research
2 Profit
3 A direct cost
4 Motion
5 Choose two from:
   – certain materials or components are no longer available
   – a competitor has made a better, more popular product
   – new materials or technologies have become available
   – customers no longer want or need the product
   – new regulations have come into force
   – materials or components have increased in price.
6 A company might end up with products left in stock which it may need to sell off cheaply or even throw away.
7 A warranty is a written promise by a company to repair or replace a product that develops a fault within a specified timeframe.
8 Shareholders invest in the hope that the value of the company will go up, allowing them to sell their shareholding for more than they bought it.
9 A team of personnel is chosen from different areas of the business to look at the project from all angles. The project leader will be selected according to their relevant skills and experience, rather than their seniority.
10 A policy is a set of rules or guidelines for the organisation. For example, it could be a policy that all staff meet with the human resources (HR) department once a year to review progress and decide on training or progression. Another policy could be that the fire alarm is tested on the first Friday of every month.

Chapter 14

Test yourself

Page 219

1 Supervisor, team leader, production manager, other qualified colleagues, mentor
2 Arrange a suitable time to meet or talk, arrange a meeting, ask a question at a training session

3 Questions you might ask include: How do I use the equipment safely? How can the equipment be isolated? How do I set up the machine?

Page 220 (left)

Working in an ethical manner is important to ensure customers are not offended. This could include showing respect for different cultures/beliefs/laws and overall public good, for example through the use of different colours that have different meanings in different cultures. Working in an ethical manner is also important to create a positive, professional reputation for the company, for example by conducting work in a safe, accurate and trustworthy manner.

Page 220 (right)

Accountability is important to ensure responsibility is taken for actions and to ensure work is of an appropriate standard. Appropriate consequences for substandard work/performance should be implemented through disciplinary procedures.

Page 221

Age, disability, gender reassignment, marriage/civil partnership, pregnancy/maternity, race, religion or belief, sex or sexual orientation.

Page 222 (left)

CPD in the workplace ensures that people can continually update their knowledge and skills. It can maintain workforce morale and motivate staff to perform as well as possible. For example, training can lead to greater efficiency and productivity. CPD also allows workers to keep up to date with changes to legislation, quality standards and company standard operating procedures (SOPs).

Page 222 (right)

Professional recognition motivates people to achieve high standards of work and ensures that outstanding performance is rewarded, for example through the Engineering and Manufacturing Awards (EMA).

Page 223

Physical characteristics define features of a person's body, whereas mental characteristics are about how a person thinks.

Page 224 (left)

Good design: safe/efficient layout of machines/equipment, design that allows for effective workflow, design that allows effective management oversight, meeting/breakout areas built into the working environment, good disability access.

Bad design: any examples where the conditions mentioned above are not met.

Page 224 (right)

Stressed workers will not produce their best work and will potentially lack focus, leading to a greater risk of errors, lack of productivity etc.

Assessment practice

1 To ensure employees have a clear understanding of their role within an organisation and know exactly what is expected of them.

2 Any two from:
- good punctuality/attendance
- demonstrating a positive attitude to work
- treating others with respect
- ensuring a professional appearance.

3 To give confidence to potential customers/clients that a job will be done well. If a company has a negative reputation it can be difficult/time consuming to repair.

4 Any two from:
- face-to-face/remote training
- training courses
- observation of colleagues
- coaching/mentoring
- obtaining further qualifications.

5 Ethical responsibilities are the moral principles that govern how an individual/organisation behaves.

6 Physical characteristics: strength, appearance or any characteristics that define features of a person's body.

7 Mental characteristics: confidence, the ability to deal with change or any characteristics that are about how a person thinks.

8 Good workplace design allows work to be carried out safely and efficiently, with good oversight built into the design.

9 Excessive workload, stress, any other relevant cause

10 Ensure workers are not tired when using equipment by placing appropriate limits on working hours and ensuring regular breaks are taken.

Chapter 15

Test yourself

Page 227

A full answer will mention all aspects of transport, handling and storage. Consider the following questions:
▶ How are the carrots packaged for transportation from the farm?
▶ What happens if the carrots are stored in an environment that is too warm or too humid?
▶ What happens if the carrots are stored in an environment that is too cold or below freezing?
▶ How long can the carrots be kept?
▶ What happens if the carrots are knocked or stored in large piles?
▶ Will the carrots keep after they have been peeled and chopped? Will they require different storage conditions?

Page 229

	Advantages	Disadvantages	Key issues
JIT	No wasted movement in production Low stock levels, so low risk Workforce can focus on tasks and not on finding what they need	Requires good, knowledgeable management Late equipment or materials have a large knock-on effect	Efficient and avoids wasteful practices
Made to order	No finished items kept in stock Only making what is needed Customers get exactly what they ask for	Long lead times, so customers have to wait Orders must come in steadily to keep production running	Reactive rather than proactive Usually only used for high-value products
Made to stock	Large quantity of products available to be sold at the best time for the market	Large stock to be carried with all the costs entailed Needs a clear idea of future market requirements and trends	Usually seasonal

Assessment practice

1 Inventory
2 Refrigeration
3 Raw materials
4 Costs to store, keep, move and dispose of stock
5 Automatic updating of stock levels, instant access to a stock database, notification of low stock levels
6 Barcodes are a way of showing a multi-digit number in a form that can be read by a scanner and be entered directly into a computer system. They can be used to simplify the entry of stock into a catalogue or database, avoiding the need for a human to enter the numbers manually, therefore making the process quicker, therefore more accurate.
7 Place to light or pick to light
8 The WEEE Directive
9 Workforce capacity planning
10 7 years

Chapter 16

Test yourself

Page 238

Transportation: unnecessary movement of materials and products

Inventory: having materials and products that are not currently being processed

Motion: wasted time or effort related to unnecessary movements of people or products

Waiting: time spent waiting for the next process to be carried out

Over-production: making more products than customers have asked for (but to the quality required)

Over-processing: making products to a higher specification than needed (but in the quantity required)

Defects: products or services failing to meet customer expectations

Skills (unused talent): not using the skills and knowledge of the workforce

Page 239

Throughput: number of products manufactured per unit of time

Changeover time: time taken to switch processes between making different products

Total processing time: sum of the actual machining time of the different processes needed to manufacture a product

Total cycle time: time to manufacture a customer order from start to finish

Scrap: (total number of products scrapped/total number of products produced) × 100

Yield: (total number of good products produced/total number of products produced) × 100

Capacity utilisation: (output achieved/potential output) × 100

Operation effective efficiency (OEE): (total number of good products produced × ideal processing time/planned production time) × 100

On-time delivery: (total number of products delivered on time/total number of products produced) × 100

Page 242

Traffic lights can be used to indicate when it is safe to enter an area, in order to prevent accidents and collisions.

Health and safety signs can be used to indicate hazards and/or required control measures, in order to reduce the risk of injury.

Floor markings around machines can be used to show areas that may be unsafe to enter when machines are operating, in order to prevent accidents and injuries.

Page 244

1 External elements in a machine changeover process are things that can be carried out while the process is running. For example, the tools and documentation needed for a changeover could be found while the previous batch is being processed, not when the process is being changed over.

Internal elements are things which currently have to be done when the process is stopped, for example, sharpening a cutting tool when changing between different products.

2 The steps involved in carrying out a SMED activity are as follows:

   1 Identify all the steps in the current changeover activity, for example, by recording the process and documenting it in detail.
   2 Remove all external elements from the changeover process.
   3 Convert internal elements to external elements.
   4 Standardise the process by documenting the new, fastest, most efficient method of carrying out the changeover, for example using a standard operating procedure.
   5 Train staff to ensure all workers follow the improved standardised process.

Improve your maths

Scrap = (total number of products scrapped/total number of products produced) × 100 = (4/(4+16)) × 100 = 20%

OEE = (total number of good products produced × ideal processing time/planned production time) × 100 = ((16 × 2.4)/60) × 100 = 64%

Assessment practice

1 Any activity which does not add value to a product
2 Waiting increases the cost of operating the business by increasing the quantity of materials and work in progress held by the company, and the space required to store them.
3 Over-production means making more products than customers have asked for (but to the quality required) whereas over-processing means making products to a higher specification than needed (but in the quantity required).
4 The eight wastes can increase both the cost of making products and the cost of operating the business.
5 To diagnose where waste is occurring and identify areas where there is the most potential for improvement
6 OEE = (total number of good products produced × ideal processing time/planned production time) × 100 = (33 × 10/(8 × 60)) × 100 = 68.75%

7   Choose two from: throughput, changeover time, total processing time, total cycle time, scrap, yield, capacity utilisation and on-time delivery.

8   Value stream mapping involves 'walking the process'. In the case of materials, this means actually following the route that they take from arriving at the factory, to leaving as finished products. Whenever there is an activity involving the materials, it is recorded as either value-added or as one of the eight wastes.

9   Choose two from: progress/countdown boards, shadow boards, standard operating procedures (SOPs), traffic lights, health and safety signs and floor markings around machines.

10 Benefit: Kanban reduces the amount of inventory, which in turn reduces the space required for work in progress and the amount of money that the business needs to operate. It also quickly identifies quality defects in batches, thereby minimising waste and scrap. Limitation: time is required to schedule each process, which may increase labour time and cost of the scheduling activity.

Chapter 17

Test yourself

Read what is needed carefully – the fact that you cannot use the common name is mentioned several times. Choose a simple product that you can describe easily and consider what, where, who, why, how and when. Your fellow learner or tutor should recognise your product using only the information you provide.

Improve your maths

E (estimated time) = (O + 4M + P)/6
= (2 + 6 × 4 + 12)/6 = six hours and 20 minutes

Assessment practice

1   A project brief

2   A typical project will start when one of three things happens:
    –   the opportunity for a new product is identified by a company (internal client), often the research and development department (technology push)
    –   the marketing department (internal client) identifies that the company needs to design a new product in order to keep up with the current market (market pull)
    –   an external client makes an enquiry (market pull).

3   Choose three from: available time, costs of the project, customer or client requirements, available staffing/skills, available machinery and benefit to the business.

4   There are costs attributable to both the time taken to undertake the training and the delivery of the training.

5   FMEA allows all stakeholders to be involved at the early planning stage to try to identify any possible risks to the project from several points of view.

6   An agenda is sent out in advance of the meeting, often by email, and lists everything that needs to be talked about.

7   More people working remotely has increased the use of video conferencing and collaborative online working.

8   E (estimated time) = (O + 4M + P)/6

9   Path 1

10 Task H

Index